Low-Dimensional Geometry

From Euclidean Surfaces to Hyperbolic Knots

STUDENT MATHEMATICAL LIBRARY
Ω IAS/PARK CITY MATHEMATICAL SUBSERIES
Volume 49

Low-Dimensional Geometry

From Euclidean Surfaces to Hyperbolic Knots

Francis Bonahon

American Mathematical Society, Providence, Rhode Island
Institute for Advanced Study, Princeton, New Jersey

The photo on the back cover of the author is reprinted with the permission of the USC Department of Mathematics.

2000 *Mathematics Subject Classification.* Primary 51M05, 51M10, 30F40, 57M25.

For additional information and updates on this book, visit
www.ams.org/bookpages/stml-49

Library of Congress Cataloging-in-Publication Data

Bonahon, Francis, 1955–
  Low-dimensional geometry : from euclidean surfaces to hyperbolic knots / Francis Bonahon.
    p. cm. – (Student mathematical library ; v. 49. IAS/Park City mathematical subseries)
  Includes bibliographical references and index.
  ISBN 978-0-8218-4816-6 (alk. paper)
  1. Manifolds (Mathematics)   2. Geometry, Hyperbolic.   3. Geometry, Plane.
4. Knot theory.   I. Title.
QA613.B66   2009
516′.07—dc22   2009005856

Table of contents

Table of contents   vii

IAS/Park City Mathematics Institute   xi

Preface   xiii

Chapter 1.   The euclidean plane   1

   §1.1.   Euclidean length and distance   1

   §1.2.   Shortest curves   3

   §1.3.   Metric spaces   3

   §1.4.   Isometries   5

   Exercises for Chapter 1   7

Chapter 2.   The hyperbolic plane   11

   §2.1.   The hyperbolic plane   11

   §2.2.   Some isometries of the hyperbolic plane   14

   §2.3.   Shortest curves in the hyperbolic plane   17

   §2.4.   All isometries of the hyperbolic plane   23

   §2.5.   Linear and antilinear fractional maps   27

   §2.6.   The hyperbolic norm   33

   §2.7.   The disk model for the hyperbolic plane   36

Exercises for Chapter 2   39

Chapter 3.   The 2-dimensional sphere   47

   §3.1.   The 2-dimensional sphere   47

   §3.2.   Shortest curves   48

   §3.3.   Isometries   49

   Exercises for Chapter 3   50

Chapter 4.   Gluing constructions   55

   §4.1.   Informal examples: the cylinder and the torus   55

   §4.2.   Mathematical definition of gluings and quotient spaces   58

   §4.3.   Gluing the edges of a euclidean polygon   61

   §4.4.   Proofs of Theorems 4.3 and 4.4   67

   §4.5.   Gluing hyperbolic and spherical polygons   79

   Exercises for Chapter 4   84

Chapter 5.   Gluing examples   89

   §5.1.   Some euclidean surfaces   89

   §5.2.   The surface of genus 2   97

   §5.3.   The projective plane   102

   §5.4.   The cylinder and the Möbius strip   103

   §5.5.   The once-punctured torus   114

   §5.6.   Triangular pillowcases   125

   Exercises for Chapter 5   126

Chapter 6.   Tessellations   133

   §6.1.   Tessellations   133

   §6.2.   Complete metric spaces   134

   §6.3.   From gluing polygon edges to tessellations   135

   §6.4.   Completeness and compactness properties   147

   §6.5.   Tessellations by bounded polygons   155

   §6.6.   Tessellations by unbounded polygons   161

   §6.7.   Incomplete hyperbolic surfaces   163

§6.8.   Poincaré's polygon theorem   169

Exercises for Chapter 6   182

Chapter 7.   Group actions and fundamental domains   185

   §7.1.   Transformation groups   185

   §7.2.   Group actions and quotient spaces   187

   §7.3.   Fundamental domains   192

   §7.4.   Dirichlet domains   197

Exercises for Chapter 7   202

Chapter 8.   The Farey tessellation and circle packing   207

   §8.1.   The Farey circle packing and tessellation   207

   §8.2.   The Farey tessellation and the once-punctured torus   212

   §8.3.   Horocircles and the Farey circle packing   214

   §8.4.   Shearing the Farey tessellation   217

Exercises for Chapter 8   222

Chapter 9.   The 3-dimensional hyperbolic space   227

   §9.1.   The hyperbolic space   227

   §9.2.   Shortest curves in the hyperbolic space   230

   §9.3.   Isometries of the hyperbolic space   231

   §9.4.   Hyperbolic planes and horospheres   234

Exercises for Chapter 9   235

Chapter 10.   Kleinian groups   241

   §10.1.   Bending the Farey tessellation   241

   §10.2.   Kleinian groups and their limit sets   247

   §10.3.   First rigorous example: fuchsian groups   252

   §10.4.   Poincaré's Polyhedron Theorem   257

   §10.5.   More examples of kleinian groups   265

   §10.6.   Poincaré, Fuchs and Klein   283

Exercises for Chapter 10   286

Chapter 11.   The figure-eight knot complement   293

§11.1.   Another crooked Farey tessellation   293

§11.2.   Enlarging the group Γ_8 294

§11.3.   Limit sets   299

§11.4.   The figure-eight knot   303

Exercises for Chapter 11   312

Chapter 12.   Geometrization theorems in dimension 3   315

§12.1.   Knots   315

§12.2.   The Geometrization Theorem for knot complements   319

§12.3.   Mostow's Rigidity Theorem   324

§12.4.   Ford domains   326

§12.5.   The general Geometrization Theorem   340

Exercises for Chapter 12   351

Appendix.   Tool Kit   355

§T.1.   Elementary set theory   355

§T.2.   Maximum, minimum, supremum, and infimum   358

§T.3.   Limits and continuity. Limits involving infinity   360

§T.4.   Complex numbers   361

Supplemental bibliography and references   365

Supplemental bibliography   365

References   369

Index   377

IAS/Park City Mathematics Institute

The IAS/Park City Mathematics Institute (PCMI) was founded in 1991 as part of the "Regional Geometry Institute" initiative of the National Science Foundation. In mid-1993 the program found an institutional home at the Institute for Advanced Study (IAS) in Princeton, New Jersey. The PCMI continues to hold summer programs in Park City, Utah.

The IAS/Park City Mathematics Institute encourages both research and education in mathematics and fosters interaction between the two. The three-week summer institute offers programs for researchers and postdoctoral scholars, graduate students, undergraduate students, high school teachers, mathematics education researchers, and undergraduate faculty. One of PCMI's main goals is to make all of the participants aware of the total spectrum of activities that occur in mathematics education and research: we wish to involve professional mathematicians in education and to bring modern concepts in mathematics to the attention of educators. To that end the summer institute features general sessions designed to encourage interaction among the various groups. In-year activities at sites around the country form an integral part of the High School Teacher Program.

Each summer a different topic is chosen as the focus of the Research Program and Graduate Summer School. Activities in the Undergraduate Program deal with this topic as well. Lecture notes from the Graduate Summer School are published each year in the IAS/Park City Mathematics Series. Course materials from the Undergraduate Program, such as the current volume, are now being published as part of the IAS/Park City Mathematical Subseries in the Student Mathematical Library. We are happy to make available more of the excellent resources which have been developed as part of the PCMI.

John Polking, Series Editor

April 13, 2009

Preface

About 30 years ago, the field of 3-dimensional topology was revolutionized by Thurston's Geometrization Theorem and by the unexpected appearance of hyperbolic geometry in purely topological problems. This book aims at introducing undergraduate students to some of these striking developments. It grew out of notes prepared by the author for a three-week course for undergraduates that he taught at the Park City Mathematical Institute in June–July 2006. It covers much more material than these lectures, but the written version intends to preserve the overall spirit of the course. The ultimate goal, attained in the last chapter, is to bring the students to a level where they can understand the statements of Thurston's Geometrization Theorem for knot complements and, more generally, of the general Geometrization Theorem for 3-dimensional manifolds recently proved by G. Perelman. Another leading theme is the intrinsic beauty of some of the mathematical objects involved, not just mathematically but visually as well.

The first two-thirds of the book are devoted to 2-dimensional geometry. After a brief discussion of the geometry of the euclidean plane $\mathbb{R}^2$, the hyperbolic plane $\mathbb{H}^2$, and the sphere $\mathbb{S}^2$, we discuss the construction of locally homogeneous spaces by gluing the sides of a polygon. This leads to the investigation of the tessellations that are associated to such constructions, with a special focus on one of the most beautiful objects of mathematics, the Farey tessellation of the hyperbolic plane. At this point, the deformations of the Farey tessellation by shearing lead us to jump to one dimension higher, in order to allow bending. After a few generalities on the 3-dimensional hyperbolic space $\mathbb{H}^3$, we consider the crooked tessellations obtained by bending the Farey tessellation, which naturally leads us to discussing kleinian groups and quasi-fuchsian groups. Pushing the bending of the Farey tessellation to the edge of kleinian groups, we reach the famous example associated to the complement of the figure-eight knot. At this point, we are ready to explain that this example is a manifestation of a general phenomenon. We state Thurston's Geometrization Theorem for knot complements, and illustrate how it has revolutionized knot theory in particular through the use of Ford domains. The book concludes with a discussion of the very recently proved Geometrization Theorem for 3-dimensional manifolds.

We tried to strike a balance between mathematical intuition and rigor. Much of the material is unapologetically "picture driven", as we intended to share our own enthusiasm for the beauty of some of the mathematical objects involved. However, we did not want to sacrifice the other foundation of mathematics, namely, the level of certainty provided by careful mathematical proofs. One drawback of this compromise is that the exposition is occasionally interrupted with a few proofs which are more lengthy than difficult, but can somewhat break the flow of the discourse. When this occurs, the reader is encouraged, on a preliminary reading, to first glance at the executive summary of the argument that is usually present at its beginning, and then to grab the remote control ⏬ and press the "fast forward" button until the first occurrence of the closing symbol ⏫. The reader may later need to return to some of the parts that have thus been zapped through, for the sake of mathematical rigor or because subsequent parts of the book may refer to specific arguments or definitions in these sections. For the same reason, the book is not intended to be read in a linear way. The reader is strongly advised to generously skip, at first, much of the early material in order to reach the parts with pretty pictures, such as Chapters 5, 6, 8, 10 or 11, as quickly as possible, and then to backtrack when specific definitions or arguments are needed.

The book also has its idiosyncrasies. From a mathematical point of view, the main one involves quotients of metric spaces. It is traditional here to focus only on topological spaces, to introduce the quotient topology by *fiat*, and then to claim that it accurately describes the intuitive notion of gluing in cut-and-paste constructions; this is not always very convincing. A slightly less well-trodden road involves quotient metric spaces, but only in the case of quotients under discontinuous group actions. We decided to follow a different strategy, by discussing quotient (semi-)metrics very early on and in their full generality. This approach is, in our view, much more intuitive but it comes with a price: Some proofs become somewhat technical. On the one hand, these can serve as a good introduction to the techniques of rigorous proofs in mathematics. On the other hand, the reader pressed for time can also take advantage of the fast-forward commands where indicated, and zap through these proofs in a first reading.

From a purely technical point of view, the text is written in such a way that, in theory, it does not require much mathematical knowledge beyond multivariable calculus. An appendix at the end provides a "tool kit" summarizing some of the main concepts that will be needed. In practice, however, the mathematical rigor of many arguments is likely to require a somewhat higher level of mathematical sophistication. The reader will also notice that the level of difficulty progressively increases as one proceeds from early to later chapters. Each chapter ends with a selection of exercises, a few of which can be somewhat challenging. The idea was to provide material suitable for an independent study by a dedicated undergraduate student, or for a topics course. Such a course might cover the main sections of Chapters 1–7, 9, 12, and whichever parts of the remaining chapters would be suitable for both the time available and the tastes of the instructor.

The author is delighted to thank Roger Howe for tricking him into believing that the PCMI course would not require that much work (which turned out to be wrong), and Ed Dunne for encouraging him to turn the original lecture notes into a book and for warning him that the task would be very labor intensive (which turned out to be right). The general form of the book owes much to the feedback received from the students and faculty who attended the PCMI lectures, and who were used as "guinea pigs"; this includes Chris Hiatt, who was the teaching assistant for the course. Dave Futer provided numerous and invaluable comments on an earlier draft of the manuscript, Roland van der Veen contributed a few more, and Jennifer Wright Sharp polished the final version with her excellent copy-editing. Finally, the mathematical content of the book was greatly influenced by the author's own research in this area of mathematics, which in recent years was partially supported by Grants 0103511 and 0604866 from the National Science Foundation.

Chapter 1

The euclidean plane

We are all very familiar with the geometry of the euclidean plane $\mathbb{R}^2$. We will encounter a new type of 2-dimensional geometry in the next chapter, that of the hyperbolic plane $\mathbb{H}^2$. In this chapter, we first list a series of well-known properties of the euclidean plane which, in the next chapter, will enable us to develop the properties of the hyperbolic plane in very close analogy.

Before proceeding, you are advised to briefly consult the TOOL KIT in the appendix for a succinct review of the basic definitions and notation concerning set theory, infima and suprema of sets of real numbers, and complex numbers.

1.1. Euclidean length and distance

The **euclidean plane** is the set

$$\mathbb{R}^2 = \{(x, y); x, y \in \mathbb{R}\}$$

consisting of all ordered pairs (x, y) of real numbers x and y.

If γ is a curve in $\mathbb{R}^2$, parametrized by the differentiable vector-valued function

$$t \mapsto \big(x(t), y(t)\big), \quad a \leqslant t \leqslant b,$$

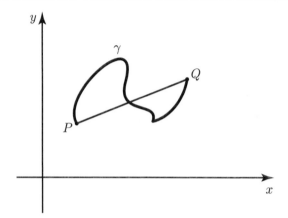

Figure 1.1. The euclidean plane

its *euclidean length* $\ell_{\mathrm{euc}}(\gamma)$ is the arc length given by

$$(1.1) \qquad \ell_{\mathrm{euc}}(\gamma) = \int_a^b \sqrt{x'(t)^2 + y'(t)^2}\, dt.$$

This length is independent of the parametrization by a well-known consequence of the chain rule.

It will be convenient to consider *piecewise differentiable* curves γ made up of finitely many differentiable curves $\gamma_1, \gamma_2, \ldots, \gamma_n$ such that the initial point of each γ_{i+1} is equal to the terminal point of γ_i. In other words, such a curve γ is differentiable everywhere except at finitely many points, corresponding to the endpoints of the γ_i, where it is allowed to have a "corner" (but no discontinuity). In this case, the length $\ell_{\mathrm{euc}}(\gamma)$ of the piecewise differentiable curve γ is defined as the sum of the lengths $\ell_{\mathrm{euc}}(\gamma_i)$ of its differentiable pieces γ_i. This is equivalent to allowing the integrand in (1.1) to be undefined at finitely many values of t where, however, it has finite left-hand and right-hand limits.

The *euclidean distance* $d_{\mathrm{euc}}(P, Q)$ between two points P and Q is the infimum of the lengths of all piecewise differentiable curves γ going from P to Q, namely

$$(1.2) \qquad d_{\mathrm{euc}}(P, Q) = \inf \left\{ \ell_{\mathrm{euc}}(\gamma); \gamma \text{ goes from } P \text{ to } Q \right\}.$$

See the TOOL KIT in the appendix for basic facts about the infimum of a set of real numbers. By definition of the infimum, the above definition means that every piecewise differentiable curve γ going from P to Q must have length greater than or equal to $d_{\text{euc}}(P, Q)$, and that there are curves whose length is arbitrarily close to $d_{\text{euc}}(P, Q)$.

1.2. Shortest curves

It is well known and easily proved (see Exercise 1.2) that the straight line provides the shortest route between two points.

Proposition 1.1. *The distance $d_{\text{euc}}(P, Q)$ is equal to the euclidean length $\ell_{\text{euc}}([P, Q])$ of the line segment $[P, Q]$ going from P to Q. In other words, $[P, Q]$ is the shortest curve going from P to Q.*   □

In particular, computing the length of a line segment by using formula (1.1) for arc length (see Exercise 1.1), we obtain the following.

Corollary 1.2. *The euclidean distance from $P_0 = (x_0, y_0)$ to $P_1 = (x_1, y_1)$ is equal to*

$$(1.3) \qquad d_{\text{euc}}(P_0, P_1) = \sqrt{(x_1 - x_0)^2 + (y_1 - y_0)^2} \qquad \square$$

1.3. Metric spaces

The euclidean plane $\mathbb{R}^2$, with its distance function d_{euc}, is a fundamental example of a metric space. A ***metric space*** is a pair (X, d) consisting of a set X together with a function $d\colon X \times X \to \mathbb{R}$ such that

    (1) $d(P, Q) \geqslant 0$ and $d(P, P) = 0$ for every $P,\ Q \in X$;

    (2) $d(P, Q) = 0$ if and only if $P = Q$;

    (3) $d(Q, P) = d(P, Q)$ for every $P,\ Q \in X$;

    (4) $d(P, R) \leqslant d(P, Q) + d(Q, R)$ for every $P,\ Q,\ R \in X$.

The fourth condition is the ***Triangle Inequality***. The function d is called the ***distance function***, the ***metric function***, or just the ***metric*** of the metric space X.

A function d that satisfies only conditions (1), (3) and (4) above is called a ***semi-distance function*** or a ***semi-metric***.

Elementary and classical properties of euclidean geometry show that $(\mathbb{R}^2, d_{\text{euc}})$ is a metric space. In particular, this explains the terminology for the Triangle Inequality. In fact, $(\mathbb{R}^2, d_{\text{euc}})$ and its higher-dimensional analogs are typical examples of metric spaces. See Exercise 1.3 for a proof that d_{euc} is a distance function which, instead of prior knowledge about euclidean geometry, uses only the definition of the euclidean distance by equation (1.2).

The main point of the definition of metric spaces is that the notions about limits and continuity that one encounters in calculus (see Section T.3 in the TOOL KIT) immediately extend to the wider context of a metric space (X, d).

For instance, a sequence of points $P_1, P_2, \ldots, P_n, \ldots$ in X *converges* to the point P_∞ if, for every $\varepsilon > 0$, there exists an integer n_0 such that $d(P_n, P_\infty) < \varepsilon$ for every $n \geqslant n_0$. This is equivalent to the property that the sequence $\big(d(P_n, P_\infty)\big)_{n \in \mathbb{N}}$ converges to 0 as a sequence of real numbers. The point P_∞ is the *limit* of the sequence $(P_n)_{n \in \mathbb{N}}$.

Similarly, a function $\varphi \colon X \to X'$ from a metric space (X, d) to a metric space (X', d') is *continuous* at $P_0 \in X$ if, for every number $\varepsilon > 0$, there exists a $\delta > 0$ such that $d'\big(\varphi(P), \varphi(P_0)\big) < \varepsilon$ for every $P \in X$ with $d(P, P_0) < \delta$. The function is *continuous* if it is continuous at every $P_0 \in X$.

We will make extensive use of the notion of a ball in a metric space (X, d). The (open) *ball* with center $P_0 \in X$ and radius $r > 0$ in (X, d) is the subset

$$B_d(P_0, r) = \{P \in X; d(P, P_0) < r\}.$$

The terminology is motivated by the case where X is the 3-dimensional euclidean space $\mathbb{R}^3$, and where d is the euclidean metric d_{euc} defined by the property that

$$d_{\text{euc}}(P_0, P_1) = \sqrt{(x_1 - x_0)^2 + (y_1 - y_0)^2 + (z_1 - z_0)^2}$$

when $P_0 = (x_0, y_0, z_0)$ and $P_1 = (x_1, y_1, z_1)$. In this case, $B_{d_{\text{euc}}}(P_0, r)$ is of course a geometric ball of radius r centered at P_0, without its boundary.

When (X, d) is the euclidean plane $(\mathbb{R}^2, d_{\text{euc}})$, a ball $B_{d_{\text{euc}}}(P_0, r)$ is an open disk of radius r centered at P_0. When (X, d) is the real line $(\mathbb{R}, d)$ with its usual metric $d(x, y) = |x, y|$, the ball $B_{d_{\text{euc}}}(x_0, r)$ is just the open interval $(x_0 - r, x_0 + r)$.

Incidentally, this may be a good spot to remind the reader of a few definitions which are often confused. In the euclidean plane $\mathbb{R}^2$, the open **disk** $B_{d_{\text{euc}}}(P_0, r) = \{P \in \mathbb{R}^2; d_{\text{euc}}(P, P_0) < r\}$ is not the same thing as the **circle** $\{P \in \mathbb{R}^2; d_{\text{euc}}(P, P_0) = r\}$ with center P_0 and radius r that bounds it. Similarly, in dimension 3, the open **ball** $B_{d_{\text{euc}}}(P_0, r) = \{P \in \mathbb{R}^3; d_{\text{euc}}(P, P_0) < r\}$ should not be confused with the **sphere** $\{P \in \mathbb{R}^3; d_{\text{euc}}(P, P_0) = r\}$ with the same center and radius.

1.4. Isometries

The euclidean plane has many symmetries. In a metric space, these are called isometries. An *isometry* between two metric spaces (X, d) and (X', d') is a bijection $\varphi \colon X \to X'$ which respects distances, namely, such that

$$d'\big(\varphi(P), \varphi(Q)\big) = d(P, Q)$$

for every $P, Q \in X$.

Recall the statement that φ is a bijection means that φ is one-to-one (or injective) and onto (or surjective), so that it has a well-defined inverse $\varphi^{-1} \colon X' \to X$. It immediately follows from definitions that the inverse φ^{-1} of an isometry φ is also an isometry.

It is also immediate that an isometry is continuous.

When there exists an isometry φ between two metric spaces (X, d) and (X', d'), then these two spaces have exactly the same properties. Indeed, φ can be used to translate any property of (X, d) to the same property for (X, d').

We are here interested in the case where $(X, d) = (X', d') = (\mathbb{R}^2, d_{\text{euc}})$. Isometries of $(\mathbb{R}^2, d_{\text{euc}})$ include:

- *translations* along a vector (x_0, y_0), defined by

$$\varphi(x, y) = (x + x_0, y + y_0);$$

- **rotations** of angle θ around the origin,

$$\varphi(x,y) = (x\cos\theta - y\sin\theta, x\sin\theta + y\cos\theta);$$

- **reflections** across a line passing through the origin and making an angle of θ with the x-axis,

$$\varphi(x,y) = (x\cos2\theta + y\sin2\theta, x\sin2\theta - y\cos2\theta);$$

- more generally, any composition of the above isometries, namely, any map φ of the form

(1.4) $$\varphi(x,y) = (x\cos\theta - y\sin\theta + x_0, x\sin\theta + y\cos\theta + y_0)$$

or

(1.5) $$\varphi(x,y) = (x\cos2\theta + y\sin2\theta + x_0, x\sin2\theta - y\cos2\theta + y_0).$$

For the last item, recall that the **composition** of two maps $\varphi\colon X \to Y$ and $\psi\colon Y \to Z$ is the map $\psi \circ \varphi\colon X \to Z$ defined by $\psi \circ \varphi(P) = \psi(\varphi(P))$ for every $P \in X$.

The above isometries are better expressed in terms of complex numbers, identifying the point $(x,y) \in \mathbb{R}^2$ with the complex number $z = x + iy \in \mathbb{C}$. See Section T.4 in the TOOL KIT for a brief summary of the main properties of complex numbers.

Using Euler's exponential notation (see Section T.4)

$$e^{i\theta} = \cos\theta + i\sin\theta,$$

the isometries listed in equations (1.4) and (1.5) can then be written as

$$\varphi(z) = e^{i\theta}z + z_0$$

and

$$\varphi(z) = e^{2i\theta}\bar{z} + z_0,$$

where $z_0 = x_0 + iy_0$ and where $\bar{z} = x - iy$ is the complex conjugate of $z = x + iy$.

Proposition 1.3. *If φ is an isometry of $(\mathbb{R}^2, d_{\mathrm{euc}}) = (\mathbb{C}, d_{\mathrm{euc}})$, then there exists a point $z_0 \in \mathbb{C}$ and an angle $\theta \in \mathbb{R}$ such that*

$$\varphi(z) = e^{i\theta}z + z_0 \quad or \quad \varphi(z) = e^{2i\theta}\bar{z} + z_0$$

for every $z \in \mathbb{C}$.

Proof. See Exercise 2.3 (and compare Theorem 2.11) for a proof of this well-known result in euclidean geometry.   □

A fundamental consequence of the abundance of isometries of the euclidean plane $(\mathbb{R}^2, d_{\mathrm{euc}})$ is the homogeneity of this metric space. A metric space (X, d) is **homogeneous** if, for any two points $P, Q \in X$, there exists an isometry $\varphi \colon X \to X$ such that $\varphi(P) = Q$. In other words, a homogeneous metric space looks the same at every point, since any property of (X, d) involving the point P also holds at any other point Q, by translating this property through the isometry φ sending P to Q.

Actually, the euclidean plane is not just homogeneous, it is **isotropic** in the sense that for any two points P_1 and $P_2 \in \mathbb{R}^2$, and for any unit vectors $\vec{v}_1$ at P_1 and $\vec{v}_2$ at P_2, there is an isometry φ of $(\mathbb{R}^2, d_{\mathrm{euc}})$ which sends P_1 to P_2 and $\vec{v}_1$ to $\vec{v}_2$. Here we are assuming the statement that φ sends the vector $\vec{v}_1$ to the vector $\vec{v}_2$ is intuitively clear; a more precise definition, using the differential $D_{P_1}\varphi$ of φ at P_1, will be given in Section 2.5.2.

As a consequence of the isotropy property, not only does the euclidean plane look the same at every point, it also looks the same in every direction.

Exercises for Chapter 1

Exercise 1.1. Using the expression given in equation (1.1) and a suitable parametrization of the line segment $[P_0, P_1]$ going from $P_0 = (x_0, y_0)$ to $P_1 = (x_1, y_1)$, show that the euclidean length of $[P_0, P_1]$ is equal to

$$\ell_{\mathrm{euc}}([P_0, P_1]) = \sqrt{(x_1 - x_0)^2 + (y_1 - y_0)^2}.$$

Exercise 1.2. The goal of this exercise is to rigorously prove that the line segment $[P, Q]$ is the shortest curve going from P to Q. Namely, consider a piecewise differentiable curve γ going from P to Q. We want to show that the euclidean length $\ell_{\mathrm{euc}}(\gamma)$ defined by equation (1.1) is greater than or equal to the length $\ell_{\mathrm{euc}}([P, Q])$ of the line segment $[P, Q]$.

a. First consider the case where $P = (x_0, y_0)$ and $Q = (x_0, y_1)$ sit on the same vertical line of equation $x = x_0$. Show that the euclidean length $\ell_{\mathrm{euc}}(\gamma)$ is greater than or equal to $|y_1 - y_0| = \ell_{\mathrm{euc}}([P, Q])$.

b. In the general case, let $\varphi \colon (x, y) \mapsto (x \cos \theta - y \sin \theta, x \sin \theta + y \cos \theta)$ be a rotation such that $\varphi(P)$ and $\varphi(Q)$ sit on the same vertical line. Show that the curve $\varphi(\gamma)$, going from $\varphi(P)$ to $\varphi(Q)$, has the same euclidean length as γ.

c. Combine parts a and b above to conclude that $\ell_{\text{euc}}(\gamma) \geqslant \ell_{\text{euc}}([P, Q])$.

Exercise 1.3. Rigorously prove that the euclidean distance function d_{euc}, as defined by equation (1.2), is a distance function on $\mathbb{R}^2$. You may need to use the result of Exercise 1.2 to show that $d_{\text{euc}}(P, Q) = 0$ only when $P = Q$. Note that the proof of the Triangle Inequality (for which you may find it useful to consult the proof of Lemma 2.1) is greatly simplified by our use of piecewise differentiable curves in the definition of d_{euc}.

Exercise 1.4. Let (X, d) be a metric space.

a. Show that $d(P, Q) - d(P, Q') \leqslant d(Q, Q')$ for every P, Q, $Q' \in X$.

b. Conclude that $\big| d(P, Q) - d(P, Q') \big| \leqslant d(Q, Q')$ for every P, Q, $Q' \in X$.

c. Use the above inequality to show that for every $P \in X$, the function $d_P \colon X \to \mathbb{R}$ defined by $d_P(Q) = d(P, Q)$ is continuous if we endow the real line $\mathbb{R}$ with the usual metric for which the distance between a and $b \in \mathbb{R}$ is equal to the absolute value $|a - b|$.

Exercise 1.5. Let $\varphi \colon X \to X'$ be a map from the metric space (X, d) to the metric space (X', d'). Show that φ is continuous at $P_0 \in X$ if and only if, for every $\varepsilon > 0$, there exists a $\delta > 0$ such that the image $\varphi\big(B_d(P_0, \delta)\big)$ of the ball $B_d(P_0, \delta) \subset X$ is contained in the ball $B_{d'}\big(\varphi(P_0), \varepsilon\big) \subset X'$.

Exercise 1.6 (Product of metric spaces). Let (X, d) and (X', d') be two metric spaces. On the product $X \times X' = \{(x, x'); x \in X, x' \in X'\}$, define

$$D \colon (X \times X') \times (X \times X') \to \mathbb{R}$$

by the property that $D\big((x, x'), (y, y')\big) = \max\{d(x, y), d'(x', y')\}$ for every (x, x'), $(y, y') \in X \times X'$. Show that D is a metric function on $X \times X'$.

Exercise 1.7. On $\mathbb{R}^2 = \mathbb{R} \times \mathbb{R}$, consider the metric function D provided by Exercise 1.6. Namely, $D\big((x, y), (x', y')\big) = \max\{|x - x'|, |y - y'|\}$ for every (x, y), $(x', y') \in \mathbb{R}^2$.

a. Show that $\frac{1}{\sqrt{2}} d_{\text{euc}}(P, P') \leqslant D(P, P') \leqslant d_{\text{euc}}(P, P')$ for every P, $P' \in \mathbb{R}^2$.

b. Let $(P_n)_{n \in \mathbb{N}}$ be a sequence in $\mathbb{R}^2$. Show that $(P_n)_{n \in \mathbb{N}}$ converges to a point $P_\infty \in \mathbb{R}^2$ for the metric D if and only if it converges to P_∞ for the metric d_{euc}.

c. Let $\varphi \colon \mathbb{R}^2 \to X$ be a map from $\mathbb{R}^2$ to a metric space (X, d). Show that φ is continuous for the metric D on $\mathbb{R}^2$ if and only if it is continuous for the metric d_{euc}.

Exercise 1.8 (Continuity and sequences). Let $\varphi \colon X \to X'$ be a map from the metric space (X, d) to the metric space (X', d').

a. Suppose that φ is continuous at P_0. Show that if P_1, P_2, ..., P_n, ... is a sequence which converges to P_0 in (X, d), then $\varphi(P_1)$, $\varphi(P_2)$, ..., $\varphi(P_n)$, ... is a sequence which converges to $\varphi(P_0)$ in (X', d').

b. Suppose that φ is not continuous at P_0. Construct a number $\varepsilon > 0$ and a sequence P_1, P_2, ..., P_n, ... in X such that $d(P_n, P_0) < \frac{1}{n}$ and $d\big(\varphi(P_n), \varphi(P_0)\big) \geqslant \varepsilon$ for every $n \geqslant 1$.

c. Combine parts a and b to show that φ is continuous at P_0 if and only if, for every sequence P_1, P_2, ..., P_n, ... converging to P_0 in (X, d), the sequence $\varphi(P_1)$, $\varphi(P_2)$, ..., $\varphi(P_n)$, ... converges to $\varphi(P_0)$ in (X', d').

Exercise 1.9. Let d and d' be two metrics on the same set X. Show that the identity map $\mathrm{Id}_X \colon (X, d) \to (X, d')$ is continuous if and only if every sequence $(P_n)_{n \in \mathbb{N}}$ that converges to some $P_\infty \in X$ for the metric d also converges to P_∞ for the metric d'. Possible hint: Compare Exercise 1.8.

Exercise 1.10. The euclidean metric of the euclidean plane is an example of a **path metric**, where the distance between two points P and Q is the infimum of the lengths of all curves joining P to Q. In the plane $\mathbb{R}^2$, let U be the U–shaped region enclosed by the polygonal curve with vertices $(0, 0)$, $(0, 2)$, $(1, 2)$, $(1, 1)$, $(2, 1)$, $(2, 2)$, $(3, 2)$, $(3, 0)$, $(0, 0)$ occurring in this order. Endow U with the metric d_U defined by the property that $d_U(P, Q)$ is the infimum of the euclidean lengths of all piecewise differentiable curves joining P to Q and completely contained in U.

a. Draw a picture of U.

b. Show that $d_U(P, Q) \geqslant d_{\mathrm{euc}}(P, Q)$ for every P, $Q \in U$.

c. Show that d_U is a metric function on U. It may be convenient to use part b above at some point of the proof.

d. If P_0 is the point $(0, 2)$, give a formula for the distance $d_U(P, P_0)$ as a function of the coordinates of $P = (x, y)$. This formula will involve several cases according to where P sits in U.

Exercise 1.11 (Lengths in metric spaces). In an arbitrary metric space (X, d), the **length** $\ell_d(\gamma)$ of a curve γ is defined as

$$\ell_d(\gamma) = \sup\left\{ \sum_{i=1}^{n} d(P_{i-1}, P_i); P_0, P_1, \ldots, P_n \in \gamma \text{ occur in this order along } \gamma \right\}.$$

In particular, the length may be infinite. For a differentiable curve γ in the euclidean plane $(\mathbb{R}^2, d_{\mathrm{euc}})$, we want to show that this length $\ell_{d_{\mathrm{euc}}}(\gamma)$ coincides with the euclidean length $\ell_{\mathrm{euc}}(\gamma)$ given by equation (1.1). For this, suppose that γ is parametrized by the differentiable function $t \mapsto \gamma(t) = \big(x(t), y(t)\big)$, $a \leqslant t \leqslant b$.

a. Show that $\ell_{d_{\mathrm{euc}}}(\gamma) \leqslant \ell_{\mathrm{euc}}(\gamma)$.

b. Cut the interval $[a, b]$ into n intervals $[t_{i-1}, t_i]$ of length $\Delta t = (b-a)/n$. Set $P_i = \gamma(t_i) = \gamma(a + i\Delta t)$ for $i = 0, 1, \ldots, n$. Show that

$$d_{\text{euc}}(P_{i-1}, P_i) \geqslant \left\| \gamma'(t_{i-1}) \right\| \Delta t - \tfrac{1}{2} K (\Delta t)^2,$$

where $K = \max\limits_{a \leqslant t \leqslant b} \left\| \gamma''(t) \right\|$ denotes the maximum length of the second derivative vector $\gamma''(t) = \big(x''(t), y''(t) \big)$ and where the length $\|(u, v)\|$ of a vector (u, v) is defined by the usual formula $\|(u, v)\| = \sqrt{u^2 + v^2}$. You may need to use the Taylor formula from multivariable calculus, which says that for every t, h,

$$\gamma(t + h) = \gamma(t) + h\gamma'(t) + h^2 R_1(t, h),$$

where the remainder $R_1(t, h)$ is such that $\|R_1(t, h)\| \leqslant \tfrac{1}{2} K$.

c. Use part b above to show that $\ell_{d_{\text{euc}}}(\gamma) \geqslant \ell_{\text{euc}}(\gamma)$.

d. Combine parts a and c above to conclude that $\ell_{d_{\text{euc}}}(\gamma) = \ell_{\text{euc}}(\gamma)$.

Exercise 1.12. Consider the length $\ell_D(\gamma)$ of a curve γ in a metric space (X, d) defined as in Exercise 1.11, in the special case where $(X, d) = (\mathbb{R}^2, D)$ is the plane $\mathbb{R}^2 = \mathbb{R} \times \mathbb{R}$ endowed with the product metric D of Exercise 1.6, defined by the property that $D\big((x, y), (x', y') \big) = \max\{|x - x'|, |y - y'|\}$ for every (x, y), $(x', y') \in \mathbb{R}^2$.

a. Show that $\ell_D(\gamma) \geqslant D(P, Q)$ for every curve γ going from P to Q.

b. Show that the length $\ell_D([P, Q])$ of the line segment $[P, Q]$ is equal to $D(P, Q)$, so that $[P, Q]$ consequently has minimum length among all curves going from P to Q.

c. Give an example where there is another curve γ going from P to Q which has minimum length $\ell_D(\gamma) = D(P, Q)$, and which is not the line segment $[P, Q]$.

d. If γ is differentiably parametrized by $t \mapsto \big(x(t), y(t) \big)$, $a \leqslant t \leqslant b$, give a condition on the derivatives $x'(t)$ and $y'(t)$ which is equivalent to the property that γ has minimum length over all curves going from $P = \big(x(a), y(a) \big)$ to $Q = \big(x(b), y(b) \big)$. (The answer depends on the relative position of P and Q with respect to each other.)

Chapter 2

The hyperbolic plane

The hyperbolic plane is a metric space which is much less familiar than the euclidean plane that we discussed in the previous chapter. We introduce its basic properties, by proceeding in very close analogy with the euclidean plane.

2.1. The hyperbolic plane

The **hyperbolic plane** is the metric space consisting of the open half-plane

$$\mathbb{H}^2 = \{(x, y) \in \mathbb{R}^2; y > 0\} = \{z \in \mathbb{C}; \operatorname{Im}(z) > 0\}$$

endowed with a new metric d_{hyp} defined below. Recall that the **imaginary part** $\operatorname{Im}(z)$ of a complex number $z = x + iy$ is just the coordinate y, while its **real part** $\operatorname{Re}(z)$ is the coordinate x.

To define the hyperbolic metric d_{hyp}, we first define the **hyperbolic length** of a curve γ parametrized by the differentiable vector-valued function

$$t \mapsto \big(x(t), y(t)\big), \quad a \leqslant t \leqslant b,$$

as

(2.1) $$\ell_{\mathrm{hyp}}(\gamma) = \int_a^b \frac{\sqrt{x'(t)^2 + y'(t)^2}}{y(t)} \, dt.$$

Again, an application of the chain rule shows that this hyperbolic length is independent of the parametrization of γ. The definition of the hyperbolic length also immediately extends to piecewise differentiable curves, by taking the sum of the hyperbolic length of the differentiable pieces, or by allowing finitely many jump discontinuities in the integrand of (2.1).

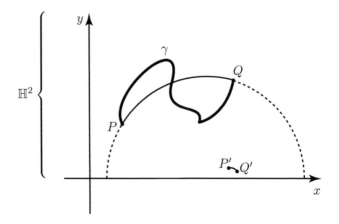

Figure 2.1. The hyperbolic plane

The *hyperbolic distance* between two points P and Q is the infimum of the hyperbolic lengths of all piecewise differentiable curves γ going from P to Q, namely

$$(2.2) \qquad d_{\text{hyp}}(P, Q) = \inf \left\{ \ell_{\text{hyp}}(\gamma); \gamma \text{ goes from } P \text{ to } Q \right\}.$$

Note the analogy with our definition of the euclidean distance in Chapter 1.

The hyperbolic distance d_{hyp} is at first somewhat unintuitive. For instance, we will see in later sections that the hyperbolic distance between the points P' and Q' indicated in Figure 2.1 is the same as the hyperbolic distance from P to Q. Also, among the curves joining P to Q, the one with the shortest hyperbolic length is the circle arc represented. With practice, we will become more comfortable with the geometry of the hyperbolic plane and see that it actually shares many important features with the euclidean plane.

But first, let us prove that the hyperbolic distance d_{hyp} is really a distance function.

Lemma 2.1. *The function*

$$d_{\mathrm{hyp}}\colon \mathbb{H}^2 \times \mathbb{H}^2 \to \mathbb{R}$$

defined by (2.2) is a distance function.

Proof. We have to check the four conditions in the definition of a distance function. The condition $d_{\mathrm{hyp}}(P, Q) \geqslant 0$ is immediate, as is the symmetry condition $d_{\mathrm{hyp}}(Q, P) = d_{\mathrm{hyp}}(P, Q)$.

To prove the Triangle Inequality, consider three points P, Q, $R \in \mathbb{H}^2$. Pick an arbitrary $\varepsilon > 0$. By definition of the hyperbolic distance as an infimum of hyperbolic lengths, there exists a piecewise differentiable curve γ going from P to Q such that $\ell_{\mathrm{hyp}}(\gamma) < d_{\mathrm{hyp}}(P, Q) + \frac{1}{2}\varepsilon$, and a piecewise differentiable curve γ' going from Q to R such that $\ell_{\mathrm{hyp}}(\gamma') < d_{\mathrm{hyp}}(Q, R) + \frac{1}{2}\varepsilon$. Chaining together these two curves γ and γ', one obtains a piecewise differentiable curve γ'' joining P to R whose length is

$$\ell_{\mathrm{hyp}}(\gamma'') = \ell_{\mathrm{hyp}}(\gamma) + \ell_{\mathrm{hyp}}(\gamma') < d_{\mathrm{hyp}}(P, Q) + d_{\mathrm{hyp}}(Q, R) + \varepsilon.$$

As a consequence,

$$d_{\mathrm{hyp}}(P, R) < d_{\mathrm{hyp}}(P, Q) + d_{\mathrm{hyp}}(Q, R) + \varepsilon.$$

Since this property holds for every $\varepsilon > 0$, we conclude that $d_{\mathrm{hyp}}(P, R) \leqslant d_{\mathrm{hyp}}(P, Q) + d_{\mathrm{hyp}}(Q, R)$ as required.

Note that our use of piecewise differentiable curves, instead of just differentiable curves, has greatly simplified this proof of the Triangle Inequality. (When γ and γ' are differentiable, the same is usually not true for γ'' since it may have a "corner" at the junction of γ and γ').

The only condition which requires some serious thought is the fact that $d_{\mathrm{hyp}}(P, Q) > 0$ if $P \neq Q$. Namely, we need to make sure that we cannot go from P to Q by curves whose hyperbolic lengths are arbitrarily small.

Consider a piecewise differentiable curve γ going from P to Q, parametrized by the piecewise differentiable function

$$t \mapsto \bigl(x(t), y(t)\bigr) \quad a \leqslant t \leqslant b,$$

with $P = \big(x(a), y(a)\big)$ and $Q = \big(x(b), y(b)\big)$. We will split the argument into two cases.

If γ does not go too high, so that $y(t) \leqslant 2y(a)$ for every $t \in [a, b]$,

$$
\begin{aligned}
\ell_{\text{hyp}}(\gamma) &= \int_a^b \frac{\sqrt{x'(t)^2 + y'(t)^2}}{y(t)}\, dt \\
&\geqslant \int_a^b \frac{\sqrt{x'(t)^2 + y'(t)^2}}{2y(a)}\, dt = \frac{1}{2y(a)} \ell_{\text{euc}}(\gamma) \\
&\geqslant \frac{1}{2y(a)} d_{\text{euc}}(P, Q).
\end{aligned}
$$

Otherwise, γ crosses the horizontal line L of equation $y = 2y(a)$. Let t_0 be the first value of t for which this happens; namely, $y(t_0) = 2y(a)$ and $y(t) < 2y(a)$ for every $t < t_0$. Let γ' denote the part of γ corresponding to the values of t with $a \leqslant t \leqslant t_0$. This curve γ' joins P to the point $\big(x(t_0), y(t_0)\big) \in L$, so that its euclidean length $\ell_{\text{euc}}(\gamma')$ is greater than or equal to the euclidean distance from P to the line L, which itself is equal to $y(a)$. Therefore,

$$
\begin{aligned}
\ell_{\text{hyp}}(\gamma) \geqslant \ell_{\text{hyp}}(\gamma') &= \int_a^{t_0} \frac{\sqrt{x'(t)^2 + y'(t)^2}}{y(t)}\, dt \\
&\geqslant \int_a^{t_0} \frac{\sqrt{x'(t)^2 + y'(t)^2}}{2y(a)}\, dt = \frac{1}{2y(a)} \ell_{\text{euc}}(\gamma') \\
&\geqslant \frac{1}{2y(a)} y(a) = \tfrac{1}{2}.
\end{aligned}
$$

In both cases, we found that $\ell_{\text{hyp}}(\gamma) \geqslant C$ for a positive constant

$$
C = \min \left\{ \tfrac{1}{2y(a)} d_{\text{euc}}(P, Q), \tfrac{1}{2} \right\} > 0,
$$

which depends only on P and Q (remember that $y(a)$ is the y-coordinate of P). If follows that $d_{\text{hyp}}(P, Q) \geqslant C > 0$ cannot be 0 if $P \neq Q$.   □

2.2. Some isometries of the hyperbolic plane

The hyperbolic plane $(\mathbb{H}^2, d_{\text{hyp}})$ has many symmetries. Actually, we will see that it is as symmetric as the euclidean plane.

2.2.1. Homotheties and horizontal translations. Some of these isometries are surprising at first. These include the **homotheties** defined by $\varphi(x, y) = (\lambda x, \lambda x)$ for some $\lambda > 0$. Indeed, if the piecewise differentiable curve γ is parametrized by

$$t \mapsto \big(x(t), y(t)\big), \quad a \leqslant t \leqslant b,$$

its image $\varphi(\gamma)$ under φ is parametrized by

$$t \mapsto \big(\lambda x(t), \lambda y(t)\big), \quad a \leqslant t \leqslant b.$$

Therefore,

$$
\begin{aligned}
\ell_{\mathrm{hyp}}\big(\varphi(\gamma)\big) &= \int_a^b \frac{\sqrt{\lambda^2 x'(t)^2 + \lambda^2 y'(t)^2}}{\lambda y(t)} \, dt \\
&= \int_a^b \frac{\sqrt{x'(t)^2 + y'(t)^2}}{y(t)} \, dt \\
&= \ell_{\mathrm{hyp}}(\gamma).
\end{aligned}
$$

Since φ establishes a one-to-one correspondence between curves joining P to Q and curves joining $\varphi(P)$ to $\varphi(Q)$, it follows from the definition of the hyperbolic metric that $d_{\mathrm{hyp}}\big(\varphi(P), \varphi(Q)\big) = d_{\mathrm{hyp}}(P, Q)$ for every $P, Q \in \mathbb{H}^2$. This proves that the homothety φ is indeed an isometry of $(\mathbb{H}^2, d_{\mathrm{hyp}})$.

The **horizontal translations** defined by $\varphi(x, y) = (x + x_0, y)$ for some $x_0 \in \mathbb{R}$ are more obvious isometries of $(\mathbb{H}^2, d_{\mathrm{hyp}})$, as is the **reflection** $\varphi(x, y) = (-x, y)$ across the y-axis.

2.2.2. The homogeneity property of the hyperbolic plane. The isometries obtained by composing homotheties and horizontal translations are enough to prove that the hyperbolic plane is homogeneous. Recall that the **composition** of two maps $\varphi \colon X \to X'$ and $\psi \colon X' \to X''$ is the map $\psi \circ \varphi \colon X \to X''$ defined by $\psi \circ \varphi(P) = \psi\big(\varphi(P)\big)$ for every $P \in X$. If, in addition, X, X' and X'' are metric spaces, if φ is an isometry from (X, d) to (X', d') and if ψ is an isometry from (X', d') to (X'', d''), then $\psi \circ \varphi$ is an isometry from (X, d) to (X'', d'') since

$$
\begin{aligned}
d''\big(\psi \circ \varphi(P), \psi \circ \varphi(Q)\big) &= d''\big(\psi(\varphi(P)), \psi(\varphi(Q))\big) \\
&= d'\big(\varphi(P), \varphi(Q)\big) \\
&= d(P, Q).
\end{aligned}
$$

Proposition 2.2. *The hyperbolic plane* $(\mathbb{H}^2, d_{\text{hyp}})$ *is homogeneous. Namely, for every* P, $Q \in \mathbb{H}^2$, *there exists an isometry* φ *of* $(\mathbb{H}^2, d_{\text{hyp}})$ *such that* $\varphi(P) = Q$.

Proof. If $P = (a, b)$ and $Q = (c, d) \in \mathbb{H}^2$, with b, $d > 0$, the homothety φ of ratio $\lambda = \frac{d}{b}$ sends P to the point $R = (\frac{ad}{b}, d)$ with the same y-coordinate d as Q. Then the horizontal translation $\psi(x, y) = (x + c - \frac{ad}{b}, y)$ sends R to Q. The composition $\psi \circ \varphi$ now provides an isometry sending P to Q. $\qquad\qquad\qquad\square$

2.2.3. The standard inversion. We now consider an even less obvious isometry of $(\mathbb{H}^2, d_{\text{hyp}})$. The **standard inversion**, or **inversion across the unit circle**, or **inversion** for short, is defined by

$$\varphi(x, y) = \left(\frac{x}{x^2 + y^2}, \frac{y}{x^2 + y^2} \right).$$

This map is better understood in polar coordinates, as it sends the point with polar coordinates $[r, \theta]$ to the point with polar coordinates $[\frac{1}{r}, \theta]$. See Figure 2.2.

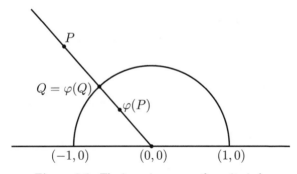

Figure 2.2. The inversion across the unit circle

Lemma 2.3. *The inversion across the unit circle is an isometry of the hyperbolic plane* $(\mathbb{H}^2, d_{\text{hyp}})$.

Proof. If γ is a piecewise differentiable curve parametrized by

$$t \mapsto (x(t), y(t)), \quad a \leqslant t \leqslant b,$$

its image $\varphi(\gamma)$ under the inversion φ is parametrized by

$$t \mapsto \big(x_1(t), y_1(t)\big), \quad a \leqslant t \leqslant b,$$

with

$$x_1(t) = \frac{x(t)}{x(t)^2 + y(t)^2} \quad \text{and} \quad y_1(t) = \frac{y(t)}{x(t)^2 + y(t)^2} .$$

Then

$$x_1'(t) = \frac{\big(y(t)^2 - x(t)^2\big)x'(t) - 2x(t)y(t)y'(t)}{\big(x(t)^2 + y(t)^2\big)^2}$$

and

$$y_1'(t) = \frac{\big(x(t)^2 - y(t)^2\big)y'(t) - 2x(t)y(t)x'(t)}{\big(x(t)^2 + y(t)^2\big)^2}$$

so that after simplifications,

$$x_1'(t)^2 + y_1'(t)^2 = \frac{x'(t)^2 + y'(t)^2}{\big(x(t)^2 + y(t)^2\big)^2} .$$

It follows that

$$
\begin{aligned}
\ell_{\text{hyp}}\big(\varphi(\gamma)\big) &= \int_a^b \frac{\sqrt{x_1'(t)^2 + y_1'(t)^2}}{y_1(t)} dt \\
&= \int_a^b \frac{\sqrt{x'(t)^2 + y'(t)^2}}{y(t)} dt \\
&= \ell_{\text{hyp}}(\gamma).
\end{aligned}
$$

As before, this shows that the inversion φ is an isometry of the hyperbolic plane. □

2.3. Shortest curves in the hyperbolic plane

In euclidean geometry, the shortest curve joining two points is the line segment with these two points as endpoints. We want to identify the shortest curve between two points in the hyperbolic plane.

We begin with a special case.

Lemma 2.4. *If $P_0 = (x_0, y_0)$, $P_1 = (x_0, y_1) \in \mathbb{H}^2$ are located on the same vertical line, then the line segment $[P_0, P_1]$ has the shortest hyperbolic length among all piecewise differentiable curves going from*

P_0 to P_1. *In addition, the hyperbolic length of any other curve joining P_0 to P_1 has strictly larger hyperbolic length, and*

$$d_{\mathrm{hyp}}(P_0, P_1) = \ell_{\mathrm{hyp}}([P_0, P_1]) = \ln \left| \frac{y_1}{y_0} \right|.$$

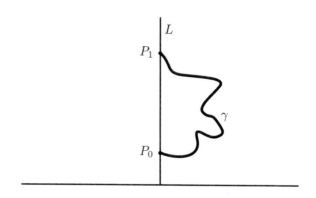

Figure 2.3. Vertical lines are shortest

Proof. Assuming $y_0 \leqslant y_1$ without loss of generality, let us first compute the hyperbolic length of $[P_0, P_1]$. Parametrize this line segment by

$$t \mapsto (x_0, t), \quad y_0 \leqslant t \leqslant y_1.$$

Then,

$$\ell_{\mathrm{hyp}}([P_0, P_1]) = \int_{y_0}^{y_1} \frac{\sqrt{0^2 + 1^2}}{t}\, dt = \ln \frac{y_1}{y_0}.$$

Now, consider a piecewise differentiable curve γ going from P_0 to P_1, which is parametrized by

$$t \mapsto \big(x(t), y(t)\big), \quad a \leqslant t \leqslant b.$$

Its hyperbolic length is

$$\ell_{\mathrm{hyp}}(\gamma) = \int_a^b \frac{\sqrt{x'(t)^2 + y'(t)^2}}{y(t)}\, dt \geqslant \int_a^b \frac{|y'(t)|}{y(t)}\, dt$$

$$\geqslant \int_a^b \frac{y'(t)}{y(t)}\, dt = \ln \frac{y(b)}{y(a)} = \ln \frac{y_1}{y_0} = \ell_{\mathrm{hyp}}([P_0, P_1]).$$

In addition, for the first term to be equal to the last one, the above two inequalities must be equalities. Equality in the first inequality requires that the function $x(t)$ be constant, while equality in the second one implies that $y(t)$ is weakly increasing. This shows that the curve γ is equal to the line segment $[P_0, P_1]$ if $\ell_{\mathrm{hyp}}(\gamma) = \ell_{\mathrm{hyp}}([P_0, P_1])$. $\square$

For future reference, we note the following estimate, which is proved by the same argument as the second half of the proof of Lemma 2.4.

Lemma 2.5. *For any two points $P_0 = (x_0, y_0)$, $P_1 = (x_1, y_1) \in \mathbb{H}^2$,*

$$d_{\mathrm{hyp}}(P_0, P_1) \geqslant \left| \ln \frac{y_1}{y_0} \right|. \qquad \square$$

In our determination of shortest curves in the hyperbolic plane, the next step is the following.

Lemma 2.6. *For any $P, Q \in \mathbb{H}^2$ that are not on the same vertical line, there exists an isometry of the hyperbolic plane $(\mathbb{H}^2, d_{\mathrm{hyp}})$ such that $\varphi(P)$ and $\varphi(Q)$ are on the same vertical line. In addition, the line segment $[\varphi(P), \varphi(Q)]$ is the image under φ of the unique circle arc joining P to Q and centered on the x-axis.*

Proof. Since P and Q are not on the same vertical line, the perpendicular bisector line of P and Q intersects the x-axis at some point R. The point R is equidistant from P and Q for the euclidean metric, so that there is a circle C centered at R and passing through P and Q. Note that C is the only circle passing through P and Q that is centered on the x-axis.

The circle C intersects the x-axis in two points. Let φ_1 be a horizontal translation sending one of these points to $(0,0)$. Then $C' = \varphi_1(C)$ is a circle passing through the origin and centered at some point $(a, 0)$.

In particular, the equation of the circle C' in polar coordinates is $r = 2a \cos\theta$. Its image under the inversion φ_2 is the curve with polar coordinate equation $r = \dfrac{1}{2a \cos\theta}$, namely, the vertical line L whose equation in cartesian coordinates is $x = \dfrac{1}{2a}$.

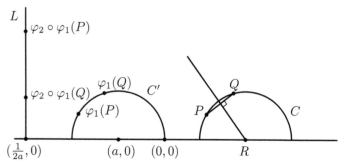

Figure 2.4. Circle arcs centered on the x-axis are shortest.

The composition $\varphi_2 \circ \varphi_1$ sends the circle C to the vertical line L. In particular, it sends the points P and Q to two points on the vertical line L. Restricting $\varphi_2 \circ \varphi_1$ to points in $\mathbb{H}^2$ then provides the isometry φ of $(\mathbb{H}^2, d_{\text{hyp}})$ that we were looking for.   □

Lemma 2.6 can be extended immediately to the case where P and Q sit on the same vertical line L by interpreting L as a circle of infinite radius whose center is located at infinity on the x-axis. Indeed, the vertical line L of equation $x = a$ can be seen as the limit as x tends to $+\infty$ or to $-\infty$ of the circle of radius $|x - a|$ centered at the point $(x, 0)$. With this convention, any two P, $Q \in \mathbb{H}^2$ can be joined by a unique circle arc centered on the x-axis.

Theorem 2.7. *Among all curves joining P to Q in $\mathbb{H}^2$, the circle arc centered on the x-axis (possibly a vertical line segment) is the unique one that is shortest for the hyperbolic length ℓ_{hyp}.*

Proof. If P and Q are on the same vertical line, this is proved by Lemma 2.4.

Otherwise, Lemma 2.6 provides an isometry φ sending P and Q to two points P' and Q' on the same vertical line L. By Lemma 2.4, the shortest curve from P' to Q' is the line segment $[P', Q']$. Since an isometry sends shortest curves to shortest curves, the shortest curve from P to Q is the image of the line segment $[P', Q']$ under the inverse isometry φ^{-1}. By the second statement of Lemma 2.6, this image is the circle arc joining P to Q and centered on the x-axis.   □

In a metric space where the distance function is defined by taking the infimum of the arc lengths of certain curves, such as the euclidean plane and the hyperbolic plane, there is a technical term for "shortest curve". More precisely, a **geodesic** is a curve γ such that for every $P \in \gamma$ and for every $Q \in \gamma$ sufficiently close to P, the section of γ joining P to Q is the shortest curve joining P to Q (for the arc length considered).

For instance, Proposition 1.1 says that geodesics in the euclidean plane $(\mathbb{R}^2, d_{\text{euc}})$ are line segments, whereas Theorem 2.7 shows that geodesics in the hyperbolic plane $(\mathbb{H}^2, d_{\text{hyp}})$ are circle arcs centered on the x-axis. By convention, line segments and circle arcs may include some, all, or none of their endpoints (in much the same way as an interval in the number line $\mathbb{R}$ may be open, closed or semi-open).

A **complete geodesic** is a geodesic which cannot be extended to a larger geodesic. From the above observations, complete geodesics of the euclidean plane are straight lines. Complete geodesics of the hyperbolic plane are open semi-circles centered on the x-axis and delimited by two points of the x-axis (including vertical half-lines going from a point on the x-axis to infinity).

For future reference, we now prove the following technical result.

Lemma 2.8. *Let $P_0 = (0, y_0)$ and $P_1 = (0, y_1)$ be two points of the upper half $L = \{(0, y); y > 0\} \subset \mathbb{H}^2$ of the y-axis, with $y_1 > y_0$, and let g be a complete hyperbolic geodesic passing through P_0. See Figure 2.5. Then the following are equivalent:*

(1) *P_0 is the point of g that is closest to P_1 for the hyperbolic distance d_{hyp};*

(2) *g is the complete geodesic g_0 that is orthogonal to L at P_0, namely, it is the euclidean semi-circle of radius y_0 centered at $(0, 0)$ and joining $(y_0, 0)$ to $(-y_0, 0)$.*

Proof. Lemmas 2.4 and 2.5 show that for every point $P = (u, v)$ on the geodesic g_0,

$$d_{\text{hyp}}(P_1, P) \geqslant \ln \frac{y_1}{v} \geqslant \ln \frac{y_1}{y_0} = d_{\text{hyp}}(P_1, P_0).$$

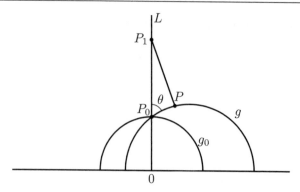

Figure 2.5

As a consequence, the point P_0 is closest to P_1 among all points of g_0.

Conversely, if g is another complete hyperbolic geodesic that passes through P_0 and makes an angle of $\theta \neq \frac{\pi}{2}$ with L at P_0, we want to find a point $P \in g$ with $d_{\mathrm{hyp}}(P_1, P) < d_{\mathrm{hyp}}(P_1, P_0)$.

For $P = (u, v) \in g$, the standard parametrization of the line segment $[P_1, P]$ gives that its hyperbolic length is equal to

$$\ell_{\mathrm{hyp}}([P_1, P]) = \int_0^1 \frac{\sqrt{u^2 + (v - y_1)^2}}{y_1 + t(v - y_1)} \, dt$$
$$= \frac{\sqrt{u^2 + (v - y_1)^2}}{y_1 - v} \ln \frac{y_1}{v}.$$

We now let the point $P = (u, v)$ vary on the geodesic g near P_0. When $u = 0$, we have that $v = y_0$ and $\dfrac{dv}{du} = \cot \theta$. Differentiating the above formula then gives that still at $u = 0$,

$$\frac{d}{du} \ell_{\mathrm{hyp}}([P_1, P]) = -\frac{1}{y_0} \cot \theta.$$

In particular, unless $\theta = \frac{\pi}{2}$, this derivative is different from 0 and there exists near $P_0 = (0, y_0)$ a point $P = (u, v)$ of g such that

$$d_{\mathrm{hyp}}(P_1, P) \leqslant \ell_{\mathrm{hyp}}([P_1, P]) < \ell_{\mathrm{hyp}}([P_1, P_0]) = d_{\mathrm{hyp}}([P_1, P_0]). \qquad \square$$

2.4. All isometries of the hyperbolic plane

So far, we have encountered three types of isometries of the hyperbolic plane $(\mathbb{H}^2, d_{\mathrm{hyp}})$: homotheties, horizontal translations and the inversion. In this section, we describe all isometries of $(\mathbb{H}^2, d_{\mathrm{hyp}})$.

It is convenient to use complex numbers. In this framework,

$$\mathbb{H}^2 = \{z \in \mathbb{C}; \mathrm{Im}(z) > 0\},$$

where the imaginary part $\mathrm{Im}(z)$ is the y-coordinate of $z = x + iy$.

In complex coordinates, a homothety is of the form $z \mapsto \lambda z$ for a real number $\lambda > 0$, a horizontal translation is of the form $z \mapsto z + x_0$ with $x_0 \in \mathbb{R}$, and the inversion is of the form $z \mapsto \frac{z}{|z|^2} = \frac{1}{\bar{z}}$, where $\bar{z} = x - iy$ is the complex conjugate of $z = x + iy$ and where $|z| = \sqrt{x^2 + y^2} = \sqrt{z\bar{z}}$ is its **modulus** (also called **absolute value**).

We can obtain more examples of isometries by composition of isometries of these types. Recall that the composition $\psi \circ \varphi$ of two maps φ and ψ is defined by $\psi \circ \varphi(P) = \psi(\varphi(P))$, and that the composition of two isometries is an isometry.

Lemma 2.9. *All maps of the form*

(2.3) $$z \mapsto \frac{az + b}{cz + d} \quad \text{with } a, b, c, d \in \mathbb{R} \text{ and } ad - bc = 1$$

or

(2.4) $$z \mapsto \frac{c\bar{z} + d}{a\bar{z} + b} \quad \text{with } a, b, c, d \in \mathbb{R} \text{ and } ad - bc = 1$$

are isometries of the hyperbolic plane $(\mathbb{H}^2, d_{\mathrm{hyp}})$.

Proof. We will show that every such map is a composition of horizontal translations $z \mapsto z + x_0$ with $x_0 \in \mathbb{R}$, of homotheties $z \mapsto \lambda z$ with $\lambda > 0$, and of inversions $z \mapsto \frac{1}{\bar{z}}$. Since a composition of isometries is an isometry, this will prove the result.

When $a \neq 0$, the map of equation (2.4) is the composition of

$$z \mapsto z + \frac{b}{a}, \quad z \mapsto \frac{1}{\bar{z}}, \quad z \mapsto \frac{1}{a^2}z \quad \text{and } z \mapsto z + \frac{c}{a}.$$

In particular, this map is the composition of several isometries of $\mathbb{H}^2$ and is therefore an isometry of $\mathbb{H}^2$.

Composing once more with $z \mapsto \frac{1}{z}$, we obtain the map of equation (2.3), thereby showing that this map is also an isometry of the hyperbolic plane when $a \neq 0$.

When $a = 0$, so that $c \neq 0$, the map of equation (2.3) is the composition of

$$z \mapsto \frac{cz + b + d}{cz + d},$$

which is an isometry of $\mathbb{H}^2$ by the previous case, and of the horizontal translation $z \mapsto z - 1$. It follows that the map of equation (2.3) is also an isometry of $\mathbb{H}^2$ when $a = 0$.

Finally, composing with $z \mapsto \frac{1}{\bar{z}}$ shows that the map of equation (2.4) is an isometry of $\mathbb{H}^2$ when $a = 0$.   □

Conversely, we will show that every isometry of the hyperbolic plane is of one the two types considered in Lemma 2.9. The proof of this fact hinges on the following property.

Lemma 2.10. *Let φ be an isometry of the hyperbolic plane $(\mathbb{H}^2, d_{\mathrm{hyp}})$ such that $\varphi(iy) = iy$ for every $y > 0$. Then either $\varphi(z) = z$ for every $z \in \mathbb{H}^2$ or $\varphi(z) = -\bar{z}$ for every z.*

Proof. Let $L = \{iy; y > 0\}$ be the upper half of the y-axis. By hypothesis, φ fixes every point of L.

For every $iy \in L$, let g_y be the unique hyperbolic complete geodesic that passes through iy and is orthogonal to L. Namely, g_y is the euclidean semi-circle of radius y centered at 0 and contained in $\mathbb{H}^2$. Since φ is an isometry and $\varphi(iy) = iy$, we know that it sends g_y to a complete geodesic g passing through iy. We will use Lemma 2.8 to prove that $g = g_y$.

Indeed, this statement characterizes the geodesic g_y by the property that for any $y_1 > y$, the point iy is the point of g_y that is closest to iy_1. As a consequence, since φ is an isometry, $\varphi(iy) = iy$ is the point of $\varphi(g_y) = g$ that is closest to $\varphi(iy_1) = iy_1$. Lemma 2.8 then shows that $g = g_y$, so that $\varphi(g_y) = g_y$.

Now, if $P = u + iv$ is a point of g_y, its image $\varphi(g_y)$ is one of the two points of g_y that are at distance $d_{\mathrm{hyp}}(P, iy)$ from iy. One of these two points is P, the other one is $-u + iv$ by symmetry.

We conclude that $\varphi(u+iv) = u+iv$ or $-u+iv$ for every $u+iv \in \mathbb{H}^2$ (since $u + iv$ belongs to some geodesic g_y). Since φ is an isometry, it is continous. It follows that either $\varphi(u + iv) = u + iv$ for every $u + iv \in \mathbb{H}^2$ or $\varphi(u+iv) = -u+iv$ for every $u+iv \in \mathbb{H}^2$. This can be rephrased as either $\varphi(z) = z$ for every $z \in \mathbb{H}^2$ or $\varphi(z) = -\bar{z}$ for every $z \in \mathbb{H}^2$. $\qquad\qquad\qquad\qquad\qquad\qquad\qquad\qquad\qquad\qquad\qquad\qquad\square$

A minor corollary of Lemma 2.9 is that $\varphi(z) = \dfrac{az + b}{cz + d}$ with a, b, c, $d \in \mathbb{R}$ and $ad - bc = 1$ sends the upper half-space $\mathbb{H}^2$ to itself; this can also be easily checked "by hand". This map is not defined at the boundary point $z = -\frac{d}{c}$. However, if we introduce a point ∞ at infinity of the real line $\mathbb{R}$ (without distinguishing between $+\infty$ and $-\infty$), the same formula defines a map

$$\varphi \colon \mathbb{R} \cup \{\infty\} \to \mathbb{R} \cup \{\infty\}$$

by setting $\varphi(-\frac{d}{c}) = \infty$ and $\varphi(\infty) = \frac{a}{c}$. This map is specially designed to be continuous. Indeed,

$$\lim_{x \to -\frac{d}{c}} \varphi(x) = \infty \quad \text{and} \quad \lim_{x \to \infty} \varphi(x) = \frac{a}{c}$$

in the "obvious" sense, which is made precise in Section T.3 of the TOOL KIT.

The same applies to a map of the form $\varphi(z) = \dfrac{c\bar{z} + d}{a\bar{z} + b}$ with a, b, c, $d \in \mathbb{R}$ and $ad - bc = 1$. These extensions are often convenient, as in the proof of the following statement.

Theorem 2.11. *The isometries of the hyperbolic plane* $(\mathbb{H}^2, d_{\mathrm{hyp}})$ *are exactly the maps of the form*

$$\varphi(z) = \frac{az + b}{cz + d} \ \textit{with } a, b, c, d \in \mathbb{R} \textit{ and } ad - bc = 1$$

or

$$\varphi(z) = \frac{c\bar{z} + d}{a\bar{z} + b} \ \textit{with } a, b, c, d \in \mathbb{R} \textit{ and } ad - bc = 1.$$

Proof. We already proved in Lemma 2.9 that all maps of these two types are isometries of the hyperbolic plane.

Conversely, let φ be an isometry of $\mathbb{H}^2$, and consider again the positive part $L = \{iy; y > 0\}$ of the y-axis. Since L is a complete geodesic of $\mathbb{H}^2$, its image under the isometry φ is also a complete geodesic of $\mathbb{H}^2$, namely, a euclidean semi-circle bounded by two distinct points u, $v \in \mathbb{R} \cup \{\infty\}$. Here u or v will be ∞ exactly when $\varphi(L)$ is a vertical half-line. In addition, if we orient L from 0 to ∞, we require without loss of generality that the corresponding orientation of $\varphi(L)$ goes from u to v.

First, consider the case where u and v are both different from ∞. The hyperbolic isometry

$$\psi(z) = \frac{az - au}{cz - cv},$$

with a and $c \in \mathbb{R}$ chosen so that $ac(u - v) = 1$, sends u to 0 and v to ∞. It follows that the composition $\psi \circ \varphi$ fixes the two points 0 and ∞. As a consequence, the isometry $\psi \circ \varphi$ sends the complete geodesic L to itself, and respects its orientation. In particular, $\psi \circ \varphi(\mathrm{i}) = \mathrm{i}t$ for some $t > 0$. Replacing a by $a/\sqrt{t}$ and c by $c\sqrt{t}$ in the definition of ψ, we can arrange that $\psi \circ \varphi(\mathrm{i}) = \mathrm{i}$. Then, $\psi \circ \varphi$ sends each $\mathrm{i}y \in L$ to a point of L that is at the same hyperbolic distance from i as $\mathrm{i}y$; since $\psi \circ \varphi$ respects the orientation of L, the only possibility is that $\psi \circ \varphi(\mathrm{i}y) = \mathrm{i}y$ for every $y > 0$.

Applying Lemma 2.10, we conclude that either $\psi \circ \varphi(z) = z$ for every z or $\psi \circ \varphi(z) = -\bar{z}$ for every z. In the first case,

$$\varphi(z) = \psi^{-1}(z) = \frac{-cvz + au}{-cz + a},$$

where the formula for the inverse function ψ^{-1} is obtained by solving the equation $\psi(z') = z$ (compare Exercise 2.10). In the second case,

$$\varphi(z) = \psi^{-1}(-\bar{z}) = \frac{cv\bar{z} + au}{c\bar{z} + a}.$$

In both cases, $\varphi(z)$ is of the type requested.

It remains to consider the cases where u or v is ∞. The argument is identical, using the isometries

$$\psi(z) = \frac{-a}{cz - cv}$$

with $ac = 1$ when $u = \infty$, and

$$\psi(z) = \frac{az - au}{c}$$

with $ac = 1$ when $v = \infty$. $\qquad\qquad\qquad\qquad\qquad\qquad\square$

The isometries of $(\mathbb{H}^2, d_{\text{hyp}})$ of the form $\varphi(z) = \dfrac{az + b}{cz + d}$ with a, b, c, $d \in \mathbb{R}$ and $ad - bc = 1$ are called **linear fractional maps** with real coefficients. Those of the form $\varphi(z) = \dfrac{c\bar{z} + d}{a\bar{z} + b}$ with a, b, c, $d \in \mathbb{R}$ and $ad - bc = 1$ are **antilinear fractional maps**.

2.5. Linear and antilinear fractional maps

In this section we establish a few fundamental properties of linear and antilinear fractional maps. Since later we will need to consider maps of this type with arbitrary complex (and not just real) coefficients, we prove these properties in this higher level of generality.

In this context, a **linear fractional map** is a nonconstant map φ of the form $\varphi(z) = \dfrac{az + b}{cz + d}$ with complex coefficients a, b, c, $d \in \mathbb{C}$. Elementary algebra shows that φ is nonconstant exactly when $ad - bc \neq 0$. Dividing all coefficients by one of the two complex square roots $\pm\sqrt{ad - bc}$, we can consequently arrange that $ad - bc = 1$ without changing the map φ. We will systematically require that the coefficients a, b, c, $d \in \mathbb{C}$ satisfy this condition $ad - bc = 1$.

So far, the map φ is not defined at $z = \frac{d}{c}$. However, this can easily be fixed by introducing a point ∞ at infinity of $\mathbb{C}$. Let the **Riemann sphere** be the union $\widehat{\mathbb{C}} = \mathbb{C} \cup \{\infty\}$. Then a linear fractional $\varphi(z) = \dfrac{az + b}{cz + d}$ with a, b, c, $d \in \mathbb{C}$ and $ad - bc = 1$ defines a map $\varphi \colon \widehat{\mathbb{C}} \to \widehat{\mathbb{C}}$ by setting $\varphi(-\frac{d}{c}) = \infty$ and $\varphi(\infty) = \frac{a}{c}$. This map is continuous for the obvious definition of continuity at infinity because

$$\lim_{z \to -\frac{d}{c}} \varphi(z) = \infty \quad \text{and} \quad \lim_{z \to \infty} \varphi(z) = \frac{a}{c},$$

where these limits involving infinity are defined exactly as in Section T.3 of the TOOL KIT, but replacing absolute values of real numbers by moduli (= absolute values) of complex numbers.

Similarly, a general **antilinear fractional map** is a map $\varphi \colon \widehat{\mathbb{C}} \to \widehat{\mathbb{C}}$ of the form $\varphi(z) = \dfrac{c\bar{z} + d}{a\bar{z} + b}$ with a, b, c, $d \in \mathbb{C}$ and $ad - bc = 1$, with the convention that $\varphi(-\frac{b}{\bar{a}}) = \infty$ and $\varphi(\infty) = \frac{c}{a}$.

See Exercise 2.8 for an explanation of why the Riemann sphere $\widehat{\mathbb{C}} = \mathbb{C} \cup \{\infty\}$ can indeed be considered as a sphere. See also Exercise 2.12 for another interpretation of $\widehat{\mathbb{C}}$, which sheds a different light on linear fractional maps.

2.5.1. Some special (anti)linear fractional maps.

We already encountered the **homotheties**

$$z \mapsto \lambda z = \frac{\lambda^{\frac{1}{2}} z + 0}{0z + \lambda^{-\frac{1}{2}}}$$

with positive real ratio $\lambda > 0$. If we allow complex coefficients, we can also consider the **rotations**

$$z \mapsto e^{i\theta} z = \frac{e^{i\frac{\theta}{2}} z + 0}{0z + e^{-i\frac{\theta}{2}}}$$

of angle $\theta \in \mathbb{R}$ around the origin, and the **translations**

$$z \mapsto z + z_0 = \frac{z + z_0}{0z + 1}$$

for arbitrary complex numbers $z_0 \in \mathbb{C}$.

We also considered the **inversion across the unit circle**

$$z \mapsto \frac{z}{|z|^2} = \frac{1}{\bar{z}} = \frac{0\bar{z} + 1}{\bar{z} + 0}.$$

Lemma 2.12. *Every linear or antilinear fractional map* $\varphi \colon \widehat{\mathbb{C}} \to \widehat{\mathbb{C}}$ *is a composition of homotheties, translations, rotations, and inversions across the unit circle.*

Proof. The proof is identical to the purely algebraic argument that we already used in the proof of Lemma 2.9 for linear and linear fractional maps with real coefficients. $\square$

Actually, there is no reason to prefer the unit circle to any other circle. If C is the circle of radius R centered at the point $z_0 \in \mathbb{C}$, the **inversion** across the circle C is the antilinear fractional map φ defined by the property that

$$\varphi(z) - z_0 = R^2 \frac{z - z_0}{|z - z_0|^2}$$

or, equivalently, that

$$\varphi(z) = \frac{\frac{z_0}{R}\bar{z} + \frac{R^2-|z_0|^2}{R}}{\frac{1}{R}\bar{z} - \frac{\bar{z}_0}{R}}.$$

Namely, φ sends z to the point that is on the same ray issued from z_0 as z, and it is at the euclidean distance $R^2/d_{\text{euc}}(z, z_0)$ from z_0. This inversion fixes every point of the circle C, and exchanges the inside and the outside of C.

There is an interesting limit case of inversions as we let the center and the radius of the circle go to infinity. For given t, t_0 and $\theta_0 \in \mathbb{R}$, set $z_0 = te^{i\theta_0}$ and $R = t - t_0$. If we let t tend to $+\infty$, the circle C converges to the line L that passes through the point $t_0e^{i\theta_0}$ and makes an angle of $\theta_0 + \frac{\pi}{2}$ with the x-axis. On the other hand, the inversion φ across C converges to the map $z \mapsto -e^{2i\theta_0}\bar{z} + 2t_0e^{i\theta_0}$, which is just the reflection across the line L.

In this way, if we interpret the line L as a circle of infinite radius centered at infinity, we can also consider the euclidean reflection across L as an inversion across this circle. Note that every line L can be obtained in this way.

2.5.2. Differentials. Recall that if $\varphi \colon U \to \mathbb{R}^2$ is a differentiable function defined on a region $U \subset \mathbb{R}^2$ by $\varphi(x, y) = \big(f(x, y), g(x, y)\big)$, the *differential* or *tangent map* of φ at a point $P_0 = (x_0, y_0)$ in the interior of U is the linear map $D_{P_0}\varphi \colon \mathbb{R}^2 \to \mathbb{R}^2$ with matrix

$$\begin{pmatrix} \frac{\partial f}{\partial x}(P_0) & \frac{\partial f}{\partial y}(P_0) \\ \frac{\partial g}{\partial x}(P_0) & \frac{\partial g}{\partial y}(P_0) \end{pmatrix}.$$

Namely,

$$D_{P_0}\varphi(\vec{v}) = D_{P_0}\varphi(a, b) = \left(a\frac{\partial f}{\partial x}(P_0) + b\frac{\partial f}{\partial y}(P_0), a\frac{\partial g}{\partial x}(P_0) + b\frac{\partial g}{\partial y}(P_0)\right)$$

for every vector $\vec{v} = (a, b) \in \mathbb{R}^2$.

The differential map $D_{P_0}\varphi$ also has the following geometric interpretation.

Lemma 2.13. *Let the differentiable map $\varphi \colon U \to \mathbb{R}^2$ be defined over a region U containing the point P_0 in its interior. Then, for every parametrized curve γ in U which passes through P_0 and is tangent to*

*the vector $\vec{v}$ there, its image under φ is tangent to the vector $D_{P_0}\varphi(\vec{v})$
at the point $\varphi(P_0)$.*

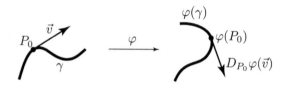

Figure 2.6. The geometry of the differential map

Proof. Suppose that the map φ is given by $\varphi(x,y) = (f(x,y), g(x,y))$,
and that the curve γ is parametrized by $t \mapsto (x(t), y(t))$. If the
point P_0 corresponds to $t = t_0$, namely, if $P_0 = (x(t_0), y(t_0))$, then
$\vec{v} = (x'(t_0), y'(t_0))$.

The image of the curve γ under φ is parametrized by

$$t \mapsto \varphi(x(t), y(t)) = \Big(f(x(t), y(t)), g(x(t), y(t))\Big).$$

Applying the chain rule for functions of several variables, its tangent
vector at $\varphi(P_0)$ is equal to

$$\frac{d}{dt}\Big(f(x(t), y(t)), g(x(t), y(t))\Big)_{t=t_0}$$

$$= \Big(\frac{d}{dt}f(x(t), y(t))_{t=t_0}, \frac{d}{dt}g(x(t), y(t))_{t=t_0}\Big)$$

$$= \Big(\frac{\partial f}{\partial x}(P_0)\,x'(t_0) + \frac{\partial f}{\partial y}(P_0)\,y'(t_0), \frac{\partial g}{\partial x}(P_0)\,x'(t_0) + \frac{\partial g}{\partial y}(P_0)\,y'(t_0)\Big)$$

$$= D_{P_0}\varphi(\vec{v}).$$

$\square$

An immediate consequence of this geometric interpretation is the
following property.

Corollary 2.14.

$$D_{P_0}(\psi \circ \varphi) = (D_{\varphi(P_0)}\psi) \circ (D_{P_0}\varphi).$$

$\square$

The differential maps of linear and antilinear fractional maps have
a particularly nice expression in complex coordinates.

Proposition 2.15. *If the linear fractional map φ is defined by $\varphi(z) = \dfrac{az+b}{cz+d}$ where a, b, c, $d \in \mathbb{C}$ with $ad - bc = 1$, its differential map $D_{z_0}\varphi \colon \mathbb{C} \to \mathbb{C}$ at $z_0 \in \mathbb{C}$ with $z_0 \neq -\frac{d}{c}$ is such that*

$$D_{z_0}\varphi(v) = \frac{1}{(cz_0 + d)^2}\, v$$

for every $v \in \mathbb{C}$.

If the antilinear fractional map ψ is defined by $\psi(z) = \dfrac{c\bar{z}+d}{a\bar{z}+b}$ where a, b, c, $d \in \mathbb{C}$ with $ad - bc = 1$, its differential map $D_{z_0}\psi \colon \mathbb{C} \to \mathbb{C}$ at $z_0 \in \mathbb{C}$ with $z_0 \neq -\frac{b}{a}$ is such that

$$D_{z_0}\psi(v) = \frac{1}{(a\bar{z}_0 + b)^2}\, \bar{v}$$

for every $v \in \mathbb{C}$.

Proof. We will use Lemma 2.13. Given $v \in \mathbb{C}$ interpreted as a vector, consider the line segment γ parametrized by $t \mapsto z(t) = z_0 + tv$. Note that $z(0) = z_0$ and that $z'(0) = v$.

Lemma 2.13 then implies that

$$
\begin{aligned}
D_{z_0}\varphi(v) &= \frac{d}{dt}\varphi\bigl(z(t)\bigr)\Big|_{t=0} = \lim_{h\to 0} \frac{1}{h}\Bigl(\varphi\bigl(z(h)\bigr) - \varphi\bigl(z(0)\bigr)\Bigr) \\
&= \lim_{h\to 0} \frac{1}{h}\left(\frac{az_0 + ahv + b}{cz_0 + chv + d} - \frac{az_0 + b}{cz_0 + d}\right) \\
&= \lim_{h\to 0} \frac{v}{(cz_0 + chv + d)(cz_0 + d)} \\
&= \frac{1}{(cz_0 + d)^2}\, v,
\end{aligned}
$$

using the property that $ad - bc = 1$.

The argument is identical for the antilinear fractional map ψ. $\quad\square$

For future reference, we note that the same computation yields:

Complement 2.16. *If $\varphi(z) = \dfrac{az+b}{cz+d}$ where $ad - bc$ is not necessarily equal to 1, then*

$$D_{z_0}\varphi(v) = \frac{ad - bc}{(cz_0 + d)^2}\, v. \qquad\qquad \square$$

A consequence of Proposition 2.15 is that the differential map of a linear fractional map is the composition of a homothety with a rotation, and the differential map of an antilinear fractional map is the composition of a homothety with a reflection. This has the following important consequence.

Corollary 2.17. *The differential map $D_{z_0}\varphi$ of a linear fractional map φ respects angles and orientation in the sense that for any two nonzero vectors $\vec{v}_1$, $\vec{v}_2 \in \mathbb{C}$, the oriented angle from $D_{z_0}\varphi(\vec{v}_1)$ to $D_{z_0}\varphi(\vec{v}_2)$ is the same as the oriented angle from $\vec{v}_1$ to $\vec{v}_2$, measuring oriented angles counterclockwise in $\mathbb{C}$.*

The differential map $D_{z_0}\psi$ of an antilinear fractional linear map ψ respects angles and reverses orientation in the sense that for any two nonzero vectors $\vec{v}_1$, $\vec{v}_2 \in \mathbb{C}$, the oriented angle from $D_{z_0}\psi(\vec{v}_1)$ to $D_{z_0}\psi(\vec{v}_2)$ is the opposite of the oriented angle from $\vec{v}_1$ to $\vec{v}_2$. $\square$

Incidentally, Corollary 2.17 shows that a linear fractional map cannot coincide with an antilinear fractional map.

2.5.3. (Anti)linear fractional maps and circles. Another fundamental property of linear and antilinear fractional maps is that they send circles to circles. For this, we have to include all lines as circles of infinite radius centered at infinity. More precisely, let a **circle** in the Riemann sphere $\widehat{\mathbb{C}} = \mathbb{C} \cup \{\infty\}$ be either a euclidean circle in $\mathbb{C}$ or the union $L \cup \{\infty\}$ of a line $L \subset \mathbb{C}$ and the point ∞.

Proposition 2.18. *A linear or antilinear fractional map $\varphi\colon \widehat{\mathbb{C}} \to \widehat{\mathbb{C}}$ sends each circle of $\widehat{\mathbb{C}}$ to a circle of $\widehat{\mathbb{C}}$.*

Proof. By Lemma 2.12, φ is a composition of homotheties, rotations, translations and inversions across the unit circle. Since homotheties, rotations and translations clearly send circles to circles, it suffices to consider the case where φ is the inversion across the unit circle.

It is convenient to use polar coordinates. In polar coordinates r and θ, the circle C of radius R centered at $z_0 = r_0 e^{i\theta_0}$ has equation

$$r^2 - 2r\,r_0 \cos(\theta - \theta_0) + r_0^2 - R^2 = 0.$$

The inversion φ sends the point with polar coordinates $[r, \theta]$ to the point of coordinates $[\frac{1}{r}, \theta]$. The image of the circle C under φ is therefore the curve of equation

$$\frac{1}{r^2} - \frac{2r_0}{r}\cos(\theta - \theta_0) + r_0^2 - R^2 = 0.$$

If $|r_0| \neq R$ or, equivalently, if the circle C does not contain the origin 0, simplifying the above equation shows that this curve is the circle of radius $\dfrac{R}{|r_0^2 - R^2|}$ centered at $\dfrac{z_0}{r_0^2 - R^2}$.

If $|z_0| = R$, we get the curve of polar equation $r = \dfrac{1}{2r_0 \cos(\theta - \theta_0)}$, which of course is a line.

Finally, we need to consider the case where C is a line. In polar coordinates, its equation is of the form $r = \dfrac{1}{2r_0 \cos(\theta - \theta_0)}$ for some r_0 and θ_0. Then its image under φ has equation $r = 2r_0 \cos(\theta - \theta_0)$, and consequently it is a circle passing through the origin. $\qquad \square$

2.6. The hyperbolic norm

If $\vec{v} = (a, b)$ is a vector in $\mathbb{R}^2$, its **euclidean magnitude** or **euclidean norm** is its usual length

$$\|\vec{v}\|_{\mathrm{euc}} = \sqrt{a^2 + b^2}.$$

For instance, if $\vec{v}$ is the velocity of a particle moving in the euclidean plane, $\|\vec{v}\|_{\mathrm{euc}}$ describes the speed of this particle.

In the hyperbolic plane, distances are measured differently according to where we are in the plane, and consequently so are speeds. If $\vec{v}$ is a vector based at the point $z \in \mathbb{H}^2 \subset \mathbb{C}$, its **hyperbolic norm** is

$$\|\vec{v}\|_{\mathrm{hyp}} = \frac{1}{\mathrm{Im}(z)} \|\vec{v}\|_{\mathrm{euc}}.$$

To justify this definition, let γ be a curve in $\mathbb{H}^2$, parametrized by $t \mapsto z(t)$, $a \leqslant t \leqslant b$. In particular, the tangent vector of γ at the point $z(t)$ is the derivative $z'(t)$, and must be considered as a vector based at $z(t)$. Then the euclidean and hyperbolic lengths of γ are given by the very similar formulas

$$\ell_{\mathrm{euc}}(\gamma) = \int_a^b \|z'(t)\|_{\mathrm{euc}} \, dt$$

and

$$\ell_{\mathrm{hyp}}(\gamma) = \int_a^b \|z'(t)\|_{\mathrm{hyp}} \, dt.$$

If φ is a differentiable map and $\vec{v}$ is a vector based at P, its image $D_P\varphi(\vec{v})$ under the differential map is a vector based at $\varphi(P)$. Indeed, see the geometric interpretation of the differential $D_P\varphi$ given by Lemma 2.13.

Lemma 2.19. *If φ is an isometry of $(\mathbb{H}^2, d_{\mathrm{hyp}})$, then $\|D_{z_0}\varphi(\vec{v})\|_{\mathrm{hyp}} = \|\vec{v}\|_{\mathrm{hyp}}$ for every vector $\vec{v}$ based at $z_0 \in \mathbb{H}^2$.*

Proof. We could go back to basic principles about the metric d_{hyp}, but it is easier to use a straight computation.

Consider the case where φ is a linear fractional $\varphi(z) = \dfrac{az+b}{cz+d}$ with a, b, c, $d \in \mathbb{R}$ and $ad - bc = 1$. By Proposition 2.15, if $\vec{v}$ is a vector based at $z_0 \in \mathbb{H}^2$,

$$\|D_{z_0}\varphi(\vec{v})\|_{\mathrm{euc}} = \frac{1}{|cz_0+d|^2} \|\vec{v}\|_{\mathrm{euc}}.$$

On the other hand,

$$\mathrm{Im}(\varphi(z_0)) = \frac{1}{2\mathrm{i}} \left(\varphi(z_0) - \overline{\varphi(z_0)} \right) = \frac{1}{2\mathrm{i}} \left(\frac{az_0+b}{cz_0+d} - \frac{a\bar{z}_0+b}{c\bar{z}_0+d} \right)$$

$$= \frac{1}{2\mathrm{i}} \frac{z_0 - \bar{z}_0}{|cz_0+d|^2} = \frac{1}{|cz_0+d|^2} \mathrm{Im}(z_0).$$

Therefore,

$$\|D_{z_0}\varphi(\vec{v})\|_{\mathrm{hyp}} = \frac{1}{\mathrm{Im}(\varphi(z_0))} \|D_{z_0}\varphi(\vec{v})\|_{\mathrm{euc}}$$

$$= \frac{1}{\mathrm{Im}(z_0)} \|\vec{v}\|_{\mathrm{euc}} = \|\vec{v}\|_{\mathrm{hyp}}.$$

The argument is essentially identical for an antilinear fractional map $\varphi(z) = \dfrac{c\bar{z}+d}{a\bar{z}+b}$. □

2.6.1. The isotropy property of the hyperbolic plane. We now show that like the euclidean plane, the hyperbolic plane $\mathbb{H}^2$ is isotropic. Recall that this means that not only can we send any point $z_1 \in \mathbb{H}^2$ to any other point $z_2 \in \mathbb{H}^2$ by an isometry φ of $(\mathbb{H}^2, d_{\mathrm{hyp}})$, but we can even arrange that φ sends any given direction at z_1 to any arbitrary direction at z_2. As a consequence, the hyperbolic plane looks the same at every point and in every possible direction.

Proposition 2.20. *Let $\vec{v}_1$ be a vector based at $z_1 \in \mathbb{H}^2$, and let $\vec{v}_2$ be a vector based at $z_2 \in \mathbb{H}^2$ with $\|\vec{v}_1\|_{\mathrm{hyp}} = \|\vec{v}_2\|_{\mathrm{hyp}}$. Then there is an isometry φ of $(\mathbb{H}^2, d_{\mathrm{hyp}})$ which sends z_1 to z_2 and whose differential map $D_{z_1}\varphi$ sends $\vec{v}_1$ to $\vec{v}_2$.*

Proof. Let $\theta \in \mathbb{R}$ be the angle from $\vec{v}_1$ to $\vec{v}_2$ measured in the usual euclidean way, namely, after moving $\vec{v}_1$ to the point z_2 by a euclidean translation of $\mathbb{R}^2$.

There exists $c, d \in \mathbb{R}$ such that $cz_1 + d = e^{-i\frac{\theta}{2}}$. Indeed, finding c and d amounts to solving a linear system of two equations. If $z_1 = x_1 + iy_1$, one finds $c = \frac{1}{y_1}\sin\frac{\theta}{2}$ and $d = \cos\frac{\theta}{2} - \frac{x_1}{y_1}\sin\frac{\theta}{2}$, but the precise value is really irrelevant.

Then one can find (many) $a, b \in \mathbb{R}$ such that $ad - bc = 1$. This is again a simple linear equation problem after observing that c and d cannot be both equal to 0.

Let φ_1 be the linear fractional defined by $\varphi_1(z) = \frac{az+b}{cz+d}$. Because of our choice of a, b, c, d, Proposition 2.15 shows that $D_{z_1}\varphi_1$ is the complex multiplication by $e^{i\theta}$, namely, the rotation of angle θ. As a consequence, still comparing angles and directions in the usual euclidean way, $\vec{v}_3 = D_{z_1}\varphi_1(\vec{v}_1)$ is parallel to $\vec{v}_2$ and points in the same direction.

Let $z_3 = \varphi_1(z_1)$. Let φ_2 be an isometry of $\mathbb{H}^2$ sending z_3 to z_2. As in our proof of the homogeneity of $\mathbb{H}^2$ in Proposition 2.2, we can even arrange that φ_2 is the composition of a homothety with a horizontal translation, so that $D_{z_3}\varphi_2$ is a homothety. In particular, $D_{z_3}\varphi_2$ sends each vector to one which is parallel to it.

Then $\varphi = \varphi_2 \circ \varphi_1$ sends z_1 to z_2, and its differential $D_{z_1}\varphi = D_{z_3}\varphi_2 \circ D_{z_1}\varphi_1$ sends $\vec{v}_1$ to a vector $\vec{v}_2' = D_{z_1}\varphi(\vec{v}_1) = D_{z_3}\varphi_2(\vec{v}_3)$, which is based at z_2 and is parallel to $\vec{v}_2$.

By Lemma 2.19,

$$\|\vec{v}_2'\|_{\mathrm{hyp}} = \|D_{z_1}\varphi(\vec{v}_1)\|_{\mathrm{hyp}} = \|\vec{v}_1\|_{\mathrm{hyp}} = \|\vec{v}_2\|_{\mathrm{hyp}}.$$

The vectors $\vec{v}_2$ and $\vec{v}_2'$ are based at the same point z_2, they are parallel, they point in the same direction, and they have the same hyperbolic norm. Consequently, they must be equal.

We therefore have found an isometry φ of $(\mathbb{H}^2, d_{\text{hyp}})$ such that $\varphi(z_1) = z_2$ and $D_{z_1}\varphi(\vec{v}_1) = \vec{v}_2$. $\square$

2.7. The disk model for the hyperbolic plane

We now describe a new model for the hyperbolic plane, namely, another metric space (X, d) which is isometric to $(\mathbb{H}^2, d_{\text{hyp}})$. This model is sometimes more convenient for performing computations. Another side benefit, mathematically less important but not negligible, is that it often leads to prettier pictures, as we will have the opportunity to observe in later chapters.

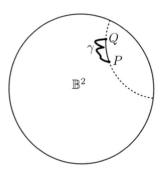

Figure 2.7. The disk model for the hyperbolic plane

Let $\mathbb{B}^2$ be the open disk of radius 1 centered at 0 in the complex plane $\mathbb{C}$, namely, in the sense of metric spaces introduced in Section 1.3, the ball $B_{d_{\text{euc}}}((0,0), 1)$ in the euclidean plane $(\mathbb{R}^2, d_{\text{euc}})$.

For a vector $\vec{v}$ based at the point $z \in \mathbb{B}^2$, define its $\mathbb{B}^2$-norm as

$$\|\vec{v}\|_{\mathbb{B}^2} = \frac{2}{1 - |z|^2} \|\vec{v}\|_{\text{euc}},$$

where $\|\vec{v}\|_{\text{euc}}$ is the euclidean norm of V. Then, as for the euclidean and hyperbolic plane, define the $\mathbb{B}^2$-length of a piecewise differentiable curve γ in $\mathbb{B}^2$ parametrized by $t \mapsto z(t)$, $a \leqslant t \leqslant b$, as $\ell_{\mathbb{B}^2}(\gamma) =$

$\int_a^b \|z'(t)\|_{\mathbb{B}^2} \, dt$. Finally, given two points $P, Q \in \mathbb{B}^2$, define their $\mathbb{B}^2$-distance $d_{\mathbb{B}^2}(P, Q)$ as the infimum of the lengths $\ell_{\mathbb{B}^2}$ as γ ranges over all piecewise differentiable curves going from P to Q.

Let Φ be the fractional linear map defined by $\Phi(z) = \dfrac{-z + i}{z + i}$. Beware that the coefficients of Φ do not satisfy the usual relation $ad - bc = 1$. This could be achieved by dividing all the coefficients by one of the complex square roots $\pm\sqrt{-2i}$, but the resulting expression would be clumsy and cumbersome.

Proposition 2.21. *The linear fractional map* $\Phi(z) = \dfrac{-z + i}{z + i}$ *induces an isometry from* $(\mathbb{H}^2, d_{\mathrm{hyp}})$ *to* $(\mathbb{B}^2, d_{\mathbb{B}^2})$.

Proof. Note that $|\Phi(z)| = 1$ when $z \in \mathbb{R}$, so that Φ sends $\mathbb{R} \cup \{\infty\}$ to the unit circle. As a consequence, Φ sends the upper half-plane $\mathbb{H}^2$ to either the inside or the outside of the unit circle in $\mathbb{C} \cup \{\infty\}$. Since $\Phi(i) = 0$, we conclude that $\Phi(\mathbb{H}^2)$ is equal to the inside $\mathbb{B}^2$ of the unit circle.

Consider the differential $D_z\Phi \colon \mathbb{C} \to \mathbb{C}$ of Φ at $z \in \mathbb{H}^2$. By Proposition 2.15 and Complement 2.16,

$$
\begin{aligned}
\|D_z\Phi(v)\|_{\mathbb{B}^2} &= \frac{2}{1 - |\Phi(z)|^2} \|D_z\Phi(v)\|_{\mathrm{euc}} \\
&= \frac{2}{1 - \left|\frac{-z+i}{z+i}\right|^2} \left| -\frac{2i}{(z+i)^2} v \right| \\
&= \frac{4}{|z+i|^2 - |-z+i|^2} |v| \\
&= \frac{4}{(z+i)(\bar{z}-i) - (-z+i)(-\bar{z}-i)} |v| \\
&= \frac{2}{i(\bar{z}-z)} |v| = \frac{1}{\mathrm{Im}(z)} |v| = \|v\|_{\mathrm{hyp}}.
\end{aligned}
$$

From this computation, we conclude that Φ sends a curve γ in $\mathbb{H}^2$ to a curve $\Phi(\gamma)$ in $\mathbb{B}^2$ such that $\ell_{\mathbb{B}^2}(\Phi(\gamma)) = \ell_{\mathrm{hyp}}(\gamma)$.

Taking the infimum of the lengths of such curves, it follows that $d_{\mathbb{B}^2}(\Phi(P), \Phi(Q)) = d_{\mathrm{hyp}}(P, Q)$ for every $P, Q \in \mathbb{H}^2$. In other words, Φ defines an isometry from $(\mathbb{H}^2, d_{\mathrm{hyp}})$ to $(\mathbb{B}^2, d_{\mathbb{B}^2})$. $\square$

In particular, this proves that $d_{\mathbb{B}^2}$ is a metric and not just a semi-metric (namely that $d_{\mathbb{B}^2}(P, Q) = 0$ only when $P = Q$), which is a property that we had implicitly assumed so far.

Proposition 2.22. *The geodesics of $(\mathbb{B}^2, d_{\mathbb{B}^2})$ are the arcs contained in euclidean circles that are orthogonal to the circle $\mathbb{S}^1$ bounding $\mathbb{B}^2$, including straight lines passing through the origin.*

Proof. Since Φ is an isometry from $(\mathbb{H}^2, d_{\text{hyp}})$ to $(\mathbb{B}^2, d_{\mathbb{B}^2})$, the geodesics of $(\mathbb{B}^2, d_{\mathbb{B}^2})$ are just the images under Φ of the geodesics of $(\mathbb{H}^2, d_{\text{hyp}})$.

Because linear fractionals send circles to circles (Proposition 2.18) and respect angles (Corollary 2.17), the result follows immediately from the fact that geodesics of $(\mathbb{H}^2, d_{\text{hyp}})$ are exactly circle arcs in euclidean circles centered on the x-axis or, equivalently, orthogonal to this x-axis. $\qquad\square$

Proposition 2.23. *The isometries of $(\mathbb{B}^2, d_{\mathbb{B}^2})$ are exactly the restrictions to $\mathbb{B}^2$ of all linear and antilinear fractional maps of the form*

$$\varphi(z) = \frac{\alpha z + \beta}{\bar{\beta} z + \bar{\alpha}} \quad or \quad \varphi(z) = \frac{\alpha \bar{z} + \beta}{\bar{\beta} \bar{z} + \bar{\alpha}}$$

with $|\alpha|^2 - |\beta|^2 = 1$.

Proof. Since Φ is an isometry from $(\mathbb{H}^2, d_{\text{hyp}})$ to $(\mathbb{B}^2, d_{\mathbb{B}^2})$, the isometries of $(\mathbb{B}^2, d_{\mathbb{B}^2})$ are exactly those maps of the form $\Phi \circ \psi \circ \Phi^{-1}$ where ψ is an isometry of $(\mathbb{H}^2, d_{\text{hyp}})$.

If ψ is a linear fractional map of the form $\psi(z) = \dfrac{az + b}{cz + d}$ with a, b, c, $d \in \mathbb{R}$ and $ad - bc = 1$, then

$$\Phi \circ \psi \circ \Phi^{-1}(z) = \frac{(ai - b + c + di)z + (-ai - b - c + di)}{(-ai + b + c + di)z + (ai + b - c + di)}$$

is of the form indicated for

$$\alpha = \frac{1}{2}(a + bi - ci + d)$$

$$\text{and} \quad \beta = \frac{1}{2}(-a + bi + ci + d).$$

Conversely, writing $\alpha + \beta = b\mathrm{i} + d$ and $\alpha - \beta = a - c\mathrm{i}$ with a, b, c, $d \in \mathbb{R}$, any map $z \mapsto \dfrac{\alpha z + \beta}{\bar{\beta} z + \bar{\alpha} z}$ with $|\alpha|^2 - |\beta|^2 = 1$ is of the form $\Phi \circ \psi \circ \Phi^{-1}$ for some a, b, c, $d \in \mathbb{R}$ with $ad - bc = 1$.

The argument is identical for antilinear fractional maps.   □

Exercises for Chapter 2

Exercise 2.1. Rigorously prove that a horizontal translation $\varphi \colon \mathbb{H}^2 \to \mathbb{H}^2$, defined by the propery that $\varphi(x, y) = (x + x_0, y)$ for a given $x_0 \in \mathbb{R}$, is an isometry of the hyperbolic plane $(\mathbb{H}^2, d_{\mathrm{hyp}})$.

Exercise 2.2 (An explicit formula for the hyperbolic distance). The goal of this exercise is to show that the hyperbolic distance $d_{\mathrm{hyp}}(z, z')$ from z to $z' \in \mathbb{H}^2 \subset \mathbb{C}$ is equal to

$$D(z, z') = \log \frac{|z - \bar{z}'| + |z - z'|}{|z - \bar{z}'| - |z - z'|}.$$

a. Show that $d_{\mathrm{hyp}}(z, z') = D(z, z')$ when z and z' are on the same vertical line.

b. Show that $D(\varphi(z), \varphi(z')) = D(z, z')$ for every z, $z' \in \mathbb{H}^2$ when $\varphi \colon \mathbb{H}^2 \to \mathbb{H}^2$ is a horizontal translation, a homothety or the inversion across the unit circle.

c. Use the proof of Lemma 2.6 to show that $d_{\mathrm{hyp}}(z, z') = D(z, z')$ for every z, $z' \in \mathbb{H}^2$.

Exercise 2.3. Adapt the proof of Theorem 2.11 to prove that every isometry of the euclidean plane $(\mathbb{R}^2, d_{\mathrm{euc}})$ is of the form $\varphi(z) = z_0 + z\mathrm{e}^{\mathrm{i}\theta}$ or $\varphi(z) = z_0 + \bar{z}\mathrm{e}^{\mathrm{i}2\theta}$ for some $z_0 \in \mathbb{C}$ and $\theta \in \mathbb{R}$.

Exercise 2.4 (Perpendicular bisector). The **perpendicular bisector** of the two distinct points P and $Q \in \mathbb{H}^2$ is the geodesic b_{PQ} defined as follows. Let M be the midpoint of the geodesic g joining P to Q. Then b_{PQ} is the complete geodesic that passes through M and is orthogonal to g.

a. Let ρ be the inversion across the euclidean circle that contains b_{PQ}. Show that ρ sends the geodesic g to itself and exchanges P and Q.

b. Show that $d_{\mathrm{hyp}}(P, R) = d_{\mathrm{hyp}}(Q, R)$ for every $R \in b_{PQ}$. Possible hint: Use part a.

c. Suppose that P and R are on opposite sides of g_{PQ}, in the sense that the geodesic k joining P to R meets b_{PQ} in a point S. Combine pieces of k and $\rho(k)$ to construct a piecewise differentiable curve k' which goes from Q to R, which has the same hyperbolic length as k, and which is not geodesic. Conclude that $d_{\mathrm{hyp}}(Q, R) < d_{\mathrm{hyp}}(P, R)$.

d. As a converse to part b, show that $d_{\text{hyp}}(P, R) \neq d_{\text{hyp}}(Q, R)$ whenever $R \notin b_{PQ}$. Possible hint: Use part c.

Exercise 2.5 (Orthogonal projection). Let g be a complete geodesic of $\mathbb{H}^2$, and consider a point $P \in \mathbb{H}^2$.

a. First consider the case where g is a vertical half-line $g = \{(x_0, y) \in \mathbb{R}^2; y > 0\}$. Show that there exists a unique complete geodesic h containing P and orthogonally cutting g at some point Q (namely, h and g meet in Q and form an angle of $\frac{\pi}{2}$ there).

b. In the case of a general complete geodesic g, show that there exists a unique complete geodesic h containing P and orthogonally cutting g at some point Q. Possible hint: Use Lemma 2.6.

c. Show that Q is the point of g that is closest to P, in the sense that $d_{\text{hyp}}(P, Q') > d_{\text{hyp}}(P, Q)$ for every $Q' \in g$ different from Q. Possible hint: First consider the case where the geodesic h is equal to a vertical half-line and where the point P lies above Q on this half-line, then apply Lemma 2.6.

Exercise 2.6 (Hyperbolic rotation around i). For $\theta \in \mathbb{R}$, consider the fractional linear map defined by

$$\varphi(z) = \frac{z \cos \frac{\theta}{2} + \sin \frac{\theta}{2}}{-z \sin \frac{\theta}{2} + \cos \frac{\theta}{2}}.$$

a. Show that φ fixes the point $i \in \mathbb{H}^2$, and that its differential $D_i\varphi$ at i is just the rotation of angle θ. Hint: Use Proposition 2.15 to compute $D_i\varphi$.

b. For an arbitrary $z_0 \in \mathbb{H}^2$, give a similar formula for the hyperbolic rotation of angle θ around the point z_0, namely, for the isometry φ: $\mathbb{H}^2 \to \mathbb{H}^2$ for which $\varphi(z_0) = z_0$ and $D_{z_0}\varphi$ is the rotation of angle θ.

Exercise 2.7 (Classification of hyperbolic isometries). Consider an isometry φ of the hyperbolic plane $(\mathbb{H}^2, d_{\text{hyp}})$, defined by the linear fractional map $\varphi(z) = \frac{az+b}{cz+d}$ with a, b, c, $d \in \mathbb{R}$ and $ad - bc = 1$.

a. Show that if $(a + d)^2 > 4$, φ has no fixed point in $\mathbb{H}^2$ but fixes exactly two points of $\mathbb{R} \cup \{\infty\}$. Conclude that in this case there exists an isometry ψ of $(\mathbb{H}^2, d_{\text{hyp}})$ such that $\psi \circ \varphi \circ \psi^{-1}$ is a homothety $z \mapsto \lambda z$ with $\lambda > 0$. (Hint: Choose ψ so that it sends the fixed points to 0 and ∞). Find a relationship between λ and $(a + d)^2$. A hyperbolic isometry of this type is said to be **loxodromic**.

b. Show that if $(a + d)^2 < 4$, φ has a unique fixed point in $\mathbb{H}^2$. Conclude that in this case there is an isometry ψ of $(\mathbb{H}^2, d_{\text{hyp}})$ such that $\psi \circ \varphi \circ \psi^{-1}$ is the linear fractional map of Exercise 2.6 for some $\theta \in \mathbb{R}$. (Hint: Choose ψ so that it sends the fixed point to i). Find a relationship between θ and $(a + d)^2$. A hyperbolic isometry of this type is said to be **elliptic**.

c. Show that if $(a + d)^2 = 4$ and if φ is not the identity map defined by $\varphi(z) = z$, then φ has a unique fixed point in $\mathbb{R} \cup \{\infty\}$. Conclude that in this case there is an isometry ψ of $(\mathbb{H}^2, d_{\mathrm{hyp}})$ such that $\psi \circ \varphi \circ \psi^{-1}$ is the horizontal translation $z \mapsto z + 1$. (Hint: Choose ψ so that it sends the fixed point to ∞). A hyperbolic isometry of this type is said to be **parabolic**.

Exercise 2.8 (Stereographic projection). Let $\mathbb{S}^2$ be the unit sphere in the 3-dimensional euclidean space $\mathbb{R}^3$, consisting of those points $(x, y, z) \in \mathbb{R}^3$ such that $x^2 + y^2 + z^2 = 1$. Consider the map $\rho \colon \mathbb{S}^2 \to \mathbb{R}^2 \cup \{\infty\}$ defined as follows. If $(x, y, z) \neq (0, 0, 1)$, then

$$\rho(x, y, z) = \left(\frac{x}{1 - z}, \frac{y}{1 - z} \right) \in \mathbb{R}^2;$$

otherwise, $\rho(0, 0, 1) = \infty$.

a. Show that when $P = (x, y, z)$ is not the "North Pole" $N = (0, 0, 1)$, its image $\rho(P)$ is just the point where the line NP crosses the xy-plane in $\mathbb{R}^3$.

b. Show that $\rho \colon \mathbb{S}^2 \to \mathbb{R}^2 \cup \{\infty\}$ is continuous at every $P_0 \in \mathbb{S}^2$. When $P_0 = N$ so that $\varphi(P_0) = \infty$, this means that for every large $\eta > 0$ there exists a small $\delta > 0$ such that $d_{\mathrm{euc}}(\rho(P), O) > \eta$ for every $P \in \mathbb{S}^2$ with $d_{\mathrm{euc}}(P, P_0) < \delta$, where O is the origin in $\mathbb{R}^2$. (Compare the calculus definition of infinite limits, as reviewed in Section T.3 of the TOOL KIT.)

c. Show that the inverse function $\rho^{-1} \colon \mathbb{R}^2 \cup \{\infty\} \to \mathbb{S}^2$ is continuous at every $Q_0 \in \mathbb{R}^2 \cup \{\infty\}$. When $Q_0 = \infty$ so that $\rho^{-1}(Q_0) = N$, this means that for every small $\varepsilon > 0$, there exists a large $\eta > 0$ such that $d_{\mathrm{euc}}(\rho^{-1}(Q), N) < \varepsilon$ for every $Q \in \mathbb{R}^2$ with $d_{\mathrm{euc}}(Q, O) > \eta$.

In other words, ρ is a homeomorphism from $\mathbb{S}^2$ to $\mathbb{R}^2 \cup \{\infty\}$. (See Section 5.1 for a definition of homeomorphisms).

Exercise 2.9. Let z_0, z_1 and z_∞ be three distinct points in the Riemann sphere $\widehat{\mathbb{C}} = \mathbb{C} \cup \{\infty\}$. Show that there exist a unique linear fractional map φ and a unique antilinear fractional map ψ such that $\varphi(0) = \psi(0) = z_0$, $\varphi(1) = \psi(1) = z_1$ and $\varphi(\infty) = \psi(\infty) = z_\infty$.

Exercise 2.10.

a. Show that the linear fractional map $\varphi \colon \widehat{\mathbb{C}} \to \widehat{\mathbb{C}}$ defined by $\varphi(z) = \frac{az+b}{cz+d}$, with a, b, c, $d \in \mathbb{C}$ such that $ad - bc = 1$, is bijective and that its inverse φ^{-1} is the linear fractional map $\varphi^{-1}(z) = \frac{dz-b}{-cz+a}$. Hint: Remember that $\varphi^{-1}(z)$ is the number u such that $\varphi(u) = z$.

b. Give a similar formula for the inverse of the antilinear fractional map $\psi(z) = \frac{c\bar{z}+d}{a\bar{z}+b}$ with $ad - bc = 1$.

Exercise 2.11.

a. Show that any linear or antilinear fractional map can be written as the composition of finitely many inversions across circles.

b. Show that when a linear fractional map is written as the composition of finitely many inversions across circles, the number of inversions is even. (Hint: Corollary 2.17.) Similarly, show that when an antilinear fractional map is written as the composition of finitely many inversions across circles, the number of inversions is odd.

Exercise 2.12 (Linear fractional maps and projective lines). Let the **real projective line** $\mathbb{RP}^1$ consist of all 1-dimensional linear subspaces of the vector space $\mathbb{R}^2$. Namely, $\mathbb{RP}^1$ is the set of all lines L through the origin in $\mathbb{R}^2$. Since such a line L is determined by its slope $s \in \mathbb{R} \cup \{\infty\}$, this provides an identification $\mathbb{RP}^1 \cong \mathbb{R} \cup \{\infty\}$.

a. Let $\Phi_A : \mathbb{R}^2 \to \mathbb{R}^2$ be the linear map defined by the matrix $A = \begin{pmatrix} a & b \\ c & d \end{pmatrix}$ with determinant $ad - bc$ equal to 1. Similarly, consider the linear fractional map $\varphi_A : \mathbb{R} \cup \{\infty\} \to \mathbb{R} \cup \{\infty\}$ defined by $\varphi_A(s) = \frac{as + b}{cs + d}$. Show that Φ_A sends the line $L \in \mathbb{RP}^1$ with slope $s \in \mathbb{R} \cup \{\infty\}$ to the line $\Phi_A(L)$ with slope $\varphi_A(s)$.

b. Use part a to show that $\varphi_{AA'} = \varphi_A \circ \varphi_{A'}$, where AA' denotes the product of the matrices A and A'.

c. Similarly, consider $\mathbb{C}^2 = \mathbb{C} \times \mathbb{C}$ as a vector space over the field $\mathbb{C}$, and let the **complex projective line** $\mathbb{CP}^1$ consist of all 1–dimensional linear subspaces $L \subset \mathbb{C}^2$. Such a complex line L is determined by its **complex slope** $s \in \mathbb{C} \cup \{\infty\}$ defined by the following property. If L is not the line $\{0\} \times \mathbb{C}$, it intersects $\{1\} \times \mathbb{C}$ at the point $(1, s)$; if $L = \{0\} \times \mathbb{C}$, its complex slope is $s = \infty$. Let $\Phi_A : \mathbb{C}^2 \to \mathbb{C}^2$ be the complex linear map defined by the matrix $A = \begin{pmatrix} a & b \\ c & d \end{pmatrix}$ with complex entries $a, b, c, d \in \mathbb{C}$ and with determinant 1. Show that for the above identification $\mathbb{CP}^1 \cong \mathbb{C} \cup \{\infty\}$, the map $\mathbb{CP}^1 \to \mathbb{CP}^1$ induced by Φ_A corresponds to the linear fractional map $\varphi_A : \mathbb{C} \cup \{\infty\} \to \mathbb{C} \cup \{\infty\}$ defined by $\varphi_A(s) = \frac{as + b}{cs + d}$.

Exercise 2.13 (Hyperbolic disks). Recall from Section 1.3 that in a metric space (X, d) the ball of radius r centered at $P \in X$ is $B_d(P, r) = \{Q \in X; d(P, Q) < r\}$.

a. Let O be the center of the disk model $\mathbb{B}^2$ of Section 2.7. Show that the ball $B_{d_{\mathbb{B}^2}}(O, r)$ in $\mathbb{B}^2$ coincides with the euclidean open disk of radius $\tanh \frac{r}{2}$ centered at O.

b. Show that every hyperbolic ball $B_{d_{\text{hyp}}}(P, r)$ in the hyperbolic plane $\mathbb{H}^2$ is a euclidean open disk. Possible hint: Use part a, the isometry $\Phi \colon (\mathbb{H}^2, d_{\text{hyp}}) \to (\mathbb{B}^2, d_{\mathbb{B}^2})$ of Proposition 2.21, Proposition 2.2 and Proposition 2.18.

c. Show that the ball $B_{d_{\text{hyp}}}(P, r)$ centered at $P = (x, y) \in \mathbb{H}^2$ is the euclidean open disk with euclidean radius $2y \sinh r$ and with euclidean center $(x, 2y \cosh r)$. Possible hint: Look at the two points where the boundary of this disk meets the vertical line passing through P.

Exercise 2.14 (Hyperbolic area). If D is a region in $\mathbb{H}^2$, define its **hyperbolic area** as

$$\text{Area}_{\text{hyp}}(D) = \iint_D \frac{1}{y^2} \, dx \, dy.$$

a. Let $\rho \colon \mathbb{H}^2 \to \mathbb{H}^2$ be the standard inversion. Show that $\rho(D)$ has the same hyperbolic area as D. Possible hints: Polar coordinates may be convenient; alternatively, one can use the change of variables formula for double integrals (see part c).

b. Show that an isometry of $\mathbb{H}^2$ sends each region $D \subset \mathbb{H}^2$ to a region of the same hyperbolic area.

c. Let Φ be the isometry from $(\mathbb{H}^2, d_{\text{hyp}})$ to $(\mathbb{B}^2, d_{\mathbb{B}^2})$ provided by Proposition 2.21. Show that for every region D in $\mathbb{H}^2$

$$\text{Area}_{\text{hyp}}(D) = \iint_{\Phi(D)} \frac{1}{(1 - x^2 - y^2)^2} \, dx \, dy.$$

It may be convenient to use the change of variables formula for double integrals, which in this case says that

$$\iint_{\Phi(D)} f(u, v) \, du \, dv = \iint_D f\big(\Phi(x, y)\big) \, |\det D_\Phi| \, dx \, dy$$

for every function $f \colon \Phi(D) \to \mathbb{R}$, and to borrow computations from the proof of Proposition 2.21 to evaluate the determinant $\det D_\Phi$ of the differential map D_φ.

d. Show that a ball $B_{d_{\text{hyp}}}(P, r)$ of radius r in $\mathbb{H}^2$ has hyperbolic area

$$\text{Area}_{\text{hyp}}\big(B_{d_{\text{hyp}}}(P, r)\big) = 2\pi(\cosh r - 1) = 4\pi \sinh^2 \frac{r}{2}.$$

Hint: First consider the case where $P = (1, 0) = \Phi^{-1}(O)$, and use part c and the result of Exercise 2.13a.

Exercise 2.15 (Area of hyperbolic triangles). For every θ with $0 < \theta \leqslant \frac{\pi}{2}$, let T_θ be the (infinite) hyperbolic triangle with vertices i, $e^{i\theta} = \cos\theta + i\sin\theta$

and ∞. Namely, T_θ is the region of $\mathbb{H}^2$ bounded below by the euclidean circle of radius 1 centered at the origin, on the left by the y-axis, and on the right by the line $x = \cos\theta$.

   a. Show that the hyperbolic area $\mathrm{Area}_{\mathrm{hyp}}(T_\theta)$, as defined in Exercise 2.14, is finite.

   b. Show that $\dfrac{d}{d\theta}\,\mathrm{Area}_{\mathrm{hyp}}(T_\theta) = -1$. Conclude that $\mathrm{Area}_{\mathrm{hyp}}(T_\theta) = \dfrac{\pi}{2} - \theta$.

   c. Let T be a finite triangle in the hyperbolic plane, namely, the region of $\mathbb{H}^2$ bounded by the three geodesics joining any two of three distinct points P, Q, R. Let α, β, $\gamma \in [0, \pi]$ be the respective angles of T at its three vertices. Show that

$$\mathrm{Area}_{\mathrm{hyp}}(T_\theta) = \pi - \alpha - \beta - \gamma.$$

Hint: Express this area as a linear combination of the hyperbolic areas of six suitably chosen infinite hyperbolic triangles, each isometric to a triangle T_θ as in parts a and b.

Exercise 2.16 (Crossratio). The *crossratio* of four distinct points z_1, z_2, z_3, $z_4 \in \widehat{\mathbb{C}} = \mathbb{C} \cup \{\infty\}$ is

$$K(z_1, z_2, z_3, z_4) = \frac{(z_1 - z_3)(z_2 - z_4)}{(z_1 - z_4)(z_2 - z_3)} \in \mathbb{C},$$

using the straightforward extension by continuity of this formula when one of the z_i is equal to ∞. For instance, $K(z_1, z_2, z_3, \infty) = \frac{z_1 - z_3}{z_2 - z_3}$. Show that $K\big(\varphi(z_1), \varphi(z_2), \varphi(z_3), \varphi(z_4)\big)$ is equal to $K(z_1, z_2, z_3, z_4)$ for every linear fractional map $\varphi \colon \widehat{\mathbb{C}} \to \widehat{\mathbb{C}}$, and that it is equal to the complex conjugate of $K(z_1, z_2, z_3, z_4)$ for every antilinear fractional map $\varphi \colon \widehat{\mathbb{C}} \to \widehat{\mathbb{C}}$. Possible hint: First check the property for a few simple (anti)linear fractional maps, and then apply Lemma 2.12.

Exercise 2.17 (Another formula for the hyperbolic distance). Given two distinct points z_1, z_2 of the hyperbolic plane $\mathbb{H}^2 = \{z \in \mathbb{C}; \mathrm{Im}(z) > 0\}$, let x_1, $x_2 \in \widehat{\mathbb{R}} = \mathbb{R} \cup \{\infty\}$ be the two endpoints of the complete geodesic g passing through z_1 and z_2, in such a way that x_1, z_1, z_2 and x_2 occur in this order on g. Show that

$$d_{\mathrm{hyp}}(z_1, z_2) = \log\frac{(z_1 - x_2)(z_2 - x_1)}{(z_1 - x_1)(z_2 - x_2)}.$$

Possible hint: First consider the case where $x_1 = 0$ and $x_2 = \infty$, and then use the invariance property for the crossratio proved in Exercise 2.16.

Exercise 2.18. Show that a crossratio formula similar to that to Exercise 2.17 holds in the disk model $(\mathbb{B}^2, d_{\mathbb{B}^2})$ of Section 2.7. Hint: Use the invariance property for the crossratio proved in Exercise 2.16.

Exercise 2.19 (The projective model for the hyperbolic plane). Let $\mathbb{B}^2$ be the open unit disk of Section 2.7. Consider the map $\Psi\colon \mathbb{B}^2 \to \mathbb{B}^2$ defined by

$$\Psi(x, y) = \left(\frac{2x}{1 + x^2 + y^2}, \frac{2y}{1 + x^2 + y^2}\right).$$

  a. Show that Ψ is bijective.

  b. Show that if g is a complete geodesic of $(\mathbb{B}^2, d_{\mathbb{B}^2})$ which is a circle arc centered on the x-axis, its image $\Psi(g)$ is the euclidean line segment with the same endpoints.

  c. Let $\rho_\theta\colon \mathbb{B}^2 \to \mathbb{B}^2$ be the euclidean rotation of angle θ around the origin O. Show that $\Psi \circ \rho_\theta = \rho_\theta \circ \Psi$.

  d. Combine parts a and b to show that Ψ sends each complete geodesic g of $(\mathbb{B}^2, d_{\mathbb{B}^2})$ to the euclidean line segment with the same endpoints.

For a vector $\vec{v}$ based at $P \in \mathbb{B}^2$, define its **projective norm** $\|\vec{v}\|_{\mathrm{proj}} = \|D_P\Psi^{-1}(\vec{v})\|_{\mathbb{B}^2}$. For every piecewise differentiable curve γ in $\mathbb{B}^2$ parametrized by $t \mapsto \gamma(t)$, $a \leqslant t \leqslant b$, define its **projective length** as $\ell_{\mathrm{proj}}(\gamma) = \int_a^b \|\gamma'(t)\|_{\mathrm{proj}}dt$. Finally, consider the new metric d_{proj} on $\mathbb{B}^2$ by the property that $d_{\mathrm{proj}}(P, Q) = d_{\mathbb{B}^2}\left(\Psi^{-1}(P), \Psi^{-1}(Q)\right)$ for every P, $Q \in \mathbb{B}^2$. In particular, Ψ is now an isometry from $(\mathbb{B}^2, d_{\mathbb{B}^2})$ to $(\mathbb{B}^2, d_{\mathrm{proj}})$.

  e. Show that for every P, $Q \in \mathbb{B}^2$, the projective distance $d_{\mathrm{proj}}(P, Q)$ is equal to the infimum of the projective lengths of all piecewise differentiable curves going from P to Q in $\mathbb{B}^2$. Show that this infimum is equal to the projective length of the euclidean line segment from P to Q.

  f. Given a vector $\vec{v}$ based at $P \in \mathbb{B}^2$, draw the line L passing through P and parallel to $\vec{v}$, and let A and B be the two points where it meets the unit circle $\mathbb{S}^1$ bounding $\mathbb{B}^2$. Show that

$$\|\vec{v}\|_{\mathrm{proj}} = \frac{d_{\mathrm{euc}}(A, B)}{d_{\mathrm{euc}}(A, P)d_{\mathrm{euc}}(B, P)}\|\vec{v}\|_{\mathrm{euc}}.$$

The computations may be a little easier if one first restricts attention to the case where P is on the x-axis, and then use part c to deduce the general case from this one.

  g. For any two distinct P, $Q \in \mathbb{B}^2$, let A, $B \in \mathbb{S}^1$ be the points where the line PQ meets the circle $\mathbb{S}^1$ in such a way that A, P, Q, B occur in this order on the line. Combine parts e and f to show that

$$d_{\mathrm{proj}}(P, Q) = \frac{1}{2}\log\frac{d_{\mathrm{euc}}(A, Q)d_{\mathrm{euc}}(P, B)}{d_{\mathrm{euc}}(A, P)d_{\mathrm{euc}}(P, Q)}.$$

(Compare the crossratio formula of Exercise 2.17.)

The metric space $(\mathbb{B}^2, d_{\mathrm{proj}})$, which is isometric to the hyperbolic plane $(\mathbb{H}^3, d_{\mathrm{hyp}})$, is called the **projective model** or the **Cayley-Klein model** for the hyperbolic plane. It is closely related to the geometry of the 3-dimensional projective plane $\mathbb{RP}^2$, defined in close analogy with the projective line $\mathbb{RP}^1$ of Exercise 2.12 and consisting of all lines passing through the origin in $\mathbb{R}^3$. The fact that its geodesics are euclidean line segments makes this projective model quite attractive for some problems.

Chapter 3

The 2-dimensional sphere

The euclidean plane $(\mathbb{R}^2, d_{\text{euc}})$ and the hyperbolic plane $(\mathbb{H}^2, d_{\text{hyp}})$ have the fundamental property that they are both homogeneous and isotropic. There is another well-known 2-dimensional space which shares this property, namely, the 2-dimensional sphere in $\mathbb{R}^3$.

This is a relatively familiar space but, as will become apparent in later chapters, its geometry is not as fundamental as hyperbolic geometry or, to a lesser extent, euclidean geometry. For this reason, its discussion will be somewhat de-emphasized in this book. We only need a brief description of this space and of its main properties.

3.1. The 2-dimensional sphere

The **2-*dimensional sphere*** is the set

$$\mathbb{S}^2 = \{(x, y, z) \in \mathbb{R}^3; x^2 + y^2 + z^2 = 1\}$$

consisting of those points in the 3-dimensional space $\mathbb{R}^3$ which are at euclidean distance 1 from the origin. Namely, $\mathbb{S}^2$ is the euclidean sphere of radius 1 centered at the origin $O = (0, 0, 0)$.

Given two points P, $Q \in \mathbb{S}^2$, we can consider all piecewise differentiable curves γ that are completely contained in $\mathbb{S}^2$ and join P to

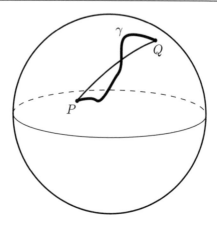

Figure 3.1. The 2-dimensional sphere

Q. The **spherical distance** from P to Q is defined as the infimum

$$d_{\mathrm{sph}}(P,Q) = \{\ell_{\mathrm{euc}}(\gamma); \gamma \text{ goes from } P \text{ to } Q \text{ in } \mathbb{S}^2\}$$

of their usual euclidean arc lengths $\ell_{\mathrm{euc}}(\gamma)$. Here, as in Chapter 1, the euclidean arc length of a piecewise differentiable curve γ parametrized by

$$t \mapsto \big(x(t), y(t), z(t)\big), \quad a \leqslant t \leqslant b,$$

is given by

$$\ell_{\mathrm{euc}}(\gamma) = \int_a^b \sqrt{x'(t)^2 + y'(t)^2 + z'(t)^2}\, dt.$$

The definition immediately shows that this spherical distance $d_{\mathrm{sph}}(P,Q)$ is greater than or equal to the usual euclidean distance $d_{\mathrm{euc}}(P,Q)$ from P to Q in $\mathbb{R}^3$. In particular, this proves that $d_{\mathrm{sph}}(P,Q) = 0$ only when $P = Q$. By the same arguments as in Lemma 2.1, d_{sph} also satisfies the Symmetry Condition and the Triangle Inequality. This proves that d_{sph} is really a metric.

3.2. Shortest curves

A **great circle** in the sphere $\mathbb{S}^2$ is the intersection of $\mathbb{S}^2$ with a plane passing through the origin. Equivalently, a great circle is a circle of radius 1 contained in $\mathbb{S}^2$. A **great circle arc** is an arc contained in a great circle.

Elementary geometric considerations show that any two P, $Q \in \mathbb{S}^2$ can be joined by a great circle arc of length $\leqslant \pi$. In addition, this circle arc is unique unless P and Q are **antipodal**, namely unless $Q = (-x, -y, -z)$ if $P = (x, y, z)$. When P and Q are antipodal, there are many great circle arcs of length π going from P to Q.

Theorem 3.1. *The geodesics of* $(\mathbb{S}^2, d_{\mathrm{sph}})$ *are exactly the great circle arcs. The shortest curves joining two points* P, $Q \in \mathbb{S}^2$ *are exactly the great circle arcs of length* $\leqslant \pi$ *going from* P *to* Q.

Proof. We sketch a proof of this result in Exercise 3.1. ☐

Note that we encounter here a new phenomenon, where a geodesic is not necessarily the shortest curve joining its endpoints; this happens for every great circle arc of length $> \pi$. Recall that in the definition of a geodesic γ the part of γ that joins P to Q is required to be the shortest curve joining P to Q *only when Q is sufficiently close to P.*

Another new phenomenon is that great circles provide **closed geodesics** of $\mathbb{S}^2$, namely, geodesics which are closed curves in the sense that they return to their initial point.

3.3. Isometries

In $\mathbb{R}^3$, a rotation φ around a line L respects euclidean distances and arc lengths. If, in addition, the rotation axis L passes through the origin O, then φ sends the sphere $\mathbb{S}^2$ to itself. Consequently, any rotation φ around a line passing through the origin induces an isometry of $(\mathbb{S}^2, d_{\mathrm{sph}})$.

These isometries are sufficient to show that the sphere is homogeneous and isotropic, as indicated by the following statement.

Proposition 3.2. *Given two points* P, $Q \in \mathbb{S}^2$, *a vector* $\vec{v}$ *tangent to* $\mathbb{S}^2$ *at* P, *and a vector* $\vec{w}$ *tangent to* $\mathbb{S}^2$ *at* Q *such that* $\|\vec{v}\|_{\mathrm{euc}} = \|\vec{w}\|_{\mathrm{euc}}$, *there exists a rotation* φ *around a line passing through the origin* O *such that* $\varphi(P) = Q$ *and* $D_P\varphi(\vec{v}) = \vec{w}$.

Proof. One easily finds a rotation φ_1 such that $\varphi_1(P) = Q$, for instance a rotation whose axis is orthogonal to the lines OP and OQ. Then, $D_P\varphi_1(\vec{v})$ is a vector tangent to $\mathbb{S}^2$ at $\varphi_1(P) = Q$ whose euclidean length is

$$\|D_P\varphi_1(\vec{v})\|_{\text{euc}} = \|\vec{v}\|_{\text{euc}} = \|\vec{w}\|_{\text{euc}}.$$

As a consequence, there exists a unique rotation φ_2 around the line OQ whose differential map $D_Q\varphi_2$ sends $D_P\varphi_1(\vec{v})$ to $\vec{w}$. Then the composition $\varphi = \varphi_2 \circ \varphi_1$ sends P to Q, and its differential $D_P\varphi = D_Q\varphi_2 \circ D_P\varphi_1$ sends $\vec{v}$ to $\vec{w}$.

By a classical property (see Exercise 3.2), the composition $\varphi = \varphi_2 \circ \varphi_1$ of two rotations is also a rotation around a line passing through the origin. $\square$

We already saw that a rotation φ around a line L passing through the origin provides an isometry of $\mathbb{S}^2$. If we compose φ with the reflection across the plane orthogonal to L at O, we obtain a **rotation-reflection**.

Note that rotations include the identity map, which is a rotation of angle 0. As a consequence, rotation-reflections also include reflections across a plane passing through the origin.

Theorem 3.3. *The isometries of $(\mathbb{S}^2, d_{\text{sph}})$ are exactly the above rotations and rotation-reflections.*

Proof. This can be proved by an argument which is very close to the proof of Theorem 2.11. See Exercise 3.4. $\square$

Exercises for Chapter 3

Exercise 3.1 (Geodesics of the sphere $\mathbb{S}^2$).

a. Let γ be a piecewise differentiable curve in $\mathbb{R}^3$ parametrized by $t \mapsto \gamma(t)$, $a \leqslant t \leqslant b$. For each t, let $\rho(t)$, $\theta(t)$ and $\varphi(t)$ be the spherical coordinates of $\gamma(t)$. Show that the euclidean length of γ is equal to

$$\ell_{\text{euc}}(\gamma) = \int_a^b \sqrt{\rho'(t)^2 + \rho(t)^2 \sin^2\varphi(t)\,\theta'(t)^2 + \rho(t)^2\varphi'(t)^2}\, dt.$$

Hint: Remember the formulas expressing rectangular coordinates in terms of spherical coordinates.

b. In the sphere $\mathbb{S}^2$, let P be the point $(0,0,1)$ and let Q be the point $(x,0,z)$ with $x \geqslant 0$. Let α be the vertical circle arc going from P to Q in $\mathbb{S}^2$, where the spherical coordinate θ is constantly equal to 0. Show that any curve γ going from P to Q has euclidean length greater than or equal to that of α.

c. Show that if P and Q are two points of the sphere $\mathbb{S}^2$, the shortest curves going from P to Q are the great circle arcs of length $\leqslant \pi$ going from P to Q. Possible hint: Use a suitable isometry of $(\mathbb{S}^2, d_{\mathrm{sph}})$ to reduce this to the case of part b.

d. Show that the geodesics of $\mathbb{S}^2$ are the great circle arcs.

Exercise 3.2. The main goal of this exercise is to show that in the euclidean space $\mathbb{R}^3$ the composition of two rotations whose axes pass through the origin $O = (0,0,0)$ is also a rotation around a line passing through the origin.

a. In $\mathbb{R}^3$, let L be a line contained in a plane Π. Let Π' be the plane obtained by rotating Π around L by an angle of $\frac{1}{2}\theta$, and let τ and τ' be the orthogonal reflections across the planes Π and Π', respectively. Show that the composition $\tau' \circ \tau$ is the rotation of angle θ around L, and that $\tau \circ \tau'$ is the rotation of angle $-\theta$ around L. Possible hint: You may find it convenient to consider the restrictions of $\tau' \circ \tau$ and $\tau \circ \tau'$ to each plane Π orthogonal to L.

b. Let L and L' be two lines passing through the origin $O = (0,0,0)$ in $\mathbb{R}^3$, and let ρ and ρ' be two rotations around L and L', respectively. Show that the composition $\rho \circ \rho'$ is a rotation around a line passing through the origin. Possible hint: Consider the plane Π containing L and L', and use the two properties of part a.

c. Show that in $\mathbb{R}^3$, the composition $\tau_1 \circ \tau_2 \circ \cdots \circ \tau_{2n}$ of an even number of orthogonal reflections τ_i across planes passing through O is a rotation (possibly the identity).

Exercise 3.3. The main goal of this exercise is to show that if τ is an orthogonal reflection across a plane Π passing through the origin $O = (0,0,0)$ and if ρ is a rotation of angle θ around a line L passing through O, then the composition $\tau \circ \rho$ is a rotation-reflection. This means that $\tau \circ \rho$ is also equal to a composition $\tau' \circ \rho'$, where ρ' is a rotation across a line L' passing through O and where τ' is the orthogonal reflection across the plane Π' *orthogonal* to L' at O.

Without loss of generality, we can assume that L is not orthogonal to Π, since otherwise we are done. Then, the plane Π_1 orthogonal to Π and containing L is uniquely determined. Let Π_2 be the image of Π_1 under the rotation of angle $-\frac{1}{2}\theta$ around L, and let Π_3 be the plane orthogonal to both Π and Π_1 at O. Let τ_1 and τ_2 be the orthogonal reflections across the planes Π_1 and Π_2, respectively.

a. Show that $\tau \circ \tau_1(P) = -P$ for every $P \in \Pi_3$ (where $-P$ denotes the point $(-x, -y, -z)$ when $P = (x, y, z)$). Drawing a picture might help.

b. Show that $\tau \circ \tau_1 \circ \tau_2(P) = -P$ for every P in the line $L' = \Pi_2 \cap \Pi_3$.

c. Let τ' be the orthogonal reflection across the plane Π' orthogonal to L' at O. Show that $\rho' = \tau' \circ \tau \circ \tau_1 \circ \tau_2$ is a rotation around the line L'. Hint: First use the result of Exercise 3.2c to show that ρ' is a rotation around some line.

d. Show that $\tau \circ \rho = \tau' \circ \rho' = \rho' \circ \tau'$, so that $\tau \circ \rho$ is a rotation-reflection. Hint: First use Exercise 3.2a to show that $\rho = \tau_1 \circ \tau_2$.

e. Show that in $\mathbb{R}^3$ the composition $\tau_1 \circ \tau_2 \circ \cdots \circ \tau_{2n+1}$ of an odd number of orthogonal reflections τ_i across planes passing through O is a rotation-reflection. Hint: Use Exercise 3.2c.

Exercise 3.4 (Isometries of the sphere $(\mathbb{S}^2, d_{\mathrm{sph}})$).

a. Adapt the proof of Theorem 2.11 to show that every isometry of the sphere $(\mathbb{S}^2, d_{\mathrm{sph}})$ is a composition of reflections across planes passing through the origin $O = (0, 0, 0)$.

b. Show that every isometry of the sphere $(\mathbb{S}^2, d_{\mathrm{sph}})$ is, either a rotation, or a rotation-reflection. Hint: Use the conclusions of Exercises 3.2c and 3.3d.

Exercise 3.5 (Spherical triangles). A **spherical triangle** is a region T of the sphere $\mathbb{S}^3$ bounded by three geodesics arcs E_1, E_2, E_3 of $\mathbb{S}^2$ meeting only at their endpoints. We also require that the angle of T at each of its three vertices is less than π, and that T is not reduced to a single point. Let α, β, and γ be three numbers in the interval $(0, \pi)$.

a. Suppose that we have found three noncoplanar vectors $\vec{u}$, $\vec{v}$, $\vec{w}$ in $\mathbb{R}^3$ such that the angle between $\vec{u}$ and $\vec{v}$ is equal to $\pi - \alpha$, the angle between $\vec{u}$ and $\vec{w}$ is equal to $\pi - \beta$, and the angle between $\vec{v}$ and $\vec{w}$ is equal to $\pi - \gamma$. Let U, V and W be the hemispheres of the sphere $\mathbb{S}^2$ consisting of all points $P \in \mathbb{S}^2$ with $\overrightarrow{OP} \cdot \vec{u} \geqslant 0$, $\overrightarrow{OP} \cdot \vec{v} \geqslant 0$, $\overrightarrow{OP} \cdot \vec{w} \geqslant 0$, respectively. Show that the intersection $T = U \cap V \cap W$ is a spherical triangle with angles α, β, γ.

b. Consider $\vec{u} = (1, 0, 0)$ and $\vec{v} = (-\cos\alpha, \sin\alpha, 0)$. Show that there exists a unit vector $\vec{w} = (a, b, c)$ with $c \neq 0$ making an angle of $\pi - \beta$ with $\vec{u}$ and an angle of $\pi - \gamma$ with $\vec{v}$ if and only if

$$\cos(\alpha + \beta) < -\cos\gamma < \cos(\alpha - \beta).$$

c. Show that the double inequality of part b is equivalent to the condition that
$$\max\{\alpha - \beta, \beta - \alpha\} < \pi - \gamma < \min\{\alpha + \beta, 2\pi - (\alpha + \beta)\},$$
which itself is equivalent to the condition that
$$\pi < \alpha + \beta + \gamma < \pi + 2\min\{\alpha, \beta, \gamma\}.$$
Hint: Note that $-\pi < \alpha - \beta < \pi$, $0 < \alpha + \beta < 2\pi$ and $0 < \pi - \gamma < \pi$.

d. Combine parts a, b and c to show that if
$$\pi < \alpha + \beta + \gamma < \pi + 2\min\{\alpha, \beta, \gamma\},$$
there exists a spherical triangle $T \subset \mathbb{S}^2$ with respective angles α, β and γ.

e. Show that if two spherical triangles T and $T' \subset \mathbb{S}^2$ have the same angles α, β, γ, there exists an isometry φ of $(\mathbb{S}^2, d_{\text{sph}})$ sending T to T'. Hint: In part b, there are only two possible unit vectors $\vec{w}$.

Exercise 3.6 (Area of spherical triangles).

a. In the sphere $\mathbb{S}^2$, consider two great semi-circles joining the "North Pole" $(0, 0, 1)$ to the "South Pole" $(0, 0, -1)$, and making an angle of α with each other at these poles. Show that the surface area of the digon bounded by these two arcs is equal to 2α. Hint: Use spherical coordinates or a proportionality argument.

b. Let T be a spherical triangle with angles α, β, γ. Let A, B and C be the great circles of $\mathbb{S}^2$ that contain each of the three edges of T. Show that these great circles subdivide the sphere $\mathbb{S}^2$ into eight spherical triangles whose angles are all of the form α, β, γ, $\pi - \alpha$, $\pi - \beta$ or $\pi - \gamma$.

c. Combine parts a and b to show that the area of the triangle T is equal to $\alpha + \beta + \gamma - \pi$. Hint: Solve a system of linear equations.

d. Show that necessarily $\pi < \alpha + \beta + \gamma < \pi + 2\min\{\alpha, \beta, \gamma\}$. Hint: First show that $0 < \alpha + \beta + \gamma - \pi < 2\alpha$.

Chapter 4

Gluing constructions

This chapter and the following one are devoted to the construction of interesting metric spaces which are locally identical to the euclidean plane, the hyperbolic plane or the sphere, but are globally very different. We start with the intuitive idea of gluing together pieces of paper, but then go on with the mathematically rigorous construction of spaces obtained by gluing together the edges of euclidean and hyperbolic polygons. This chapter is concerned with the theoretical aspects of the construction, while the next chapter will investigate various examples.

4.1. Informal examples: the cylinder and the torus

We first discuss in a very informal way the idea of creating new spaces by gluing. Precise definitions will be rigorously developed in the next section.

If one takes a rectangular piece of paper and glues the top side to the bottom side so as to respect the orientations indicated in Figure 4.1, it is well known that one gets a cylinder.

This paper cylinder can be deformed to many positions in 3-dimensional space but they all have the same metric: As long as we do not stretch the paper, the euclidean arc length of a curve drawn on

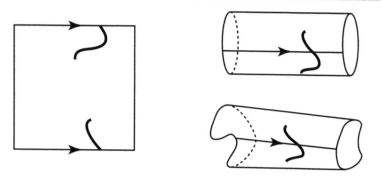

Figure 4.1. Creating a cylinder from a piece of paper

the cylinder remains constant under deformations, and it is actually equal to the arc length of the corresponding pieces of curve in the original rectangle.

One can also try, in addition to gluing the top side to the bottom side, to glue the left side to the right side of the piece of paper. Namely, after gluing the top and bottom sides together to obtain a cylinder, we can glue the left boundary curve of the cylinder to the right one. This is harder to realize physically in 3-dimensional space without crumpling the paper but if we are willing to use rubber instead of paper and to stretch the cylinder in order to put its two sides in contact, one easily sees that this creates a torus, namely, an inner tube or the surface of a donut. See Figure 4.2.

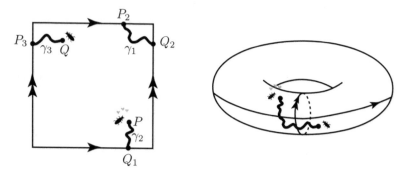

Figure 4.2. Love story on a torus

Let us try to understand the geometry of this torus from the point of view of a little bug crawling over it. For instance, suppose that in order to meet its lover the bug walks from P to Q along the curve γ indicated on the right of Figure 4.2. To measure the distance that it needs to travel, one could consider the euclidean arc length of γ for a given position of the torus in 3-dimensional space, but this will depend on the stretching that occurred when moving the torus to that position. However, if we are interested in prestretching distances, the natural thing to consider is to decompose the curve γ into pieces coming from the original piece of paper, and then take the sum of the lengths of these pieces. For instance, in the situation illustrated on Figure 4.2, the curve γ comes from three curves γ_1, γ_2 and γ_3 on the square, in such a way that each γ_i goes from a point P_i to a point Q_i, and where $P = P_1$, Q_1 is glued to P_2, Q_2 is glued to P_3, and $Q_3 = Q$. The distance traveled by our critter friend, as measured on the original piece of paper, is then the sum of the euclidean arc lengths of γ_1, γ_2 and γ_3 on this piece of paper.

In order to introduce some mathematical rigor to this discussion, let us formalize this construction. We begin with the rectangle

$$X = [a, b] \times [c, d] = \{(x, y) \in \mathbb{R}^2; a \leqslant x \leqslant b, c \leqslant y \leqslant d\}.$$

Let $\bar{X}$ be the space obtained from X by doing the gluings indicated. Some points of $\bar{X}$ correspond to exactly one point of X (located in the interior of the rectangle), some points of $\bar{X}$ correspond to two points of X (located on opposite sides of the rectangle), and one point corresponds to four points of X (namely, the corners of the rectangle). In other words, each point of $\bar{X}$ is described by a subset of X of one of the following types:

(1) the 1-element set $\{(x, y)\}$ with $a < x < b$ and $c < y < d$;

(2) the 2-element set $\{(x, c), (x, d)\}$ with $a < x < b$;

(3) the 2-element set $\{(a, y), (b, y)\}$ with $c < y < d$;

(4) the 4-element set $\{(a, c), (a, d), (b, c), (b, d)\}$.

These subsets form a **partition** of X. This means that every point of X belongs to exactly one such subset.

We could define the distance between points of $\bar{X}$ by taking the infimum of the lengths of curves joining them, as in the example of our little bug walking on the torus. The next section develops a definition that is equivalent to this idea, but is somewhat easier to state and to use.

4.2. Mathematical definition of gluings and quotient spaces

4.2.1. Partitions. Let (X, d) be a metric space, and consider a partition $\bar{X}$ of X. As indicated above, a **partition** of X is a family $\bar{X}$ of subsets $A \subset X$ such that each point $P \in X$ belongs to one and only one such subset A.

In particular, every element of the set $\bar{X}$ is a subset $A \subset X$. We can therefore consider that the set $\bar{X}$ is obtained from X by deciding that for each subset A of the partition all the points in the subset A now correspond to a single element of $\bar{X}$. In other words, all the points of A are now glued together to give a single point in $\bar{X}$. So the formalism of partitions is a good way to rigorously describe the intuitive idea of gluing points of X together. It takes a while to get used to it though since a **point** of $\bar{X}$ is also a **subset** of X.

The following notation will often be convenient. If $P \in X$ is a point of X, let $\bar{P} \in \bar{X}$ denote the corresponding point of $\bar{X}$ after the gluing. Namely, in the formalism of partitions, $\bar{P} \subset X$ is the element of the partition $\bar{X}$ such that $P \in \bar{P}$.

4.2.2. The quotient semi-metric. We now introduce a distance function on the set $\bar{X}$, along the lines of the informal discussion of the previous section.

If $\bar{P}$ and $\bar{Q}$ are two points of $\bar{X}$ corresponding to P and $Q \in X$, respectively, a **discrete walk** w from $\bar{P}$ to $\bar{Q}$ is a finite sequence $P = P_1, Q_1, P_2, Q_2, P_3, \ldots, Q_{n-1}, P_n, Q_n = Q$ of points of X such that $\bar{Q}_i = \bar{P}_{i+1}$ for every $i < n$. Namely, such a discrete walk alternates travels in X from P_i to Q_i and jumps from Q_i to a point P_{i+1} that is glued to Q_i. The d-**length** (or just the **length** if there is no ambiguity on the metric d considered) of a discrete walk w is the sum of the travel distances

$$\ell_d(w) = \sum_{i=1}^{n} d(P_i, Q_i).$$

This is the exact translation of our informal discussion of the little bug walking on the torus, except that we are now requiring the bug to follow a path that is made up of straight line segments in the rectangle. Namely, the bug should be a grasshopper instead of a snail. As it follows a discrete walk, it alternates steps where it hops in X from one point P_i to another point Q_i, and steps where it is beamed-up from Q_i to P_{i+1} by the gluing process.[1]

To make the description of a discrete walk w easier to follow, it is convenient to write $P \sim P'$ when P is glued to P', namely, when $\bar{P} = \bar{P}'$. (Compare also our discussion of equivalence relations in Exercise 4.2). Then a discrete walk w from $\bar{P}$ to $\bar{Q}$ is of the form $P = P_1$, $Q_1 \sim P_2$, $Q_2 \sim P_3$, ..., $Q_{n-1} \sim P_n$, $Q_n = Q$. This makes it a little easier to remember that the consecutive points Q_i, P_{i+1} are glued to each other, while the pairs P_i, Q_i correspond to a travel of length $d(P_i, Q_i)$ in X.

We would like to define a distance function $\bar{d}$ on $\bar{X}$ by

$$\bar{d}(\bar{P}, \bar{Q}) = \inf\{\ell_d(w); w \text{ discrete walk from } \bar{P} \text{ to } \bar{Q}\}$$

for any two points $\bar{P}$, $\bar{Q} \in \bar{X}$.

Lemma 4.1. *The above number $\bar{d}(\bar{P}, \bar{Q})$ is independent of the choice of the points P, $Q \in X$ used to represent $\bar{P}$, $\bar{Q} \in \bar{X}$. As a consequence, this defines a function $\bar{d} \colon \bar{X} \times \bar{X} \to \mathbb{R}$.*

In addition, this function $\bar{d}$ is a semi-distance on $\bar{X}$, in the sense that it satisfies the following three conditions:

(1) *$\bar{d}(\bar{P}, \bar{Q}) \geqslant 0$ and $\bar{d}(\bar{P}, \bar{P}) = 0$ for every $\bar{P}$, $\bar{Q} \in \bar{X}$ (Nonnegativity Condition);*

(2) *$\bar{d}(\bar{P}, \bar{Q}) = \bar{d}(\bar{Q}, \bar{P})$ for every $\bar{P}$, $\bar{Q} \in \bar{X}$ (Symmetry Condition);*

(3) *$\bar{d}(\bar{P}, \bar{R}) \leqslant \bar{d}(\bar{P}, \bar{Q}) + \bar{d}(\bar{Q}, \bar{R})$ for every $\bar{P}$, $\bar{Q}$, $\bar{R} \in \bar{X}$ (Triangle Inequality).*

---

[1]The original Park City lectures involved pedagogic sound effects to distinguish between these two types of moves.

Proof. To prove the first statement, consider two other points P', $Q' \in X$ such that $\bar{P}' = \bar{P}$ and $\bar{Q}' = \bar{Q}$. We need to show that $\bar{d}(\bar{P}', \bar{Q}') = \bar{d}(\bar{P}, \bar{Q})$.

If w is a discrete walk $P = P_1$, $Q_1 \sim P_2$, $Q_2 \sim P_3$, $\ldots$, $Q_{n-1} \sim P_n$, $Q_n = Q$ from $\bar{P}$ to $\bar{Q}$, starting from P and ending at Q, we can consider another discrete walk w' of the form $P' = P_0$, $Q_0 \sim P_1$, $Q_1 \sim P_2$, $Q_2 \sim P_3$, $\ldots$, $Q_{n-1} \sim P_n$, $Q_n \sim P_{n+1}$, $Q_{n+1} = Q'$ by taking $P_0 = Q_0 = P'$ and $P_{n+1} = Q_{n+1} = Q'$. This new discrete walk w' starts at P', ends at Q', and has the same length $\ell(w') = \ell(w)$ as w. Taking the infimum over all such discrete walks w, we conclude that the "distance" $\bar{d}(\bar{P}', \bar{Q}')$, defined using P' and Q', is less than or equal to $\bar{d}(\bar{P}, \bar{Q})$, defined using P and Q. Exchanging the roles of P, Q and P', Q', we similarly obtain that $\bar{d}(\bar{P}, \bar{Q}) \leqslant \bar{d}(\bar{P}', \bar{Q}')$, so that $\bar{d}(\bar{P}, \bar{Q}) = \bar{d}(\bar{P}', \bar{Q}')$.

This proves that the function $\bar{d} \colon \bar{X} \times \bar{X} \to \mathbb{R}$ is well defined.

The Nonnegativity Condition (1) is immediate.

To prove the Symmetry Condition (2), note that every discrete walk $P = P_1$, $Q_1 \sim P_2$, $Q_2 \sim P_3$, $\ldots$, $Q_{n-1} \sim P_n$, $Q_n = Q$ from $\bar{P}$ to $\bar{Q}$ provides a discrete walk $Q = Q_n$, $P_n \sim Q_{n-1}$, $P_{n-1} \sim Q_{n-2}$, $\ldots$, $P_2 \sim Q_1$, $P_1 = P$ from $\bar{Q}$ to $\bar{P}$. Since these two discrete walks have the same length, one immediately concludes that $\bar{d}(\bar{P}, \bar{Q}) = \bar{d}(\bar{Q}, \bar{P})$.

Finally, for the Triangle Inequality (3), consider a discrete walk w of the form $P = P_1$, $Q_1 \sim P_2$, $Q_2 \sim P_3$, $\ldots$, $Q_{n-1} \sim P_n$, $Q_n = Q$ going from $\bar{P}$ to $\bar{Q}$, and a discrete walk w' of the form $Q = Q'_1$, $R_1 \sim Q'_2$, $\ldots$, $R_{m-1} \sim Q'_m$, $R_m = R$ going from $\bar{Q}$ to $\bar{R}$. These two discrete walks can be chained together to give a discrete walk w'' of the form $P = P_1$, $Q_1 \sim Q_2$, $\ldots$, $Q_{n-1} \sim P_n$, $Q_n \sim Q'_1$, $R_1 \sim Q'_2$, $\ldots$, $R_{m-1} \sim Q'_m$, $R_m = R$ going from $\bar{P}$ to $\bar{R}$. Since $\ell_d(w'') = \ell_d(w) + \ell_d(w')$, taking the infimum over all such discrete walks w and w', we conclude that $\bar{d}(\bar{P}, \bar{R}) \leqslant \bar{d}(\bar{P}, \bar{Q}) + \bar{d}(\bar{Q}, \bar{R})$. $\square$

The only missing property for the semi-distance $\bar{d}$ to be a distance function is that $\bar{d}(\bar{P}, \bar{Q}) = 0$ only when $\bar{P} = \bar{Q}$.

If this property holds, we will say that the partition or gluing process is **proper**. The metric space $(\bar{X}, \bar{d})$ then is the **quotient space** of the metric space (X, d) by the partition. The distance function $\bar{d}$ is the **quotient metric** induced by d.

In the case of the torus obtained by gluing the sides of a rectangle, we will prove in Sections 4.3 and 4.4 that the gluing is proper. The same will hold for many examples that we will consider later on. However, there also exist partitions which are not proper. See Exercise 4.1 for an example where each point is glued to at most one other point.

4.2.3. The quotient map. For future reference, the following elementary observation will often be convenient.

Let $\pi\colon X \to \bar{X}$ be the **quotient map** defined by $\pi(P) = \bar{P}$.

Lemma 4.2. *For every P, $Q \in X$,*

$$\bar{d}(\bar{P}, \bar{Q}) \leqslant d(P, Q).$$

As a consequence, the quotient map $\pi\colon X \to \bar{X}$ is continuous.

Proof. The inequality is obtained by consideration of the one-step discrete walk w from $\bar{P}$ to $\bar{Q}$ defined by $P = P_1$, $Q_1 = Q$. By definition of $\bar{d}$, $\bar{d}(\bar{P}, \bar{Q}) \leqslant \ell(w) = d(P, Q)$.

The continuity of the quotient map π is an immediate consequence of this inequality. □

4.3. Gluing the edges of a euclidean polygon

This section is devoted to the special case where $\bar{X}$ is obtained by gluing together the edges of a polygon X, as in the example of the torus obtained by gluing together opposite sides of a rectangle.

4.3.1. Polygons and edge gluing data. Let X be a **polygon** in the euclidean plane $\mathbb{R}^2$. Namely, X is a region of the euclidean plane whose boundary is decomposed into finitely many line segments, lines and half-lines $E_1, E_2, \ldots, E_n$ meeting only at their endpoints. We also impose that at most two E_i can meet at any given point.

The line segments, lines and half-lines E_i bounding X are the **edges** of the polygon X. The points where two edges meet are its **vertices**.

We require in addition that X and the E_i are ***closed***, in the sense that they contain all the points of $\mathbb{R}^2$ that are in their boundary. In this specific case, this is equivalent to the property that the polygon X contains all its edges and all its vertices.

However, we allow X to go to infinity, in the sense that it may be bounded or unbounded in $(\mathbb{R}^2, d_{\text{euc}})$. A subset of a metric space is ***bounded*** when it is contained in some ball

$$B_{d_{\text{euc}}}(P, r) = \{Q \in X; d_{\text{euc}}(P, Q) < r\}$$

with finite radius $r < \infty$. Using the Triangle Inequality (and changing the radius r), one easily sees that this property does not depend on the point P chosen as center of the ball. The subset is ***unbounded*** when it is not bounded.

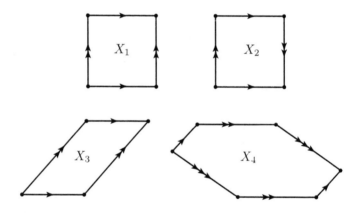

Figure 4.3. A few bounded euclidean polygons

Figures 4.3 and 4.4 illustrate a few examples of polygons.

In most cases considered, the polygon X will in addition be ***convex*** in the sense that, for every P, $Q \in X$ the line segment $[P, Q]$ joining P to Q is contained in X. We endow such a convex polygon with the restriction d_X of the euclidean metric d_{euc}, defined by the property that $d_X(P, Q) = d_{\text{euc}}(P, Q)$ for every P, $Q \in X$.

Among the polygons X_1, X_2, ..., X_7 described in Figures 4.3 and 4.4, X_7 is the only one that is not convex.

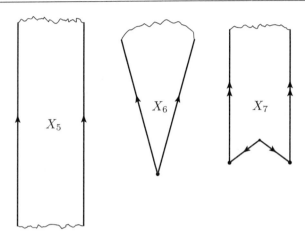

Figure 4.4. A few unbounded euclidean polygons

When X is not convex, it is more convenient to consider for P, $Q \in X$ the infimum $d_X(P, Q)$ of the euclidean length $\ell_{euc}(\gamma)$ of all piecewise differentiable curves γ joining P and Q and contained in X. The fact that the function d_X so defined is a metric on X is immediate, noting that $d_X(P, Q) \geqslant d_{euc}(P, Q)$. See Exercise 1.10 for an explicit example.

We will call this distance function d_X the ***euclidean path metric*** of the polygon X. Note that the path metric d_X coincides with the restriction of the euclidean metric d_{euc} when X is convex. In the general case, the path metric coincides *locally* with d_{euc} in the sense that every $P \in X$ is the center of a small ball $B_{d_{euc}}(P, \varepsilon)$ such that $d_X(P, Q) = d_{euc}(P, Q)$ for every $Q \in X \cap B_{d_{euc}}(P, \varepsilon)$ (because the line segment $[P, Q]$ is contained in X).

We will occasionally allow the polygon X to be made up of several disjoint pieces, so that there exist points P and $Q \in X$ which cannot be joined by a piecewise differentiable curve γ completely contained in X. In this case, $d_X(P, Q) = +\infty$ by convention. This requires that we extend the definition of metrics to allow them to take values in $[0, +\infty]$, but the extension is immediate provided we use the obvious conventions for inequalities and additions involving infinite numbers ($a \leqslant \infty$ and $a + \infty = \infty$ for every $a \in [0, \infty]$, etc. . . .).

A polygon for which this does not happen is said to be **connected**. Namely, the polygon X is connected if any two points P and $Q \in X$ can be joined by a piecewise differentiable curve γ completely contained in X. The reader who is already familiar with some notions of topology will notice that the definition we are using here is more reminiscent of that of path connectedness; however, the two definitions are equivalent for polygons.

After these preliminaries about polygons, we now describe how to glue the edges of a polygon $X \subset \mathbb{R}^2$ together. For this, we first group these edges into pairs $\{E_1, E_2\}$, $\{E_3, E_4\}$, ..., $\{E_{2p-1}, E_{2p}\}$, and then for each such pair $\{E_{2k-1}, E_{2k}\}$ we specify an isometry $\varphi_{2k-1} \colon E_{2k-1} \to E_{2k}$. Here φ_{2k-1} is an isometry for the restrictions of the metric d_X to the edges E_{2k-1} and E_{2k}, which also coincide with the restrictions of the euclidean metric d_{euc} to these edges. This is equivalent to the property that φ_{2k-1} sends each line segment in the edge E_{2k-1} to a line segment of the same length in E_{2k}.

Note that in particular, the edges E_{2k-1} and E_{2k} in a given pair must have the same length, possibly infinite. In general, the isometry φ_{2k-1} is then uniquely determined once we know how φ_{2k-1} sends what orientation of E_{2k-1} to which orientation of E_{2k}. A convenient way to describe this information is to draw arrows on E_{2k-1} and E_{2k} corresponding to these matching orientations. It is also convenient to use a different type of arrow for each pair $\{E_{2k-1}, E_{2k}\}$, so that the pairing can be readily identified on the picture. Figures 4.3 and 4.4 offer some examples.

There is one case where this arrow information is not sufficient, when E_{2k-1} and E_{2k} are both lines (of infinite length), as in the case of the infinite strip X_5 of Figure 4.4. In this case, $\varphi_{2k-1} \colon E_{2k-1} \to E_{2k}$ is only defined up to translation in one of these lines, and we will need to add extra information to describe φ_{2k-1}.

The notation will be somewhat simplified if we introduce the isometry $\varphi_{2k} \colon E_{2k} \to E_{2k-1}$ defined as the inverse $\varphi_{2k} = \varphi_{2k-1}^{-1}$ of $\varphi_{2k-1} \colon E_{2k-1} \to E_{2k}$. In this way, every edge E_i is glued to an edge $E_{i\pm 1}$ by an isometry $\varphi_i \colon E_i \to E_{i\pm 1}$, where ± 1 depends on the parity of i.

With this data of isometric edge identifications $\varphi_i \colon E_i \to E_{i\pm1}$, we can now describe the gluing of the edges of the polygon X by specifying a partition $\bar{X}$ as follows. Recall that if $P \in X$, we denote by $\bar{P} \in \bar{X}$ the corresponding element of the partition $\bar{X}$, consisting of all the points of X that are glued to P. The gluing is then defined as follows:

- if P is in the interior of the polygon X, then P is glued to no other point and $\bar{P} = \{P\}$;

- if P is in a edge E_i and is not a vertex, then $\bar{P}$ consists of the two points $P \in E_i$ and $\varphi_i(P) \in E_{i\pm1}$;

- if P is a vertex, then $\bar{P}$ consists of P and of all the vertices of X of the form $\varphi_{i_k} \circ \varphi_{i_{k-1}} \circ \cdots \circ \varphi_{i_1}(P)$, where the indices i_1, i_2, ..., i_k are such that $\varphi_{i_{j-1}} \circ \cdots \circ \varphi_{i_1}(P) \in E_{i_j}$ for every j.

The case of vertices may appear a little complicated at first, but it becomes much simpler with practice. Indeed, because each vertex belongs to exactly two edges, there is a simple method for listing all elements of $\bar{P}$ for a vertex P. The key observation is that when considering a vertex $\varphi_{i_k} \circ \varphi_{i_{k-1}} \circ \cdots \circ \varphi_{i_1}(P)$ glued to P, we can always assume that the range $\varphi_{i_j}(E_{i_j})$ of the gluing map φ_{i_j} is different from the region $E_{i_{j+1}}$ of $\varphi_{i_{j+1}}$ (since otherwise $\varphi_{i_{j+1}} = \varphi_{i_j}^{-1}$, so that these two gluing maps cancel out).

This leads to the following algorithm: Start with $P_1 = P$, and let E_{i_1} be one of the two edges containing P_1. Set $P_2 = \varphi_{i_1}(P_1)$, and let E_{i_2} be the edge containing P_2 that is different from $\varphi_{i_1}(E_{i_1})$. Iterating this process provides a sequence of vertices P_1, P_2, ..., P_j, ... and edges E_{i_1}, E_{i_2}, ..., E_{i_j}, ... such that $P_j \in E_{i_j}$, $P_{j+1} = \varphi_{i_j}(P_j)$, and $E_{i_{j+1}}$ is the edge containing P_{j+1} that is different from $\varphi_{i_j}(E_{i_j})$. Since there are only finitely many vertices, there is an index k for which $P_{k+1} = P_j$ for some $j \leqslant k$. If k is the smallest such index, one easily checks that $P_{k+1} = P_1$, and that $\bar{P} = \{P_1, P_2, \ldots, P_k\}$.

For instance, in the example of the rectangle X_1 of Figure 4.3, the four corners are glued together to form a single point in $\bar{X}_1$. For the hexagon X_4, the vertices of X_4 project to two points of $\bar{X}_4$, each of them corresponding to three vertices of the hexagon. In Figure 4.4, the infinite strip X_5 has no vertex. The quotient space $\bar{X}_7$ has two elements associated to vertices of X_7, one corresponding to exactly one vertex and another one consisting of two vertices of X_7.

4.3.2. Edge gluings are proper. Let $\bar{d}_X$ be the semi-metric on the quotient space $\bar{X}$ that is defined by the euclidean path metric d_X of the polygon $X \subset \mathbb{R}^2$.

Theorem 4.3. *If $\bar{X}$ is obtained from the euclidean polygon X by gluing together edge pairs by isometries, then the gluing is proper. Namely, the semi-distance $\bar{d}_X$ induced on $\bar{X}$ by the metric d_X of X is such that $\bar{d}_X(\bar{P}, \bar{Q}) > 0$ when $\bar{P} \neq \bar{Q}$.*

The proof of Theorem 4.3 is somewhat long and is postponed to Section 4.4.

4.3.3. Euclidean surfaces. Let us go back to our informal paper and adhesive tape discussion of the torus $\bar{X}$, obtained by gluing opposite sides of a rectangle X.

If $\bar{P}$ is the point of the torus that corresponds to the four corners of the rectangle, it should be intuitively clear (and will be rigorously proved in Lemma 4.5) that a point $\bar{Q} \in \bar{X}$ is at distance $< \varepsilon$ of $\bar{P}$ in $\bar{X}$ exactly when it corresponds to a point $Q \in X$ which is at distance $< \varepsilon$ from one of the corners of the rectangle. As a consequence, for ε small enough, the ball $B_{\bar{d}_X}(\bar{P}, \varepsilon)$ is the image in $\bar{X}$ of four quarter-disks in X. We know from experience that if we glue together four paper quarter-disks with (invisible) adhesive tape, we obtain an object which is undistinguishable from a full disk of the same radius in the euclidean plane.

The same property will hold at a point $\bar{P} \in \bar{X}$ that is the image of a point $P \in X$ located on a side of the rectangle, not at a corner. Then the ball $B_{\bar{d}}(\bar{P}, \varepsilon)$ is obtained by gluing two half-disks along their diameters, and again it has the same metric properties as a disk.

As a consequence, if our little bug crawling on the torus $\bar{X}$ is, in addition, very near sighted, it will not be able to tell that it is walking on a torus instead of a plane. This may be compared to the (hi)story of other well-known animals who thought for a long time that they were living on a plane, before progressively discovering that they were actually inhabiting a surface with the rough shape of a very large sphere.

When gluing together sectors of paper disks, a crucial property for the result to look like a full disk in the euclidean plane is that the angles of these disk sectors should add up to 2π. As is well known to anybody who has ever made a birthday hat out of cardboard, the resulting paper construction has a sharp cone point if the sum of the angles is less than 2π. Similarly, it wrinkles if the angles add up to more than 2π. Not unexpectedly, we will encounter the same condition when gluing the edges of a euclidean polygon.

Let us put this informal discussion in a more mathematical framework.

Two metric spaces (X, d) and (X', d') are **locally isometric** if for every $P \in X$ there exists an isometry between some ball $B_d(P, \varepsilon)$ centered at P and a ball $B_{d'}(P', \varepsilon)$ in X'.

Theorem 4.4. *Let $(\bar{X}, \bar{d}_X)$ be the quotient metric space obtained from a euclidean polygon (X, d_X) by gluing together pairs of edges of X by isometries. Suppose that the following additional condition holds: For every vertex P of X, the angles of X at those vertices P' of X which are glued to P add up to 2π. Then $(\bar{X}, \bar{d}_X)$ is locally isometric to the euclidean plane $(\mathbb{R}^2, d_{\mathrm{euc}})$.*

Again, although the general idea is exactly the one suggested by the above paper-and-adhesive-tape discussion, the proof of Theorem 4.4 is rather long with several cases to consider. It is given in the next section.

A metric space (X, d) which is locally isometric to the euclidean plane $(\mathbb{R}^2, d_{\mathrm{euc}})$ is a **euclidean surface**. Equivalently, the metric d is then a **euclidean metric**.

4.4. Proofs of Theorems 4.3 and 4.4

This section is devoted to the proofs of Theorems 4.3 and 4.4. These proofs are not very difficult but are a little long, with many cases to consider. They may perhaps be skipped on a first reading, which gives the reader a first opportunity to use the fast-forward command of the remote control.

To
page 79 These are the first really complex proofs that we encounter in this book. It is important to understand these arguments (and many more later) at a level which is higher than a simple manipulation of symbols. With this goal in mind, the reader is strongly encouraged to read them with a piece of paper and pencil in hand, and to draw pictures of the geometric situations involved in order to better follow the explanations.

Throughout this section, X will denote a polygon in the euclidean space $(\mathbb{R}^2, d_{\mathrm{euc}})$. We are also given isometries $\varphi_{2k-1} \colon E_{2k-1} \to E_{2k}$ and $\varphi_{2k} = \varphi_{2k-1}^{-1} \colon E_{2k} \to E_{2k-1}$ between the edges $E_1, \ldots, E_n$ of X. Then $\bar{X}$ is the space obtained from X by performing the corresponding edge gluings, and $\bar{d}_X$ is the semi-metric induced on $\bar{X}$ by the euclidean path metric d_X introduced in the previous section. Recall that in the more common case where X is convex, d_X is just the restriction of the euclidean metric d_{euc} to X.

First we need to understand the balls

$$B_{\bar{d}_X}(\bar{P}, \varepsilon) = \{\bar{Q} \in \bar{P}; \bar{d}_X(\bar{P}, \bar{Q}) < \varepsilon\},$$

at least for ε sufficiently small.

4.4.1. Small balls in the quotient space $(\bar{X}, \bar{d})$.

Lemma 4.5. *For every $\bar{P} \in \bar{X}$, there exists an $\varepsilon_0 > 0$ such that for every $\varepsilon \leqslant \varepsilon_0$ and every $Q \in X$, the point $\bar{Q} \in \bar{X}$ is in the ball $B_{\bar{d}_X}(\bar{P}, \varepsilon)$ if and only if there is a $P' \in \bar{P}$ such that $d_X(P', Q) < \varepsilon$.*

We can rephrase this in terms of the quotient map $\pi \colon X \to \bar{X}$ sending $P \in X$ to $\pi(P) = \bar{P} \in \bar{X}$. Lemma 4.5 states that for ε sufficiently small, the ball $B_{\bar{d}_X}(\bar{P}, \varepsilon)$ is exactly the union of the images under π of the balls $B_{d_X}(P', \varepsilon)$ as P' ranges over all points of $\bar{P}$.

For a better understanding of the proof, it may be useful to realize that the ball $B_{\bar{d}_X}(\bar{P}, \varepsilon)$ can be significantly larger than the union of the images $\pi\big(B_{d_X}(P', \varepsilon)\big)$ when ε is not small. This is illustrated by Figure 4.5 in the case of the torus. In these pictures, $\bar{P}$ consists of the single point P. The shaded areas represent the balls $B_{\bar{d}_X}(\bar{P}, \varepsilon)$

for various values of ε; in each shaded area, the darker area is the image of the ball $B_{d_X}(P, \varepsilon)$ under π.

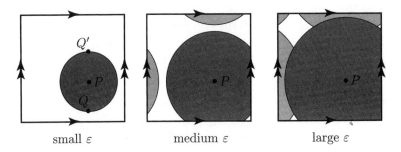

small ε medium ε large ε

Figure 4.5. Balls $B_{\bar{d}}(\bar{P}, \varepsilon)$ in $\bar{X}$

Proof of Lemma 4.5. Recall from Lemma 4.2 that $\bar{d}_X(\bar{P}, \bar{Q}) \leqslant d_X(P', Q)$ for every $P' \in \bar{P}$. Therefore, the "if" part of the statement holds without restriction on ε.

The "only if" part will take more time to prove as we will need to distinguish cases.

Let $\bar{P}$ be a point of $\bar{X}$. For a number ε_0 which will be specified later on in function of $\bar{P}$, we consider a point $Q \in X$ such that $\bar{d}_X(\bar{P}, \bar{Q}) < \varepsilon \leqslant \varepsilon_0$. We want to find a point $P' \in \bar{P}$, namely, a point of X which is glued to P, such that $d_X(P', Q) < \varepsilon$.

Since $\bar{d}_X(\bar{P}, \bar{Q}) < \varepsilon$, there exists a discrete walk w from $\bar{P}$ to $\bar{Q}$ of the form $P = P_1$, $Q_1 \sim P_2$, $Q_2 \sim P_3$, ..., $Q_{n-1} \sim P_n$, $Q_n = Q$ and whose length is such that $\ell(w) = \sum_{i=1}^{n} d_X(P_i, Q_i) < \varepsilon$.

We want to prove by induction that for every $j \leqslant n$,

(4.1) there exists $P' \in \bar{P}$ such that $d_X(P', Q_j) \leqslant \sum_{i=1}^{j} d_X(P_i, Q_i) < \varepsilon$.

We can begin the induction with $j = 1$, in which case (4.1) is trivial by taking $P' = P$.

Suppose as an induction hypothesis that (4.1) holds for j. We want to show that it holds for $j + 1$.

For this, we will distinguish cases according to the type of the point $P \in X$. We will also specify ε_0 in each case.

CASE 1. *P is in the interior of the polygon X.*

We first specify the number ε_0 needed in this case. We choose it so that the closed disk of radius ε_0 centered at P is completely contained in the interior of the polygon. Equivalently, every point on the boundary of the polygon is at distance $> \varepsilon_0$ from P.

In this case, P is the only point of $\bar{P}$.

By the induction hypothesis (4.1) and by choice of $\varepsilon_0 > \varepsilon$, the point Q_j is in the interior of the polygon. In particular, it is glued to no other point so that $P_{j+1} = Q_j$. Combining the Triangle Inequality with the induction hypothesis, we conclude that

$$d_X(P, Q_{j+1}) \leqslant d_X(P, Q_j) + d_X(P_{j+1}, Q_{j+1}) \leqslant \sum_{i=1}^{j+1} d_X(P_i, Q_i) < \varepsilon.$$

This proves (4.1) for $j + 1$.

CASE 2. *P is on an edge E_i of the polygon and not at a vertex.*

In this case, $\bar{P}$ consists of P and of exactly one other point $\varphi_i(P)$ on the edge $E_{i \pm 1}$ that is glued to E_i.

Choose $\varepsilon_1 > 0$ such that P is at distance $> \varepsilon_1$ from the other edges E_j, with $j \neq i$, of the polygon. Similarly, let ε_2 be such that $\varphi_i(P)$ is at distance $> \varepsilon_2$ from any edge other than the edge $E_{i \pm 1}$ that contains it. Choose ε_0 as the smaller of ε_1 and ε_2.

If $Q_j = P_{j+1}$, combining the induction hypothesis (4.1) with the Triangle Inequality gives, as in the case of interior points,

$$d_X(P', Q_{j+1}) \leqslant d_X(P', Q_j) + d_X(P_{j+1}, Q_{j+1}) \leqslant \sum_{i=1}^{j+1} d_X(P_i, Q_i) < \varepsilon,$$

which proves (4.1) for $j + 1$ in this case.

Otherwise, Q_j and P_{j+1} are distinct but glued together. Because $d_X(P', Q_j) < \varepsilon \leqslant \varepsilon_0$ and by choice of ε_0, these two points cannot be vertices of the polygon, so that one of them is in the edge E_i and the other one is in the edge $E_{i \pm 1}$ glued to E_i by the map φ_i. In particular, $P_{j+1} = \varphi_i^{\pm 1}(Q_j)$. Set $P'' = \varphi_i^{\pm 1}(P')$. Note that P'' is just equal to P or to $\varphi_i(P)$; in particular it is in $\bar{P}$.

We will use the crucial property that the gluing map φ_i respects distances. As a consequence, $d_X(P'', P_{j+1}) = d_X(P', Q_j)$ so that

$$d_X(P'', Q_{j+1}) \leqslant d_X(P'', P_{j+1}) + d_X(P_{j+1}, Q_{j+1})$$

$$\leqslant d_X(P', Q_j) + d_X(P_{j+1}, Q_{j+1}) \leqslant \sum_{i=1}^{j+1} d_X(P_i, Q_i) < \varepsilon$$

by the induction hypothesis. Again, this proves (4.1) for $j+1$ in this case.

CASE 3. *P is a vertex of the polygon.*

In this case, $\bar{P}$ consists of a certain number of vertices P' of the polygon. Pick ε_0 such that every point of $\bar{P}$ is at distance $> \varepsilon_0$ from the edges that do not contain it.

The proof is almost identical to that of Case 2 with only a couple of minor twists.

If $Q_j = P_{j+1}$, as in the previous two cases, the combination of the induction hypothesis (4.1) and the Triangle Inequality shows that (4.1) holds for $j+1$.

If $Q_j \neq P_{j+1}$, we distinguish cases according to whether Q_j is a vertex of X or not. If Q_j is a vertex, since $d_X(P', Q_j) < \varepsilon \leqslant \varepsilon_0$ by the induction hypothesis, this vertex must be P' by the choice of ε_0. Then, $P'' = P_{j+1}$ is also in $\bar{P}$ since it is glued to $Q_j = P'$, and therefore we found a $P'' \in \bar{P}$ such that

$$d_X(P'', Q_{j+1}) = d_X(P_{j+1}, Q_{j+1}) \leqslant \sum_{i=1}^{j+1} d_X(P_i, Q_i) < \varepsilon$$

as required.

Otherwise, Q_j is not a vertex and is glued to P_{j+1} by a gluing map $\varphi_k \colon E_k \to E_{k\pm1}$. By choice of ε_0, P' is contained in the edge E_k, so that $P'' = \varphi_k(P') \in \bar{P}$ is defined. Since the gluing map φ_k respects distances,

$$d_X(P'', Q_{j+1}) \leqslant d_X(P'', P_{j+1}) + d_X(P_{j+1}, Q_{j+1})$$

$$\leqslant d_X(P', Q_j) + d_X(P_{j+1}, Q_{j+1}) \leqslant \sum_{i=1}^{j+1} d_X(P_i, Q_i) < \varepsilon,$$

using the fact that the induction hypothesis (4.1) holds for j.

Therefore, (4.1) now holds for $j + 1$, as requested.

This completes the proof of (4.1) in all three cases, and for all j. The case $j = n$ proves Lemma 4.5 since $Q_n = Q$. $\qquad\square$

4.4.2. Proof of Theorem 4.3. We now have the tools needed to prove Theorem 4.3, namely that $\bar{d}_X(\bar{P}, \bar{Q}) > 0$ whenever $\bar{P} \neq \bar{Q}$.

Let ε_0 be associated to $\bar{P}$ by Lemma 4.5.

Because $\bar{X}$ is a partition of X, the fact that $\bar{P} \neq \bar{Q}$ implies that these two subsets $\bar{P}$ and $\bar{Q}$ of X are disjoint, namely, they have no point in common. Since these subsets are finite, it follows that there exists an $\varepsilon_1 > 0$ such that every point of $\bar{P}$ is at distance $> \varepsilon_1$ from every point of $\bar{Q}$. Set $\varepsilon = \min\{\varepsilon_0, \varepsilon_1\} > 0$.

Then $\bar{d}_X(\bar{P}, \bar{Q}) \geq \varepsilon$. Indeed, Lemma 4.5 would otherwise provide a point $P' \in \bar{P}$ such that $d_X(P', Q) < \varepsilon \leq \varepsilon_1$, thereby contradicting the definition of ε_1. $\qquad\square$

4.4.3. Proof of Theorem 4.4. For every point $\bar{P} \in \bar{X}$, we need to find an isometry ψ between a small ball $B_{\bar{d}_X}(\bar{P}, \varepsilon) \subset \bar{X}$ centered at $\bar{P}$ and a small ball $B_{d_{\mathrm{euc}}}(P', \varepsilon)$ in the euclidean plane $\mathbb{R}^2$.

Fix an ε satisfying the conclusions of Lemma 4.5. In addition, choose ε small enough that each $P' \in \bar{P}$ is at a euclidean distance $> 3\varepsilon$ from any edge that does not contain it. In particular, the balls $B_{d_X}(P', \varepsilon)$ are pairwise disjoint and are disks, half-disks or disk sectors in $\mathbb{R}^2$, according to the type of $P' \in \bar{P}$. Here, a *(euclidean) disk sector* is one of the two pieces of a euclidean disk $B_{d_{\mathrm{euc}}}(P_0, r)$ in $\mathbb{R}^2$ delimited by two half-lines issued from its center P_0, as in a slice of pie.

In addition, the Triangle Inequality shows that the balls $B_{d_X}(P', \varepsilon)$ are at a euclidean distance $> \varepsilon$ apart, in the sense that $d_{\mathrm{euc}}(Q', Q'') > \varepsilon$ if $Q' \in B_{d_X}(P', \varepsilon)$ and $Q'' \in B_{d_X}(P'', \varepsilon)$ with $P' \neq P'' \in \bar{P}$.

Lemma 4.5 says that the ball $B_{\bar{d}_X}(\bar{P}, \varepsilon)$ is obtained by gluing together the balls $B_{d_X}(P', \varepsilon)$ in X centered at the points P' that are glued to P. Let

$$B = \bigcup_{P' \in \bar{P}} B_{d_X}(P', \varepsilon)$$

denote the union of these balls.

This subset $B \subset X$ comes with two natural metrics. The first one is the restriction of the metric d_X. The second one d_B is similarly defined, but by restricting attention to curves that are contained in B. Namely, $d_B(Q, Q')$ is the infimum of the euclidean lengths $\ell_{euc}(\gamma)$ of all piecewise differentiable curves γ joining Q to Q' and contained in B. In particular, $d_B(Q, Q') = \infty$ if Q and Q' are in distinct balls $B_{d_X}(P', \varepsilon)$ and $B_{d_X}(P'', \varepsilon)$ of B.

When Q and Q' are in the same ball $B_{d_X}(P', \varepsilon)$ of B, elementary geometry shows that $d_B(Q, Q') = d_X(Q, Q')$. Indeed, $B_{d_X}(P', \varepsilon)$ is a disk, a half-disk or a disk sector. Therefore, the only case which requires some thought is that of a disk sector of angle $> \pi$ (since otherwise $d_B(Q, Q') = d_X(Q, Q') = d_{euc}(Q, Q')$ by convexity). In this case, one just needs to check that the shortest curve from Q to Q' in the polygon X is either a single line segment completely contained in $B_{d_X}(P', \varepsilon)$ or the union of two line segments meeting at the vertex P' in $B_{d_X}(P', \varepsilon)$; compare Exercise 1.10.

Since the ball $B_{\bar{d}_X}(\bar{P}, \varepsilon)$ is obtained by performing certain gluings on B it inherits a quotient semi-metric $\bar{d}_B$. The advantage of $\bar{d}_B$ is that it is entirely defined in terms of B, without reference to the rest of X.

Lemma 4.6. *The metrics $\bar{d}_X$ and $\bar{d}_B$ coincide on the ball $B_{\bar{d}_X}(\bar{P}, \frac{1}{3}\varepsilon)$.*

The restriction to the ball of radius $\frac{1}{3}\varepsilon$ is used to rule out the possibility of a "shortcut" through $\bar{X}$ making $\bar{Q}$ and $\bar{Q}'$ closer in $\bar{X}$ than in $B_{\bar{d}_X}(\bar{P}, \varepsilon) = \bar{B}$. The left-hand side of Figure 4.5 provides an example of two such points $\bar{Q}$ and $\bar{Q}' \in B_{\bar{d}_X}(\bar{P}, \varepsilon)$ such that $\bar{d}_X(\bar{Q}, \bar{Q}') < \bar{d}_B(\bar{Q}, \bar{Q}')$.

Proof of Lemma 4.6. By definition of d_X and d_B, $d_X(R, R') \leqslant d_B(R, R')$ for every R, $R' \in B$. It immediately follows that $\bar{d}_X(\bar{Q}, \bar{Q}') \leqslant \bar{d}_B(\bar{Q}, \bar{Q}')$ for every $\bar{Q}$, $\bar{Q}' \in B_{\bar{d}_X}(\bar{P}, \varepsilon)$. Incidentally, this shows that $\bar{d}_B$ is really a metric and not just a semi-metric.

To prove the reverse inequality, we need to restrict attention to $\bar{Q}$, $\bar{Q}' \in B_{\bar{d}_X}(\bar{P}, \frac{1}{3}\varepsilon)$. In particular, $\bar{d}_X(\bar{Q}, \bar{Q}') < \frac{2}{3}\varepsilon$ by the Triangle Inequality.

Let w be a discrete walk from $\bar{Q}$ to $\bar{Q}'$ in X of the form $Q = Q_1$, $Q'_1 \sim Q_2$, $Q'_2 \sim Q_3$, ..., $Q'_{n-1} \sim Q_n$, $Q'_n = Q'$, and whose d_X-length $\ell_{d_X}(w)$ is sufficiently close to $\bar{d}_X(\bar{Q}, \bar{Q}')$ that $\ell_{d_X}(w) < \frac{2}{3}\varepsilon$. Then $\bar{Q}'_i = \bar{Q}_{i+1}$ in $\bar{X}$ and, using the fact that the quotient map is distance nonincreasing (Lemma 4.2),

$$\sum_{i=1}^{n} \bar{d}_X(\bar{Q}_i, \bar{Q}_{i+1}) \leqslant \sum_{i=1}^{n} d_X(Q_i, Q'_i) < \tfrac{2}{3}\varepsilon.$$

A repeated use of the Triangle Inequality then shows that

$$\bar{d}_X(\bar{P}, \bar{Q}_i) \leqslant \bar{d}_X(\bar{P}, \bar{Q}_1) + \sum_{j=1}^{i-1} \bar{d}_X(\bar{Q}_j, \bar{Q}_{j+1}) < \tfrac{1}{3}\varepsilon + \tfrac{2}{3}\varepsilon = \varepsilon,$$

so that all $\bar{Q}_i$ are in $B_{\bar{d}_X}(\bar{P}, \varepsilon)$. Since ε satisfies the conclusions of Lemma 4.5, we conclude that all Q_i and Q'_i are in the subset B.

If $P' \neq P'' \in \bar{P}$, then $d_X(P', P'') > 3\varepsilon$ by choice of ε, and the Triangle Inequality shows that any point of the ball $B_{d_X}(P', \varepsilon)$ is at a distance $> \varepsilon$ from any point of $B_{d_X}(P'', \varepsilon)$. Since $d_X(Q_i, Q'_i) < \tfrac{1}{3}\varepsilon$, we conclude that Q_i and Q'_i are in the same ball $B_{d_X}(P', \varepsilon)$. In particular, we observed (right above the statement of Lemma 4.6) that $d_B(Q_i, Q'_i) = d_X(Q_i, Q'_i)$.

What this shows is that w is also a discrete walk from $\bar{Q}$ to $\bar{Q}'$ in B, whose d_B-length $\ell_{d_B}(w)$ is equal to its d_X-length $\ell_{d_X}(w)$. As a consequence, $\bar{d}_B(\bar{Q}, \bar{Q}') \leqslant \ell_{d_X}(w)$.

Since this holds for every discrete walk w whose length $\ell_{d_X}(w)$ is sufficiently close to $\bar{d}_X(\bar{Q}, \bar{Q}')$, we conclude that $\bar{d}_B(\bar{Q}, \bar{Q}') \leqslant \bar{d}_X(\bar{Q}, \bar{Q}')$.

Because we have already shown that the reverse inequality holds, this proves that $\bar{d}_B(\bar{Q}, \bar{Q}') = \bar{d}_X(\bar{Q}, \bar{Q}')$ for every $\bar{Q}, \bar{Q}' \in B_{\bar{d}_X}(\bar{P}, \tfrac{1}{3}\varepsilon)$.   □

We are now ready to prove Theorem 4.4. As in the proof of Lemma 4.5, we will distinguish cases according to the type of the point $P \in X$ corresponding to $\bar{P} \in \bar{X}$.

CASE 1. *P is in the interior of the polygon X.*

In particular, P is glued to no other point so that $\bar{P}$ consists only of P. Then $B = B_{d_X}(P,\varepsilon)$ and, by our choice of ε, the ball $B_{d_X}(P,\varepsilon)$ is completely contained in the interior of X. In particular, the ball $B_{d_X}(P,\varepsilon) \subset X$ is the same as the euclidean ball $B_{d_{\mathrm{euc}}}(P,\varepsilon) \subset \mathbb{R}^2$, an open disk in the euclidean plane $\mathbb{R}^2$. Also, there are no gluings between distinct points of $B = B_{d_X}(P,\varepsilon)$, so that every $\bar{Q} \in B_{\bar{d}_X}(\bar{P},\varepsilon)$ corresponds to exactly one point $Q \in B_{d_X}(P,\varepsilon)$.

Define $\psi \colon B_{\bar{d}_X}(\bar{P},\varepsilon) \to B_{d_{\mathrm{euc}}}(P,\varepsilon)$ by the property that $\psi(\bar{Q}) = Q$ for every $\bar{Q} \in B_{\bar{d}_X}(\bar{P},\varepsilon)$.

The map ψ may not be an isometry over the whole ball, but we claim that

$$d_{\mathrm{euc}}(\psi(\bar{Q}),\psi(\bar{Q}')) = \bar{d}_X(\bar{Q},\bar{Q}')$$

for every $\bar{Q}$, $\bar{Q}' \in B_{\bar{d}_X}(\bar{P}, \tfrac{1}{3}\varepsilon)$. Indeed, $\bar{d}_X(\bar{Q},\bar{Q}') = \bar{d}_B(\bar{Q},\bar{Q}')$ by Lemma 4.6. Since there are no gluings in B, one easily sees that $\bar{d}_B(\bar{Q},\bar{Q}') = d_B(Q,Q')$ (see Exercise 4.3). Finally, $d_B(Q,Q') = d_{\mathrm{euc}}(Q,Q') = d_{\mathrm{euc}}(\psi(\bar{Q}),\psi(\bar{Q}'))$ by convexity of the ball $B = B_{d_X}(P,\varepsilon)$.

This proves that the restriction of ψ to the ball $B_{\bar{d}_X}(\bar{P}, \tfrac{1}{3}\varepsilon)$ is an isometry from $\big(B_{\bar{d}_X}(\bar{P}, \tfrac{1}{3}\varepsilon), \bar{d}_X\big)$ to the euclidean disk $\big(B_{d_{\mathrm{euc}}}(P, \tfrac{1}{3}\varepsilon), d_{\mathrm{euc}}\big)$, as requested.

Having completed the analysis in Case 1, we now directly jump to the most complex case.

CASE 2. *P is a vertex of the polygon X.*

Write $\bar{P} = \{P_1, P_2, \ldots, P_k\}$ with $P = P_1$. Namely, P_1, P_2, $\ldots$, P_k are the vertices of X that are glued to P. Lemma 4.5 says that the ball $B_{\bar{d}_X}(\bar{P},\varepsilon)$ in $\bar{X}$ is the image under the quotient map $\pi \colon X \to \bar{X}$ of the union B of the balls $B_{d_X}(P_1,\varepsilon)$, $B_{d_X}(P_2,\varepsilon)$, $\ldots$, $B_{d_X}(P_k,\varepsilon)$ in X.

Because of our choice of ε, each of the balls $B_{d_X}(P_j,\varepsilon)$ in the metric space (X, d_X) is a disk sector of radius ε in $\mathbb{R}^2$, and these disk sectors are pairwise disjoint. We now need to work harder than in the previous case to rearrange these disk sectors into a full disk.

Each P_j belongs to exactly two edges E_{i_j} and $E_{i'_j}$. As in our description of vertex gluings at the end of Section 4.3.1, we can choose

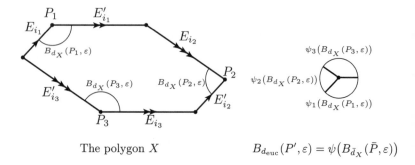

The polygon X $\qquad\qquad$ $B_{d_{\mathrm{euc}}}(P',\varepsilon) = \psi\big(B_{\bar{d}_X}(\bar{P},\varepsilon)\big)$

Figure 4.6. Gluing vertices together

the indexings so that for every j with $1 \leqslant j \leqslant k$, the gluing map φ_{i_j} sends the vertex P_j to P_{j+1} and the edge E_{i_j} to $E_{i'_{j+1}}$ with the convention that $P_{k+1} = P_1$ and $i'_{k+1} = i'_1$.

We will construct our isometry $\psi \colon B_{\bar{d}_X}(\bar{P},\varepsilon) \to B_{d_{\mathrm{euc}}}(P',\varepsilon)$ piecewise from suitable isometries ψ_j of $(\mathbb{R}^2, d_{\mathrm{euc}})$. For this, we use the following elementary property, which we list as a lemma for future reference.

Lemma 4.7. *Let $\varphi \colon g \to g'$ be an isometry between two line segments, half-lines or full-lines g and g' in $\mathbb{R}^2$. Then, φ extends to an isometry $\varphi \colon \mathbb{R}^2 \to \mathbb{R}^2$ of $(\mathbb{R}^2, d_{\mathrm{euc}})$.*

In addition, if we choose one side of g and another side for g', we can arrange that φ sends the selected side of g to the one selected for g'. The isometry φ is then uniquely determined by these properties. □

Lemma 4.7 is an immediate consequence of the classification of isometries of $(\mathbb{R}^2, d_{\mathrm{euc}})$ provided by Proposition 1.3.

In particular, for every j, we can extend the gluing map $\varphi_{i_j} \colon E_{i_j} \to E_{i'_{j+1}}$ to an isometry $\varphi_{i_j} \colon \mathbb{R}^2 \to \mathbb{R}^2$ of $(\mathbb{R}^2, d_{\mathrm{euc}})$ that sends the polygon X to the side of $E_{i'_{j+1}}$ that is opposite X.

To define the ψ_j, we begin with any isometry ψ_1 of $(\mathbb{R}^2, d_{\mathrm{euc}})$, and inductively define

$$\psi_{j+1} = \psi_j \circ \varphi_{i_j}^{-1} = \psi_1 \circ \varphi_{i_1}^{-1} \circ \varphi_{i_2}^{-1} \circ \cdots \circ \varphi_{i_j}^{-1}.$$

By induction on j and because $P_{j+1} = \varphi_j(P_j)$, the map ψ_j sends the vertex P_j to the same point $P' = \psi_1(P)$ for every j. In particular, the isometry ψ_j sends the disk sector $B_{d_X}(P_j, \varepsilon)$ to a disk sector of the disk $B_{d_{\mathrm{euc}}}(P', \varepsilon)$. Similarly, the image of the edge $E_{i'_{j+1}} = \varphi_{i_{j+1}}(E_j)$ under ψ_{j+1} is equal to the image of E_{i_j} under ψ_j. By definition of the extension of ψ_{i_j} to an isometry of $\mathbb{R}^2$, the two disk sectors $\psi_j\big(B_{d_X}(P_j, \varepsilon)\big)$ and $\psi_{j+1}\big(B_{d_X}(P_{j+1}, \varepsilon)\big)$ sit on opposite sides of $\psi_j(E_{i_j}) = \psi_{j+1}(E_{i'_{j+1}})$. It follows that the disk sectors $\psi_j\big(B_{d_X}(P_j, \varepsilon)\big)$ all fit side-by-side and in order of increasing j around their common vertex P'. See Figure 4.6.

It is now time to use the hypothesis that the internal angles of the polygon X at the vertices $P_1, P_2, \ldots, P_k \in \bar{P}$ add up to 2π. This implies that the disk sector $\psi_{k+1}\big(B_{d_X}(P_{k+1}, \varepsilon)\big) = \psi_{k+1}\big(B_{d_X}(P_1, \varepsilon)\big)$ is equal to $\psi_1\big(B_{d_X}(P_1, \varepsilon)\big)$. In particular, the two isometries ψ_{k+1} and ψ_1 of $(\mathbb{R}^2, d_{\mathrm{euc}})$ send $P_1 = P_{k+1}$ to the same point P', send the edge $E_{i_{k+1}} = E_{i'_1}$ to the same line segment or half-line issued from P', and send a side of $E_{i_{k+1}} = E_{i'_1}$ to the same side of $\psi_{k+1}(E_{i_{k+1}}) = \psi_1(E_{i'_1})$. By the uniqueness part of Lemma 4.7, it follows that $\psi_{k+1} = \psi_1$.

Finally note that when $Q \in E_{i_j}$ is glued to $Q' = \varphi_{i_j}(Q) \in E_{i'_{j+1}}$, then $\psi_j(Q) = \psi_{j+1}(Q')$. We can therefore define a map

$$\psi \colon B_{\bar{d}_X}(\bar{P}, \varepsilon) \to B_{d_{\mathrm{euc}}}(P', \varepsilon)$$

by the property that $\psi(\bar{Q})$ is equal to $\psi_j(Q)$ whenever $Q \in B_d(P_j, \varepsilon)$. The above considerations show that ψ is well defined.

We will show that ψ induces an isometry between the corresponding balls of radius $\frac{1}{3}\varepsilon$.

For this, consider two points $\bar{Q}, \bar{Q}' \in B_{\bar{d}_X}(\bar{P}, \frac{1}{3}\varepsilon)$. By Lemma 4.6 and by the Triangle Inequality, $\bar{d}_B(\bar{Q}, \bar{Q}') = \bar{d}_X(\bar{Q}, \bar{Q}') < \frac{2}{3}\varepsilon$. Let w be a discrete walk from $\bar{Q}$ to $\bar{Q}'$ in B, of the form $Q = Q_1$, $Q'_1 \sim Q_2$, $Q'_2 \sim Q_3, \ldots, Q'_{n-1} \sim Q_n$, $Q'_n = Q'$, and whose d_B-length $\ell_{d_B}(w)$ is sufficiently close to $\bar{d}_B(\bar{Q}, \bar{Q}')$ that $\ell_{d_B}(w) < \frac{2}{3}\varepsilon$. In particular, each $d_B(Q_i, Q'_i)$ is finite, so that Q_i and Q'_i belong to the same ball $B_{d_X}(P_{j_i}, \varepsilon)$. As a consequence,

$$d_{\mathrm{euc}}\big(\psi(\bar{Q}_i), \psi(\bar{Q}'_i)\big) = d_{\mathrm{euc}}\big(\psi_{j_i}(Q_i), \psi_{j_i}(Q'_i)\big)$$
$$= d_{\mathrm{euc}}(Q_i, Q'_i) \leqslant d_B(Q_i, Q'_i)$$

since each ψ_{j_i} is a euclidean isometry. Then, by iterating the Triangle Inequality and using the fact that $\bar{Q}'_i = \bar{Q}_{i+1}$,

$$d_{\mathrm{euc}}\big(\psi(\bar{Q}), \psi(\bar{Q})\big) \leqslant \sum_{i=1}^{n-1} d_{\mathrm{euc}}\big(\psi(\bar{Q}_i), \psi(\bar{Q}'_i)\big)$$

$$\leqslant \sum_{i=1}^{n-1} d_B(Q_i, Q'_i) = \ell_{d_B}(w).$$

Since this holds for every discrete walk w from $\bar{Q}$ to $\bar{Q}'$ in B whose length is sufficiently close to $\bar{d}_B(\bar{Q}, \bar{Q}')$, we conclude that

$$(4.2) \qquad\qquad d_{\mathrm{euc}}\big(\psi(\bar{Q}), \psi(\bar{Q})\big) \leqslant \bar{d}_B(\bar{Q}, \bar{Q}').$$

Conversely, let γ be the oriented line segment from $\psi(\bar{Q})$ to $\psi(\bar{Q}')$ in the disk $B_{d_{\mathrm{euc}}}(P', \frac{1}{3}\varepsilon)$. Recall that $B_{d_{\mathrm{euc}}}(P', \frac{1}{3}\varepsilon)$ is decomposed into the disk sectors $\psi_j\big(B_d(P_j, \frac{1}{3}\varepsilon)\big)$. Therefore we can split γ into line segments $\gamma_1, \gamma_2, \ldots, \gamma_n$, in this order, such that each γ_i is contained in a disk sector $\psi_{j_i}\big(B_d(P_{j_i}, \frac{1}{3}\varepsilon)\big)$.

In the disk sector $B_d(P_{j_i}, \frac{1}{3}\varepsilon) \subset X$, consider the oriented line segment $\gamma'_i = \psi_{j_i}^{-1}(\gamma_i)$ corresponding to γ_i. If the endpoints of γ'_i are labelled so that γ'_i goes from Q_i to Q'_i, we now have a discrete walk w from $\bar{Q}$ to $\bar{Q}'$ of the form $Q = Q_1, Q'_1 \sim Q_2, Q'_2 \sim Q_3, \ldots,$ $Q'_{n-1} \sim Q_n, Q'_n = Q'$, of d_B-length

$$\ell_{d_B}(w) = \sum_{i=1}^{n} d_B(Q_i, Q'_i) = \sum_{i=1}^{n} \ell_{\mathrm{euc}}(\gamma'_i) = \sum_{i=1}^{n} \ell_{\mathrm{euc}}(\gamma_i)$$

$$= \ell_{\mathrm{euc}}(\gamma) = d_{\mathrm{euc}}\big(\psi(\bar{Q}), \psi(\bar{Q})\big).$$

It follows that

$$(4.3) \qquad\qquad \bar{d}_B(\bar{Q}, \bar{Q}') \leqslant d_{\mathrm{euc}}\big(\psi(\bar{Q}), \psi(\bar{Q})\big).$$

Combining the inequalities (4.2) and (4.3), we conclude that

$$\bar{d}_X(\bar{Q}, \bar{Q}') = \bar{d}_B(\bar{Q}, \bar{Q}') = d_{\mathrm{euc}}\big(\psi(\bar{Q}), \psi(\bar{Q})\big)$$

for every $\bar{Q}, \bar{Q}' \in B_{\bar{d}_X}(\bar{P}, \frac{1}{3}\varepsilon)$. In other words, ψ induces an isometry from the ball $\big(B_{\bar{d}_X}(\bar{P}, \frac{1}{3}\varepsilon), \bar{d}_X\big)$ to the ball $\big(B_{d_{\mathrm{euc}}}(P', \frac{1}{3}\varepsilon), d_{\mathrm{euc}}\big)$.

This concludes our discussion of Case 2, where P is a vertex of X. We have one case left to consider.

CASE 3. *P is in an edge of the polygon X, but is not a vertex.*

The proof is identical to that of Case 2. Actually, it can even be considered as a special case of Case 2 by viewing P and the point P' that is glued to it as vertices of X where the internal angle is equal to π.

From page 68

This concludes the proof of Theorem 4.4. □

4.5. Gluing hyperbolic and spherical polygons

Before applying Theorems 4.3 and 4.4 to specific examples, let us look at the key ingredients of their proof in more detail. For the proof of Theorem 4.3 (and the proof of Lemma 4.5 before), in addition to standard properties of metric spaces, we mostly used the fact that the maps φ_i gluing one edge of the polygon to another preserved distances. A critical component of the proof of Theorem 4.4 was Lemma 4.7.

4.5.1. Hyperbolic polygons. All these properties have straightforward analogues in the hyperbolic plane $(\mathbb{H}^2, d_{\text{hyp}})$, provided we use the appropriate translation. For instance, the euclidean metric d_{euc} just needs to be replaced by the hyperbolic metric d_{hyp}, euclidean isometries by hyperbolic isometries, line segments and lines by geodesics, etc.... Consequently, our results automatically extend to the hyperbolic context.

The only point that requires some thought is the following property, which replaces Lemma 4.7.

Lemma 4.8. *In the hyperbolic plane $(\mathbb{H}^2, d_{\text{hyp}})$, let $\varphi\colon g \to g'$ be an isometry between two geodesics g and g'. Then, φ extends to an isometry $\varphi\colon \mathbb{H}^2 \to \mathbb{H}^2$ of $(\mathbb{H}^2, d_{\text{hyp}})$.*

In addition, if we choose one side of g and another side for g', we can arrange that φ sends the selected side of g to the one selected for g'. The isometry φ is then uniquely determined by these properties.

Proof. Pick a point $P \in g$ and a nonzero vector $\vec{v}$ tangent to g at P. In particular, $\vec{v}$ defines an orientation for g, which we can transport through φ to obtain an orientation of g'. At the point $P' = \varphi(P) \in g$, let $\vec{v}\,'$ be the vector tangent to g' in the direction of this orientation, and such that $\|\vec{v}\,'\|_{\mathrm{hyp}} = \|\vec{v}\|_{\mathrm{hyp}}$. Proposition 2.20, which shows that $(\mathbb{H}^2, d_{\mathrm{hyp}})$ is isotropic, provides a hyperbolic isometry $\psi \colon \mathbb{H}^2 \to \mathbb{H}^2$ such that $\psi(P) = P'$ and $D_P\psi(\vec{v}) = \vec{v}\,'$. In particular, ψ sends g to the geodesic which is tangent to $D_P\psi(\vec{v}) = \vec{v}\,'$ at $\psi(P) = P'$, namely, g'.

The restriction of ψ to the geodesic g is an isometry $g \to g'$ which sends P to the same point P' as the isometry φ, and sends the orientation of g to the same orientation as φ. It follows that ψ coincides with φ on g. In other words, the isometry $\psi \colon \mathbb{H}^2 \to \mathbb{H}^2$ extends $\varphi \colon g \to g'$.

If ψ sends the selected side of g to the selected side of g', we are done. Otherwise, let ρ be the hyperbolic reflection across g', namely, the isometry of $(\mathbb{H}^2, d_{\mathrm{hyp}})$ induced by the inversion across the euclidean circle containing g'. Because ρ fixes every point of g' and exchanges its two sides, the hyperbolic isometry $\varphi = \rho \circ \psi$ now has the required properties.

The uniqueness easily follows from Lemma 2.10. $\qquad\qquad$ □

We are now ready to carry out our automatic translation from euclidean to hyperbolic geometry.

Let X be a **polygon** in the hyperbolic plane $(\mathbb{H}^2, d_{\mathrm{hyp}})$. Namely, X is a region in $\mathbb{H}^2$ whose boundary in $\mathbb{H}^2$ is decomposed into finitely many hyperbolic geodesics E_1, E_2, ..., E_n meeting only at their endpoints. When we consider X as a subset of $\mathbb{R}^2$, its boundary in $\mathbb{R}^2$ may also include finitely many intervals in the real line $\mathbb{R}$ bounding the hyperbolic plane $\mathbb{H}^2$ in $\mathbb{R}^2 = \mathbb{C}$; if this is the case, note that X will be unbounded for the hyperbolic metric d_{hyp}.

In addition we require that X and the E_i contain all those points of $\mathbb{H}^2$ that are in their boundary. Namely, X and the E_i are **closed** in $\mathbb{H}^2$, although not necessarily in $\mathbb{R}^2$.

The geodesics E_i bounding X are the **edges** of the polygon X. The points where two edges meet are its **vertices**. As in the euclidean case, we require that only two edges meet at any given vertex.

Figure 4.7 offers a few examples. In this figure, X_1 is a hyperbolic octagon, with eight edges and eight vertices; it is bounded for the hyperbolic metric d_{hyp}. The hyperbolic polygon X_2 is an infinite strip, with two edges and no vertex; it touches the line $\mathbb{R}$ along two disjoint intervals, and is unbounded for the hyperbolic metric d_{hyp} (although it is bounded for the euclidean metric of $\mathbb{R}^2$). The hyperbolic quadrilateral X_3 is delimited by our edges, has no vertex in $\mathbb{H}^2$, and touches $\widehat{\mathbb{R}} = \mathbb{R} \cup \{\infty\}$ along four points, one of which is ∞. We will meet these three hyperbolic polygons again in Sections 5.2, 5.4.2 and 5.5, respectively.

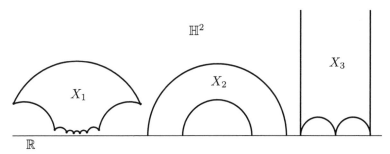

Figure 4.7. A few hyperbolic polygons

For such a hyperbolic polygon X, we can then introduce edge gluing data by, first, grouping the edges together in pairs $\{E_1, E_2\}$, $\{E_3, E_4\}, \ldots, \{E_{2p-1}, E_{2p}\}$ and then, for each such pair $\{E_{2k-1}, E_{2k}\}$, by specifying an isometry $\varphi_{2k-1} \colon E_{2k-1} \to E_{2k}$. Here φ_{2k-1} is required to be an isometry for the hyperbolic distance d_{hyp}.

As in the euclidean case, the edges E_{2k-1} and E_{2k} in the same pair must have the same hyperbolic length, possibly infinite. In general, the isometry φ_{2k-1} is then uniquely determined once we know how φ_{2k-1} sends an orientation of E_{2k-1} to an orientation of E_{2k}, and we will often describe this information by drawing matching arrows on E_{2k-1} and E_{2k}. The case where drawing arrows is not sufficient to specify φ_{2k-1} is when E_{2k-1} and E_{2k} are complete geodesics of $\mathbb{H}^2$, namely, full euclidean semi-circles centered on the real line.

As in the euclidean case, we endow X with the **path metric** d_X for which $d_X(P, Q)$ is the infimum of the hyperbolic lengths of all piecewise differentiable curves joining P to Q in X. When X is *convex*, in the sense that the geodesic arc joining any two P, $Q \in X$ is contained in X, the metric d_X clearly coincides with the restriction of the hyperbolic metric d_{hyp}.

Theorem 4.9. *If $\bar{X}$ is obtained from the hyperbolic polygon X by gluing pairs of its edges by isometries, then the gluing is proper. Namely, the semi-distance $\bar{d}_X$ induced on $\bar{X}$ by the path metric d_X of X is such that $\bar{d}_X(\bar{P}, \bar{Q}) > 0$ when $\bar{P} \neq \bar{Q}$.*

Proof. The proof is identical to that of Theorem 4.3. Just follow each step of that proof, using the appropriate translation.   □

Theorem 4.10. *Let $(\bar{X}, \bar{d}_X)$ be the quotient metric space obtained from a hyperbolic polygon (X, d_X) by gluing together pairs of edges of X by hyperbolic isometries. Suppose that the following additional condition holds: For every vertex P of X, the angles of X at those vertices P' of X which are glued to P add up to 2π. Then $(\bar{X}, \bar{d})$ is locally isometric to the hyperbolic plane $(\mathbb{H}^2, d_{\text{hyp}})$.*

Proof. The proof is identical to that of Theorem 4.4, provided that we replace Lemma 4.7 by Lemma 4.8.   □

A metric space (X, d) which is locally isometric to the hyperbolic plane $(\mathbb{H}^2, d_{\text{euc}})$ is a **hyperbolic surface**. Equivalently, the metric d is then a **hyperbolic metric**.

4.5.2. Spherical polygons. The same properties also generalize to polygons in the sphere $(\mathbb{S}^2, d_{\text{sph}})$.

A **polygon** in the sphere $(\mathbb{S}^2, d_{\text{sph}})$ is a region X of $\mathbb{S}^2$ whose boundary is decomposed into finitely many geodesics E_1, E_2, ..., E_n meeting only at their endpoints. These E_i are the **edges** of the polygon X, and the points where they meet are its **vertices**. As before, we require that X contains all its edges and vertices, and that every edge contains its endpoints. Also, exactly two edges meet at a given vertex.

We endow X with the **path metric** d_X for which $d_X(P, Q)$ is the infimum of the euclidean lengths of all piecewise differentiable curves joining P to Q in $X \subset \mathbb{S}^2 \subset \mathbb{R}^3$. When X is **convex**, in the sense that any two P, $Q \in X$ can be joined by a geodesic arc of $\mathbb{S}^2$ of length $\leqslant \pi$

which is completely contained in X, the metric d_X clearly coincides with the restriction of the spherical metric d_{sph}.

After grouping the edges together in pairs $\{E_1, E_2\}$, $\{E_3, E_4\}$, ..., $\{E_{2p-1}, E_{2p}\}$, the gluing data used consists of isometries φ_{2k-1}: $E_{2k-1} \to E_{2k}$.

As in the hyperbolic case, the key property is the following extension of Lemma 4.7.

Lemma 4.11. *In the sphere* $(\mathbb{S}^2, d_{\mathrm{sph}})$, *let* $\varphi\colon g \to g'$ *be an isometry between two geodesics* g *and* g'. *Then* φ *extends to an isometry* φ: $\mathbb{S}^2 \to \mathbb{S}^2$ *of* $(\mathbb{S}^2, d_{\mathrm{sph}})$.

In addition, if we choose one side for g *and another side for* g', *we can arrange that* φ *sends the selected side for* g *to the one selected for* g'. *The isometry* φ *is then uniquely determined by these properties.*

Proof. Since geodesics of $(\mathbb{S}^2, d_{\mathrm{sph}})$ are great circle arcs (Theorem 3.1), this is easily proved by elementary arguments in 3-dimensional euclidean geometry. □

As before, we endow the spherical polygon X with the path metric d_X for which $d_X(P, Q)$ is the infimum of the spherical lengths of all piecewise differentiable curves joining P to Q in X.

Then, by replacing Lemma 4.7 by Lemma 4.11, the proofs of Theorems 4.3 and 4.4 immediately extend to the spherical context and give the following two results.

Theorem 4.12. *If* $\bar{X}$ *is obtained from the spherical polygon* X *by gluing pairs of its edges by isometries, then the gluing is proper. Namely, the semi-distance* $\bar{d}_X$ *induced on* $\bar{X}$ *by the path metric* d_X *is really a metric.* □

Theorem 4.13. *Let* $(\bar{X}, \bar{d}_X)$ *be the quotient metric space obtained from a spherical polygon* (X, d_X) *by gluing together pairs of edges of* X *by hyperbolic isometries. Suppose that the following additional condition holds: For every vertex* P *of* X, *the angles of* X *at those vertices* P' *of* X *which are glued to* P *add up to* 2π. *Then* $(\bar{X}, \bar{d}_X)$ *is locally isometric to the sphere* $(\mathbb{S}^2, d_{\mathrm{sph}})$. □

As in the euclidean and hyperbolic cases, a metric space (X, d) which is locally isometric to the sphere $(\mathbb{S}^2, d_{\mathrm{sph}})$ is a **spherical surface**. Equivalently, the metric d is then a **spherical metric**.

Exercises for Chapter 4

Exercise 4.1. Let X be the closed interval $[0, 1]$ in $\mathbb{R}$. Let $\bar{X}$ be the partition consisting of all the subsets $\{\frac{1}{m 2^n}, \frac{m}{2^n}\}$ where m, $n \in \mathbb{N}$ are integers such that m is odd and $1 < m < 2^n$, and of all one-element subsets $\{P\}$ where $P \in [0, 1]$ is not of the form $P = \frac{1}{m 2^n}$ or $\frac{m}{2^n}$ as above. Let $\bar{d}_{\mathrm{euc}}$ be the quotient semi-metric induced on $\bar{X}$ by the usual metric $d_{\mathrm{euc}}(P, Q) = |P - Q|$ of $X = [0, 1]$.

   a. Show that for every $P \in [0, 1]$ and every $\varepsilon > 0$, there exists $Q_1, Q_2 \in [0, 1]$ such that $\bar{Q}_1 = \bar{Q}_2$ in $\bar{X}$, $d_{\mathrm{euc}}(0, Q_1) < \varepsilon$, and $d_{\mathrm{euc}}(P, Q_2) < \varepsilon$.

   b. Show that $\bar{d}_{\mathrm{euc}}(\bar{0}, \bar{P}) = 0$ for every $\bar{P} \in \bar{X}$. In particular, the semi-metric $\bar{d}_{\mathrm{euc}}$ is not a metric, and the gluing is not proper.

   c. Show that $\bar{d}_{\mathrm{euc}}(\bar{P}, \bar{Q}) = 0$ for every $\bar{P}, \bar{Q} \in \bar{X}$.

Exercise 4.2 (Equivalence relations and partitions). A **relation** on a set X is just a subset $\mathcal{R}$ of the product $X \times Y$. One way to think of this is that $\mathcal{R}$ describes a certain property involving two points of X. Namely, P and $Q \in X$ satisfy this property exactly when the pair (P, Q) is an element of $\mathcal{R}$. To emphasize this interpretation, we write $P \sim Q$ to say that $(P, Q) \in \mathcal{R}$.

   An **equivalence relation** is a relation such that

    (i) $P \sim P$ for every $P \in X$ (Reflexivity Property);
    (ii) if $P \sim Q$, then $Q \sim P$ (Symmetry Property);
    (iii) if $P \sim Q$ and $Q \sim R$, then $P \sim Q$ (Transitivity Property).

   a. Given an equivalence relation, define the **equivalence class** of $P \in X$ as

$$\bar{P} = \{Q \in X; P \sim Q\}.$$

Show that as P ranges over all points of X, the family of the equivalence classes $\bar{P}$ is a partition of X.

   b. Conversely, let $\bar{X}$ be a partition of the set X and, as usual, let $\bar{P} \in \bar{X}$ denote the subset that contains $P \in X$. Define a relation on X by the property that $P \sim Q$ exactly when P and Q belong to the same subset $\bar{P} = \bar{Q}$ of the partition. Show that $\sim$ is an equivalence relation.

Exercise 4.3 (Trivial gluing). Let (X, d) be a metric space, and consider the trivial gluing where the partition $\bar{X}$ consists only of the one-element subsets $\bar{P} = \{P\}$. In other words, no two distinct elements of X are glued together in $\bar{X}$. Let $(\bar{X}, \bar{d})$ be the resulting quotient semi-metric space.

Using the definition of the quotient semi-metric $\bar{d}$ in terms of discrete walks, rigorously prove that the quotient map $\pi\colon X \to \bar{X}$ defined by $\pi(P) = \bar{P}$ is an isometry from (X, d) to $(\bar{X}, \bar{d})$.

Exercise 4.4 (Iterated gluings). Let $\bar{X}$ be a partition of the metric space (X, d), and let $\bar{\bar{X}}$ be a partition of the quotient space $\bar{X}$. Let $\bar{d}$ be the semi-metric induced by d on $\bar{X}$, and let $\bar{\bar{d}}$ be the semi-metric induced by $\bar{d}$ on $\bar{\bar{X}}$.

a. If $P \in X$, the element $\bar{\bar{P}} = \overline{\bar{P}} \in \bar{\bar{X}}$ is a family of subsets of X, and we can consider their union $\dot{P} \subset X$. Show that the subsets $\dot{P}$ form a partition $\dot{X}$ of X.

b. Let $\varphi\colon \bar{\bar{X}} \to \dot{X}$ be the map defined by the property that $\varphi(\bar{\bar{P}}) = \dot{P}$ for every $P \in X$. Show that φ is bijective.

c. Let w be a discrete walk $P = P_1$, $Q_1 \sim P_2$, ..., $Q_{n-1} \sim P_n$, $Q_n = Q$ from $\dot{P}$ to $\dot{Q}$ in $\dot{X}$. Show that $\bar{P} = \bar{P}_1$, $\bar{Q}_1 \sim \bar{P}_2$, ..., $\bar{Q}_{n-1} \sim \bar{P}_n$, $\bar{Q}_n = \bar{Q}$ forms a discrete walk $\bar{w}$ from $\bar{P}$ to $\bar{Q}$ in $\bar{X}$, whose length is such that $\ell_{\bar{d}}(\bar{w}) \leqslant \ell_d(w)$. (Beware that the same symbol $\sim$ is used to refer to gluing with respect to the partition $\dot{X}$ in the first case, and with respect to $\bar{X}$ in the second instance.) Conclude that if $\dot{d}$ is the quotient semi-metric induced by d on $\dot{X}$, then $\bar{\bar{d}}(\bar{\bar{P}}, \bar{\bar{Q}}) \leqslant \dot{d}(\dot{P}, \dot{Q})$ for every P, $Q \in X$.

d. Given a small $\varepsilon > 0$, let $\bar{w}$ be a discrete walk $\bar{P} = \bar{P}_1$, $\bar{Q}_1 \sim \bar{P}_2$, ..., $\bar{Q}_{n-1} \sim \bar{P}_n$, $\bar{Q}_n = \bar{Q}$ from $\bar{\bar{P}}$ to $\bar{\bar{Q}}$ in $\bar{\bar{X}}$ whose length is sufficiently close to $\bar{\bar{d}}(\bar{\bar{P}}, \bar{\bar{Q}})$ that $\ell_{\bar{d}}(\bar{w}) \leqslant \bar{\bar{d}}(\bar{\bar{P}}, \bar{\bar{Q}}) + \frac{\varepsilon}{2}$. Similarly, for every i, choose a discrete walk w_i from $\bar{P}_i$ to $\bar{Q}_i$, consisting of $P_i = P_{i,1}$, $Q_{i,1} \sim P_{i,2}$, ..., $Q_{i,k_i-1} \sim P_{i,k_i}$, $Q_{i,k_i} = Q_i$ whose length is sufficiently close to $\bar{d}(\bar{P}_i, \bar{Q}_i)$ that $\ell_d(w_i) \leqslant \bar{d}(\bar{P}_i, \bar{Q}_i) + \frac{\varepsilon}{2n}$. Show that the w_i can be chained together to form a discrete walk w from $\dot{P}$ to $\dot{Q}$ in $\dot{X}$ such that $\ell_d(w) \leqslant \ell_{\bar{d}}(\bar{w}) + \frac{\varepsilon}{2}$. Conclude that $\dot{d}(\dot{P}, \dot{Q}) \leqslant \bar{\bar{d}}(\bar{\bar{P}}, \bar{\bar{Q}}) + \varepsilon$.

e. Show that φ is an isometry from $(\bar{\bar{X}}, \bar{\bar{d}})$ to $(\dot{X}, \dot{d})$.

In other words, a two-step gluing construction yields the same quotient semi-metric space as gluing everything together in one single action.

Exercise 4.5. Let X be the interval $[0, 2\pi] \subset \mathbb{R}$, and let $\bar{X}$ be the partition consisting of the two-element subset $\{0, 2\pi\}$ and of all the one-element subsets $\{P\}$ with $P \in (0, 2\pi)$. Let $\bar{d}$ be the quotient semi-metric induced on $\bar{X}$ by the usual metric $d(P, Q) = |Q - P|$ of $X = [0, 2\pi]$. In other words, $(\bar{X}, \bar{d})$ is obtained by gluing together the two endpoints of the interval $X = (0, 2\pi)$ endowed with the metric d.

Let $\mathbb{S}^1 = \{(x, y) \in \mathbb{R}^2; x^2 + y^2 = 1\}$ be the unit circle in the euclidean plane, endowed with the metric $d_{\mathbb{S}^1}$ for which $d_{\mathbb{S}^1}(P, Q)$ is the infimum of the euclidean arc lengths $\ell_{\mathrm{euc}}(\gamma)$ of all piecewise differentiable curves γ going from P to Q in $\mathbb{S}^1$. Consider the map $\varphi \colon X = [0, 2\pi] \to \mathbb{S}^1$ defined by $\varphi(t) = (\cos t, \sin t)$.

    a. Show that if $\pi \colon X \to \bar{X}$ denotes the quotient map, there exists a unique map $\bar{\varphi} \colon \bar{X} \to \mathbb{S}^1$ such that $\varphi = \bar{\varphi} \circ \pi$. Show that $\bar{\varphi}$ is bijective.

    b. Show that for every discrete walk w from $\bar{P}$ to $\bar{Q}$ in $\bar{X}$, there exists a piecewise differentiable curve γ going from $\varphi(P)$ to $\varphi(Q)$ in $\mathbb{S}^1$ whose length $\ell_{\mathrm{euc}}(\gamma)$ is equal to the length $\ell_d(w)$ of w. Conclude that $d_{\mathbb{S}^1}\big(\bar{\varphi}(\bar{P}), \bar{\varphi}(\bar{Q})\big) \leqslant \bar{d}(\bar{P}, \bar{Q})$ for every $\bar{P}, \bar{Q} \in \bar{X}$.

    c. Conversely, show that for every $P, Q \in X$, there exists a discrete walk w from $\bar{P}$ to $\bar{Q}$ in $\bar{X}$ involving at most four points of X, whose length $\ell_d(w)$ is equal to $d_{\mathbb{S}^1}\big(\varphi(P), \varphi(Q)\big)$. (Hint: Consider a shortest curve from $\varphi(P)$ to $\varphi(Q)$ in $\mathbb{S}^1$.) Conclude that $\bar{d}(\bar{P}, \bar{Q}) \leqslant d_{\mathbb{S}^1}\big(\bar{\varphi}(\bar{P}), \bar{\varphi}(\bar{Q})\big)$ for every $\bar{P}, \bar{Q} \in \bar{X}$.

    d. Combine these results to show that φ is an isometry from $(\bar{X}, \bar{d})$ to $(\mathbb{S}^1, d_{\mathbb{S}^1})$.

Exercise 4.6. In the euclidean plane $\mathbb{R}^2$, let D_1, D_2 and D_3 be three disjoint euclidean disk sectors of radius r and respective angles θ_1, θ_2 and θ_3 with $\theta_1 + \theta_2 = \theta_3 \leqslant \pi$. Let $\bar{X}$ be the quotient space obtained from $X = D_1 \cup D_2$ by isometrically gluing one edge of D_1 to an edge of D_2, sending the vertex of D_1 to the vertex of D_2. Show that if d_X and d_{D_3} are the euclidean path metrics defined as in Section 4.3.1, the quotient space $(\bar{X}, \bar{d}_X)$ is isometric to $(\bar{D}_3, d_{D_3})$. Hint: Copy parts of the proof of Theorem 4.4.

Exercise 4.7 (Euclidean cones)**.** Let D_1, D_2, ..., D_n be n disjoint euclidean disk sectors with radius r and with angles θ_1, θ_2, ..., θ_n, respectively, and let E_i and E'_i denote the two edges of D_i. Isometrically glue each edge E_i to E'_{i+1}, sending the vertex of D_i to the vertex of D_{i+1} and counting indices modulo n (so that $E'_{n+1} = E'_1$). Show that the resulting quotient space $(\bar{X}, \bar{d}_X)$ depends only on the radius r and on the angle sum $\theta = \sum_{i=1}^{n} \theta_i$. Namely, if D'_1, D'_2, ..., $D'_{n'}$ is another family of n' disjoint euclidean disk sectors of the same radius r and with respective angles θ'_1, θ'_2, ..., $\theta'_{n'}$ with $\sum_{i=1}^{n'} \theta'_i = \sum_{i=1}^{n} \theta_i$, and if these disk sectors are glued together as above, then the resulting quotient space $(\bar{X}', \bar{d}_{X'})$ is isometric to $(\bar{X}, \bar{d}_X)$. Hint: Use the results of Exercises 4.4 and 4.6 to reduce the problem to the case where $n = n'$ and $\theta_i = \theta'_i$ for every i, and to make sure that the order of the gluings does not matter.

    The space $(\bar{X}, \bar{d}_X)$ of Exercise 4.7 is a **euclidean cone** with radius r and cone angle θ. Figure 4.8 represents a few examples. Note the different shape according to whether the cone angle θ is less than, equal to, or more than 2π. When θ is equal to 2π, the cone is of course isometric to a euclidean disk.

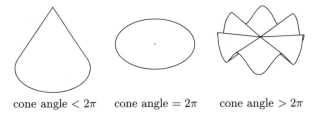

cone angle $< 2\pi$ cone angle $= 2\pi$ cone angle $> 2\pi$

Figure 4.8. Three euclidean cones

Exercise 4.8 (Surfaces with cone singularities). Let $(\bar{X}, \bar{d}_X)$ be the quotient metric space obtained from a euclidean polygon (X, d_X) by isometrically gluing together its edges. Show that for every $\bar{P} \in \bar{X}$, there exists a radius r such that the ball $B_{\bar{d}_X}(\bar{P}, r)$ is isometric to a euclidean cone, defined as in Exercise 4.7.

A space $(\bar{X}, \bar{d}_X)$ satisfying the conclusions of Exercise 4.8 is a *eu-clidean surface with cone singularities*. One can similarly define hyperbolic and spherical surfaces with cone singularities.

Exercise 4.9.

a. Let (C, d) be a euclidean cone with center P_0, radius r and cone angle θ as in Exercise 4.7. Show that for every $r' < r$, the "circle"

$$S_d(P_0, r') = \{P \in C; d(P, P_0) = r'\}$$

is a closed curve, whose length $\ell_d\big(S_d(P_0, r')\big)$ in the sense of Exercise 1.11 is equal to $\theta r'$.

b. Conclude that the angle condition of Theorem 4.4 is necessary for its conclusion to hold. Namely, if, when isometrically gluing together the sides of a euclidean polygon X, there is a vertex P such that the angles of X at those vertices P' which are glued to P do not add up to 2π, then the quotient space $(\bar{X}, \bar{d})$ is not locally isometric to the euclidean plane $(\mathbb{R}^2, d_{\text{euc}})$

Chapter 5

Gluing examples

After suffering through the long proofs of Section 4.4, we can now harvest the fruit of our labors, and apply the technology that we have built in Chapter 4 to a few examples.

5.1. Some euclidean surfaces

We begin by revisiting, in a more rigorous setting, the example of the torus that we had informally discussed in Section 4.1.

5.1.1. Euclidean tori from rectangles and parallelograms. Let X_1 be the rectangle $[a,b] \times [c,d]$, consisting of those $(x,y) \in \mathbb{R}^2$ such that $a \leqslant x \leqslant b$ and $c \leqslant y \leqslant d$. Glue the bottom edge $E_1 = [a,b] \times \{c\}$ to the top edge $E_2 = [a,b] \times \{d\}$ by the isometry $\varphi_1 \colon [a,b] \times \{c\} \to [a,b] \times \{d\}$ defined by $\varphi_1(x,c) = (x,d)$, and glue the left edge $E_3 = \{a\} \times [c,d]$ to the right edge $E_4 = \{b\} \times [c,d]$ by $\varphi_3 \colon \{a\} \times [c,d] \to \{b\} \times [c,d]$ defined by $\varphi_3(a,y) = (b,y)$. Namely, we consider the edge gluing that already appeared in Section 4.1 and which we reproduce in Figure 5.1.

With these edge identifications, the four vertices of the rectangle are glued together to form a single point $\bar{P}$ of the quotient space $\bar{X}_1$. The sum of the angles of X_1 at these vertices is

$$\frac{\pi}{2} + \frac{\pi}{2} + \frac{\pi}{2} + \frac{\pi}{2} = 2\pi.$$

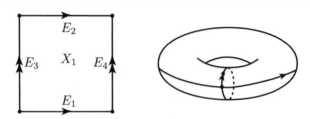

Figure 5.1. Gluing opposite sides of a rectangle

We can therefore apply Theorems 4.3 and 4.4. Note that X_1 is convex, so that the path metric d_{X_1} coincides with the resticiton of the euclidean metric d_{euc} of $\mathbb{R}^2$. Then Theorems 4.3 and 4.4 show that the metric space $(\bar{X}_1, \bar{d}_{X_1})$ is locally isometric to the euclidean metric of the euclidean plane $(\mathbb{R}^2, d_{\mathrm{euc}})$.

This is our first rigorous example of a euclidean surface.

In our informal discussion of this example in Section 4.1, we explained how $\bar{X}_1$ can be identified to the torus illustrated on the right of Figure 5.1 if we are willing to stretch the metric. The mathematically rigorous way to express this property is to use the language of topology and to say that the space $\bar{X}_1$ is homeomorphic to the torus.

A **homeomorphism** from a metric space (X, d) to another metric space (X', d') is a bijection $\varphi \colon X \to X'$ such that both φ and its inverse φ^{-1} are continuous. (See Section T.1 in the TOOL KIT for the definition of bijections and inverse maps.) The homeomorphism φ can be used as a dictionary between X and X' to translate back and forth every property involving limits and continuity. For instance, a sequence $P_1, P_2, \ldots, P_n, \ldots$ converges to the point P_∞ in X if and only if its image $\varphi(P_1), \varphi(P_2), \ldots, \varphi(P_n), \ldots$ converges to $\varphi(P_\infty)$ in X'. As a consequence, if the metric spaces (X, d) and (X', d') are **homeomorphic**, in the sense that there exists a homeomorphism $(X, d) \to (X', d')$, then (X, d) and (X', d') share exactly the same limit and continuity properties.

An example of a homeomorphism is provided by an isometry from (X, d) to (X', d'). However, a general homeomorphism $\varphi \colon X \to X'$ is much more general in the sense that the distance $d'\big(\varphi(P), \varphi(Q)\big)$ may be very different from $d(P, Q)$. The only requirement is that

$d'\big(\varphi(P), \varphi(Q)\big)$ is small exactly when $d(P, Q)$ is small (in a sense quantified with the appropriate ε and δ).

In general, we will keep our discussion of homeomorphisms at a very informal level. However, we should perhaps go through at least one example in detail.

Let the 2-**dimensional torus** be the surface $\mathbb{T}^2$ of the 3-dimensional space $\mathbb{R}^3$ obtained by revolving about the z-axis the circle in the xz-plane that is centered at the point $(R, 0, 0)$ and has radius $r < R$, as in the right-hand side of Figure 5.1. We consider $\mathbb{T}^2$ as a metric space by endowing it with the restriction of the 3-dimensional metric d_{euc} of $\mathbb{R}^3$. Different choices of r and R give different subsets of $\mathbb{R}^3$, but these are easily seen to be homeomorphic.

Lemma 5.1. *Let $(\bar{X}_1, \bar{d}_{X_1})$ be the quotient metric space obtained from the rectangle $X_1 = [a, b] \times [c, d]$ by gluing together opposite edges by euclidean translations. Then $\bar{X}_1$ is homeomorphic to the 2-dimensional torus $\mathbb{T}^2$.*

Proof. To simplify the notation, we restrict attention to the case of the square $X_1 = [-\pi, \pi] \times [-\pi, \pi]$. However, the argument straightforwardly extends to general rectangles $[a, b] \times [c, d]$ by suitable rescaling of the variables.

Let $\rho \colon X_1 \to \mathbb{T}^2$ be the map defined by

$$\rho(\theta, \varphi) = \big((R + r \cos \varphi) \cos \theta, (R + r \cos \varphi) \sin \theta, r \sin \varphi\big).$$

Geometrically, $\rho(\theta, \varphi)$ is obtained by rotating by an angle of θ around the z-axis the point $(R + r \cos \varphi, 0, r \sin \varphi)$ of the circle in the xz-plane with center $(R, 0, 0)$ and radius r. From this description it is immediate that $\rho(\theta, \varphi) = \rho(\theta', \varphi')$ exactly when (θ, φ) and (θ', φ') are glued together to form a single point of $\bar{X}_1$. It follows that ρ induces a bijection $\bar{\rho} \colon \bar{X}_1 \to \mathbb{T}^2$ defined by the property that $\bar{\rho}(\bar{P}) = \rho(P)$ for any $P \in \bar{P}$.

The map ρ is continuous by the usual calculus arguments. Using the property that $\bar{d}(\bar{P}, \bar{Q}) \leqslant d(P, Q)$ (Lemma 4.2), it easily follows that $\bar{\rho}$ is continuous.

To prove that the inverse function $\bar\rho^{-1}\colon \mathbb{T}^2 \to \bar X_1$ is continuous at the point $Q_0 \in \mathbb{T}^2$, we need to distinguish cases according to the type of the point Q_0.

Consider the most complex case, where $Q_0 = (-R+r, 0, 0)$ is the image under $\bar\rho$ of the point $\bar P_0 \in \bar X_1$ corresponding to the four vertices $(\pm\pi, \pm\pi)$ of X_1. If $Q = (x, y, z) \in \mathbb{T}^2$ is near Q_0, we can explicitly compute all (θ, φ) such that $\rho(\theta, \varphi) = Q$. Indeed, $\varphi = \pi - \arcsin \frac{z}{r}$ if $z > 0$, $\varphi = -\pi - \arcsin \frac{z}{r}$ if $z < 0$, and $\varphi = \pm\pi$ if $z = 0$. Similarly, $\theta = \pi - \arcsin \frac{y}{R + r\cos\varphi}$ if $y > 0$, $\theta = -\pi - \arcsin \frac{y}{R + r\cos\varphi}$ if $y < 0$, and $\theta = \pm\pi$ if $y = 0$. By continuity of the function arcsin, it follows that (θ, φ) will be arbitrarily close to one of the corners $(\pm\pi, \pm\pi)$ if $Q = (x, y, z)$ is sufficiently close to $Q_0 = (-R + r, 0, 0)$.

For $\varepsilon > 0$ small enough, Lemma 4.5 shows that the ball $B_{\bar d}(\bar P_0, \varepsilon)$ is just the image under the quotient map $X_1 \to \bar X_1$ of the four quarter-disks of radius ε centered at the four vertices $(\pm\pi, \pm\pi)$ of X_1. By the observations above, there exists a $\delta > 0$ such that whenever $Q \in \mathbb{T}^2$ is such that $d_{\mathrm{euc}}(Q, Q_0) < \delta$, any $(\theta, \varphi) \in X_1$ with $\rho(\theta, \varphi) = Q$ is within a distance $< \varepsilon$ of one of the vertices $(\pm\pi, \pm\pi)$. Since $\bar\rho^{-1}(Q)$ is the image in $\bar X_1$ of any $(\theta, \varphi) \in X_1$ with $\rho(\theta, \varphi) = Q$, we conclude that $\bar d(\bar\rho^{-1}(Q), P_0) < \varepsilon$.

Since $P_0 = \bar\rho^{-1}(Q_0)$, this proves that $\bar\rho^{-1}$ is continuous at $Q_0 = (-R + r, 0, 0)$.

The continuity at the other $Q_0 \in \mathbb{T}^2$ is proved by a similar case-by-case analysis. □

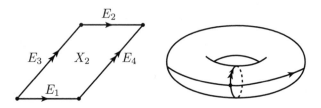

Figure 5.2. Gluing opposite sides of a parallelogram

We can consider a variation of this example by replacing the rectangle by a parallelogram X_2 and gluing again opposite sides by translations. As in the case of the rectangle, the four vertices of the parallelogram are glued to a single point. Because the angles of a euclidean parallelogram add up to 2π, the angle condition of Theorem 4.4 is satisfied, and we conclude that the quotient metric space $(\bar{X}_2, \bar{d})$ is a euclidean surface.

The parallelogram X_2 can clearly be stretched to assume the shape of a rectangle in such a way that the gluing data for X_2 gets transposed to the gluing data for X_1. See Exercise 5.1. It follows that the quotient surface $\bar{X}_2$ is again a torus.

5.1.2. Euclidean Klein Bottles. Given a rectangle $X_3 = [a, b] \times [c, d]$, we can also glue its sides together using different gluing maps. For instance, we can still glue the bottom edge $E_1 = [a, b] \times \{c\}$ to the top edge $E_2 = [a, b] \times \{d\}$ by the isometry $\varphi_1 \colon [a, b] \times \{c\} \to [a, b] \times \{d\}$ defined by $\varphi_1(x, c) = (x, d)$, but glue the left edge $E_3 = \{a\} \times [c, d]$ to the right edge $E_4 = \{b\} \times [c, d]$ by $\varphi_3 \colon \{a\} \times [c, d] \to \{b\} \times [c, d]$ defined by $\varphi_3(a, y) = (b, d - y)$. Namely, the gluing map flips the left edge upside down before sending it to the right edge by a translation.

Again, the four vertices of X_3 are glued together to form a single point of the quotient space $\bar{X}_3$. Since the angles of X_3 at these four vertices add up to 2π, the combination of Theorems 4.3 and 4.4 shows that $(\bar{X}_3, \bar{d}_{X_3})$ is locally isometric to the euclidean plane $(\mathbb{R}^2, d_{\mathrm{euc}})$.

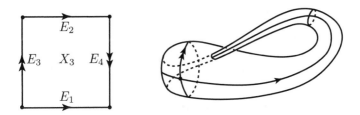

Figure 5.3. Another way of gluing opposite sides of a rectangle

To understand the global shape of $\bar{X}_3$, we first glue the bottom and top sides together, to form a cylinder as in the case of the torus. We then need to glue the left side of the cylinder to the right side by a translation *followed by a flip*. This time, the difficulty of physically realizing this in 3-dimensional space goes well beyond the need for

stretching the paper. It can actually be shown to be impossible to realize, in the sense that there is no subset of $\mathbb{R}^3$ which is homeomorphic
to $\bar{X}_3$.

The right-hand side of Figure 5.3 offers an approximation, where
the surface crosses itself along a closed curve. Each point of this
self-intersection curve has to be understood as corresponding to two
points of the surface $\bar{X}_3$. Introducing an additional space dimension,
this picture can also be used to represent an object in 4–dimensional
space. This is similar to the way a figure eight ∞ in the plane can
be deformed to a curve ∞ with no self-intersection in 3-dimensional

Photo credit: Acme Klein Bottle.

Figure 5.4. A physical Klein bottle

space, by pushing parts of the figure eight up and down near the
point where it crosses itself. In the same way, the object represented
on the right-hand side of Figure 5.3 can be deformed to a subset of
the 4–dimensional space $\mathbb{R}^4$ that is homeomorphic to $\bar{X}_3$.

The surface $\bar{X}_3$ is a **Klein bottle**.

The Klein bottle was introduced in 1882 by Felix Klein (1849–
1925), as an example of pathological surface. The "bottle" terminology is usually understood to reflect the fact that a Klein bottle
can be obtained from a regular wine bottle by stretching its neck and
connecting it to the base after passing inside of the bottle. Another interpretation (unverified, and not incompatible with the previous one)
claims that it comes from a bad pun, or a bad translation from the
German, in which the *Kleinsche Fläche* (Klein surface) became the
Kleinsche Flasche (Klein bottle). The latter version probably provides a better story, whereas the first one makes for better pictures.
This second point is well illustrated by the physical glass model of
Figure 5.4, borrowed from the web site www.kleinbottle.com of the
company *Acme Klein Bottle*, which offers many Klein bottle-shaped
products for sale.

5.1.3. Gluing opposite sides of a hexagon. Let us now go beyond quadrilaterals and consider a hexagon X_4 where we glue opposite
edges together, as indicated on Figure 5.5; compare also Figures 4.3
and 4.6. The vertices of X_4 project to two points of the quotient space
$\bar{X}_4$, each corresponding to three vertices of X_4. More precisely, if we
label the vertices P_1, P_2, ..., P_6 in this order as one goes around the
hexagon, the odd vertices P_1, P_3 and P_5 are glued together to form
one point of $\bar{X}_4$, and the even vertices P_2, P_4 and P_6 form another
point of $\bar{X}_4$. Consequently, we need the hexagon X_4 to satisfy the
following two conditions:

(1) opposite edges have the same length;

(2) the angles of X_4 at the odd vertices P_1, P_3, P_5 add up to
2π.

Recall that the sum of the angles of a euclidean hexagon is always
equal to 4π, so that condition (2) is equivalent to the property that
the angles of X_4 at its even vertices P_2, P_4, P_6 add up to 2π. A little

exercise in elementary euclidean geometry shows that in a hexagon
satisfying conditions (1) and (2) above, opposite edges are necessarily
parallel; see Exercise 5.7.

If the hexagon X_4 satisfies conditions (1) and (2), we can again
apply Theorems 4.3 and 4.4 to show that the quotient metric space
$(\bar{X}_4, \bar{d}_{X_4})$ is a euclidean surface.

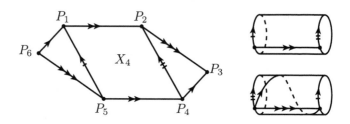

Figure 5.5. Gluing opposite sides of a hexagon

To understand the global shape of $\bar{X}_4$ up to homeomorphism,
we can consider the diagonals P_1P_5 and P_2P_4 of the hexagon X_4,
as in Figure 5.5. These two diagonals cut the hexagon into three
pieces, the parallelogram $P_1P_2P_4P_5$ and the two triangles $P_2P_3P_4$
and $P_5P_6P_1$. Gluing the edges P_1P_2 and P_4P_5 of the parallelogram
$P_1P_2P_4P_5$ provides a cylinder, whose boundary consists of the two
images of the diagonals P_1P_5 and P_2P_4. Similarly, gluing the two
triangles $P_2P_3P_4$ and $P_5P_6P_1$ together by identifying the edge P_2P_3
to the edge P_5P_6 and the edge P_3P_4 to P_6P_1 gives another cylinder,
whose boundary again corresponds to the images of the diagonals
P_1P_5 and P_2P_4. See the right-hand side of Figure 5.5. This proves
that splitting the quotient space $\bar{X}_4$ along the images of the diagonals
P_1P_5 and P_2P_4 gives two cylinders. In particular, $\bar{X}_4$ can be recovered
from these two cylinders by gluing them back together according to
the pattern described on the right of Figure 5.5. It easily follows that
$\bar{X}_4$ is homeomorphic to the torus.

As announced in an earlier disclaimer, this discussion of the construction of a homeomorphism from the quotient space $\bar{X}_4$ to the
torus is somewhat informal. However, with a little reflection, you
should be able to convince yourself that this description could be

made completely rigorous if needed. The same will apply to other informal descriptions of homeomorphisms later on.

5.2. The surface of genus 2

We saw in the last section that if we start with a euclidean rectangle or parallelogram and if we glue opposite edges by a translation, we obtain a euclidean torus. Similarly, for a euclidean hexagon satisfying appropriate conditions on its edge lengths and angles, we found out that gluing opposite edges again yields a euclidean torus.

We can go one step further and glue opposite sides of an octagon X, as on the left-hand side of Figure 5.6.

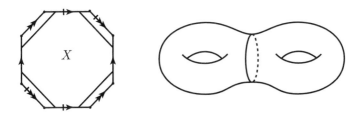

Figure 5.6. Gluing opposite edges of an octagon

We claim that the quotient space $\bar{X}$ is homeomorphic to the **surface of genus** 2 represented on the right of Figure 5.6, namely, a sphere with 2 handles. (A torus is a sphere with one handle added.)

To see this, one can cut out a smaller octagon from X as indicated in Figure 5.6. This smaller octagon X_1 can be seen as a rectangle whose corners have been cut off. Gluing opposite edges of this rectangle, we see that the image $\bar{X}_1$ in $\bar{X}$ is just a torus from which a square (corresponding to the triangles removed from the rectangle) has been removed. See Figure 5.7.

It remains to consider the four strips forming the complement X_2 of X_1 in X. Gluing these four strips together along their short edges gives a big square minus a smaller square, namely, some kind of square annulus, as in the left-hand side of Figure 5.8. Flipping this annulus inside out, as in the middle picture of Figure 5.8, and

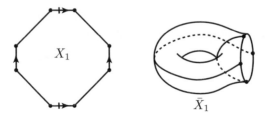

Figure 5.7. One half of Figure 5.6

Figure 5.8. The other half of Figure 5.6

then gluing the outside sides, we see that the image $\bar{X}_2$ of X_2 in the quotient space $\bar{X}$ is again a torus minus a square.

Finally, the quotient space $\bar{X}$ is obtained by gluing the two surfaces $\bar{X}_1$ and $\bar{X}_2$ along their boundaries, which gives the surface of genus 2 of Figure 5.6.

Let us now try to put a euclidean metric on the quotient space $\bar{X}$. The eight vertices of the octagon X are glued together to form a single point of $\bar{X}$. Consequently, if we want to apply Theorems 4.3 and 4.4, we need to use a euclidean octagon where opposite edges have the same length, and where the sum of the angles at the vertices is equal to 2π. Unfortunately, in euclidean geometry, the angles of an octagon add up to 6π!

Therefore, it seems impossible to put a euclidean metric on $\bar{X}$. (It can be proved that this is indeed the case, and that the surface of genus 2 admits no euclidean metric; see Exercise 5.16). However, hyperbolic geometry will provide us with a suitable octagon.

The first step is the following.

Lemma 5.2. *In the hyperbolic plane* $\mathbb{H}^2$, *there exists a triangle* T *with angles* $\frac{\pi}{2}$, $\frac{\pi}{8}$ *and* $\frac{\pi}{8}$.

See Proposition 5.13 and Exercise 5.15 for a more general construction of hyperbolic triangles with prescribed angles.

Proof. We will actually use euclidean geometry to construct this hyperbolic triangle.

We begin with the hyperbolic geodesic g with endpoints 0 and ∞. Namely g is the vertical half-line beginning at 0. Then consider the complete geodesic h that is orthogonal to g at the point i. Namely, h is a euclidean semi-circle of radius 1 centered at 0. We are looking for a third geodesic k which makes an angle of $\frac{\pi}{8}$ with both g and h.

For every $y \leqslant 1$, let k_y be the complete geodesic that passes through the point iy and makes an angle of $\frac{\pi}{8}$ with g. Namely, k_y is a euclidean semi-circle of radius $y \csc \frac{\pi}{8}$ centered at the point $y \cot \frac{\pi}{8}$, and consequently meets g as long as $\left(\sin \frac{\pi}{8}\right) / \left(1 + \cos \frac{\pi}{8}\right) < y \leqslant 1$. See Figure 5.9.

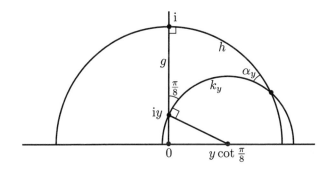

Figure 5.9. A hyperbolic triangle with angles $\frac{\pi}{2}$, $\frac{\pi}{8}$ and $\frac{\pi}{8}$

The angle α_y between k_y and g at their intersection point depends continuously on y. It is equal to $\frac{3\pi}{8}$ when $y = 1$, and approaches 0 as y tends to $\left(\sin \frac{\pi}{8}\right) / \left(1 + \cos \frac{\pi}{8}\right)$. Consequently, by the Intermediate Value Theorem there exists a value of y for which $\alpha_y = \frac{\pi}{8}$. By applying the Cosine Formula to the triangle formed by the intersection point of g and k_y and by the centers of these euclidean semi-circles, one could actually find an explicit formula for this y, but this is not necessary.

For this value of y, the hyperbolic geodesics g, h and k_y delimit the hyperbolic triangle T required.   □

Lemma 5.3. *In the hyperbolic plane $\mathbb{H}^2$, there exists an octagon X where all edges have the same length and where all angles are equal to $\frac{\pi}{4}$.*

Proof. Let T be the triangle provided by Lemma 5.2. List its vertices as P_0, P_1 and P_2 in such a way that P_0 is the vertex with angle $\frac{\pi}{2}$.

We start with 16 isometric copies T_1, T_2, ..., T_{16} of T. Namely, the T_i are hyperbolic triangles for which there exists isometries φ_i: $\mathbb{H}^2 \to \mathbb{H}^2$ such that $T_i = \varphi_i(T)$.

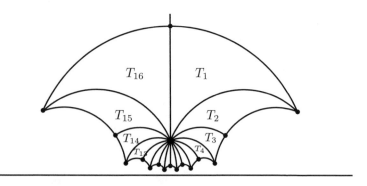

Figure 5.10. A hyperbolic octagon with all angles equal to $\frac{\pi}{4}$

Pick an arbitrary point $Q \in \mathbb{H}^2$. We can choose the isometries φ_i so that $\varphi_i(P_2) = Q$ for every i. In addition, using Lemma 4.8, we can arrange that:

(1) if i is even, φ_i and φ_{i-1} send the edge P_0P_2 to the same geodesic, so that T_i and T_{i-1} have this edge $\varphi_i(P_0P_2) = \varphi_{i-1}(P_0P_2)$ in common;

(2) if $i > 1$ is odd, φ_i and φ_{i-1} send the edge P_1P_2 to the same geodesic, so that T_i and T_{i-1} have this edge $\varphi_i(P_1P_2) = \varphi_{i-1}(P_1P_2)$ in common;

(3) for every $i > 1$, the triangles T_i and T_{i-1} sit on opposite sides of their common edge.

Since $16\frac{\pi}{8} = 2\pi$, the T_i fit nicely together around the point Q, and the last edge of T_{16} comes back to match the first edge of T_1. In particular, $\varphi_{16}(P_1P_2) = \varphi_1(P_1P_2)$. See Figure 5.10.

When i is even, the two geodesic arcs $\varphi_i(P_0P_1) = \varphi_{i-1}(P_0P_1)$ meet at $\varphi_i(P_0) = \varphi_{i-1}(P_0)$ and make an angle of $\frac{\pi}{2} + \frac{\pi}{2} = \pi$ at that point. It follows that the union of these two geodesic arcs forms a single geodesic arc.

Therefore, the union X of the 16 triangles T_i is an octagon in the hyperbolic plane $\mathbb{H}^2$. Its angles are all equal to $2\frac{\pi}{8} = \frac{\pi}{4}$. Its edges all have the same length, namely, twice the length of the edge P_0P_1 of the original triangle T. $\qquad\square$

The symmetries of this hyperbolic octagon are more apparent in Figure 5.11, which represents its image in the disk model for $\mathbb{H}^2$, introduced in Section 2.7, if we arrange that the center Q of X corresponds to the center O of $\mathbb{B}^2$.

Figure 5.11. The hyperbolic octagon of Figure 5.10 in the disk model

Let X be the hyperbolic octagon provided by Lemma 5.3, and let $\bar{X}$ be the quotient space obtained by gluing together opposite sides of X, as in Figure 5.6. Then Theorems 4.9 and 4.10 assert that the metric d_X induces a quotient metric $\bar{d}_X$ on this quotient space X, and that the metric space $(\bar{X}, \bar{d}_X)$ is locally isometric to the hyperbolic plane.

In particular, we have constructed a hyperbolic surface $(\bar{X}, \bar{d}_X)$ which is homeomorphic to the surface of genus 2.

5.3. The projective plane

We now construct a spherical surface, which is different from the sphere $\mathbb{S}^2$.

Let X be a hemisphere in $\mathbb{S}^2$. To turn X into a polygon, pick two antipodal points P_1 and $P_2 = -P_1$ on the great circle C delimiting X, which will be the vertices of the polygon. These two vertices split C into two edges E_1 and E_2. We now have a spherical polygon X, which we endow with the path metric d_X. Note that X is convex, so that d_X is just the restriction of the spherical metric d_{sph}.

Now, glue E_1 to E_2 by the antipode map $\varphi_1 \colon E_1 \to E_2$ defined by $\varphi_1(P) = -P$. This gluing data defines a quotient space $(\bar{X}, \bar{d}_X)$.

The polygon X and its gluing data are represented in Figure 5.12.

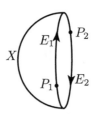

Figure 5.12. The projective plane

The angles of X at P_1 and P_2 are both equal to π, and consequently add up to 2π. Therefore, we can apply Theorems 4.12 and 4.13, and show that the quotient space $(\bar{X}, \bar{d}_X)$ is locally isometric to the sphere $(\mathbb{S}^2, d_{\mathrm{sph}})$. This quotient space $(\bar{X}, \bar{d}_X)$ is called the *projective plane*.

In Exercise 5.10, we show that the projective plane can also be interpreted as the space of lines passing through the origin in the 3-dimensional space $\mathbb{R}^3$.

5.4. The cylinder and the Möbius strip

We now consider unbounded polygons and the surfaces obtained by gluing their edges together.

The simplest case is that of an infinite strip where the two edges are glued together which provides a cylinder or a Möbius strip. These examples may appear somewhat trivial at first, but they already display many features that we will encounter in more complicated surfaces.

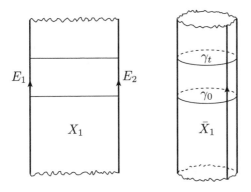

Figure 5.13. A euclidean cylinder

5.4.1. Euclidean cylinders and Möbius strips.
We can begin with an infinite strip X_1 in the euclidean plane $\mathbb{R}^2$, bounded by two parallel lines E_1 and E_2. Orient E_1 and E_2 in the same direction, and glue them by an isometry $\varphi_1 \colon E_1 \to E_2$ respecting these orientations. Because there are no vertices on E_1 and E_2, this is a situation where the gluing map φ_1 is not uniquely determined by these properties. Indeed, there are many possible choices for φ_1, all differing from each other by composition with a translation of $\mathbb{R}^2$ parallel to E_1 and E_2.

Pick any such gluing map $\varphi_1 \colon E_1 \to E_2$, and consider the corresponding quotient space $(\bar{X}_1, \bar{d}_{X_1})$. Since X_1 has no vertex, the angle hypothesis of Theorem 4.4 is automatically satisfied. Therefore, Theorems 4.3 and 4.4 show that $(\bar{X}_1, \bar{d}_{X_1})$ is a euclidean surface.

This euclidean surface is easily seen to be homeomorphic to the cylinder.

For another vertical strip X_2 in $\mathbb{R}^2$ bounded by parallel lines E_1 and E_2, we can orient E_1 and E_2 so that they now point in opposite directions and choose a gluing map $\varphi \colon E_1 \to E_2$ respecting these orientations. Again, there are many different choices for this gluing map, differing by composition with translations parallel to E_1 and E_2.

For each choice of such a gluing map φ_1, another application of Theorems 4.3 and 4.4 shows that the corresponding quotient space $(\bar{X}_2, \bar{d}_{X_2})$ is a euclidean surface.

The topology of the quotient space $(\bar{X}_2, \bar{d}_{X_2})$ is now very different. Indeed, this space is homeomorphic to the famous **Möbius strip**.

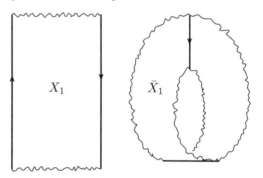

Figure 5.14. A euclidean Möbius strip

The Möbius strip is named after August Möbius (1790–1868), who conceived of this nonorientable surface in 1858 while working on geometric properties of polyhedra. He never published this work, which was only discovered after his death. Credit for the discovery of the Möbius strip should probably go to Johann Benedict Listing (1808–1882) instead, who independently described the Möbius strip in 1858. Incidentally, Listing made another important contribution to the themes of this monograph. He coined the word "topology" (or *Topologie* in the German original) in an 1836 letter, and the first printed occurrence of this term appears in his book *Vorstudien zur Topologie*, published in 1847.

5.4.2. Hyperbolic cylinders. We can make completely analogous constructions in hyperbolic geometry by replacing the euclidean strip with a strip X_3 in the hyperbolic plane $\mathbb{H}^2$, bounded by two disjoint complete geodesics E_1 and E_2. However, there are essentially two different shapes for such an infinite strip in $\mathbb{H}^2$.

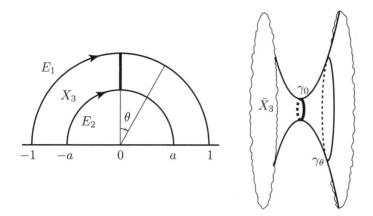

Figure 5.15. A hyperbolic cylinder

First, the endpoints of E_1 and E_2 on $\mathbb{R} \cup \{\infty\}$ may all be distinct. By an easy algebraic exercise with linear fractional maps, there exists an isometry of $\mathbb{H}^2$ sending E_1 to the complete geodesic with endpoints ± 1 and sending E_2 to the complete geodesic with endpoints $\pm a$, for some $a < 1$. Consequently, we can assume, without loss of generality, that E_1 goes from -1 to $+1$ and that E_2 goes from $-a$ to $+a$.

To glue E_1 to E_2 we need an isometry φ_1 sending E_1 to E_2. Among such isometries of $\mathbb{H}^2$, the simplest one is the homothety defined by $\varphi_1(z) = az$. Choose this specific isometry as a gluing map, and let $(\bar{X}_3, \bar{d}_{X_3})$ be the corresponding quotient metric space. The combination of Theorems 4.9 and 4.10 shows that $(\bar{X}_3, \bar{d}_{X_3})$ is locally isometric to the hyperbolic plane $(\mathbb{H}^2, d_{\mathrm{hyp}})$; namely, it is a hyperbolic surface. We are now in the situation of Figure 5.15.

This surface is easily shown to be homeomorphic to the cylinder. However, its geometry is very different from that of a euclidean cylinder.

Indeed, in the euclidean case, assume that X_1 is the vertical strip $\{(x,y) \in \mathbb{R}^2; 0 \leqslant x \leqslant a\}$ and that we glue $E_1 = \{(x,y) \in R^2; x = 0\}$ to $E_2 = \{(x,y) \in \mathbb{R}^2; x = a\}$ by the horizontal translation $\varphi_1 \colon (x,y) \mapsto (x+a, y)$. (See Exercise 7.9 in Chapter 7 for an analysis of the other possible gluing maps, which shows that the general case is essentially equivalent to the one considered here.) The quotient space $(\bar{X}_1, \bar{d}_{X_1})$ can then be decomposed as a union of closed curves γ_t where, for each t, γ_t is the image in $\bar{X}_1$ of the horizontal line segment $\{(x,y) \in R^2; 0 \leqslant t \leqslant a, y = t\}$. These closed curves all have length a, and the set of points at distance δ from the central curve γ_0 consists exactly of the two curves $\gamma_\delta \cup \gamma_{-\delta}$. In particular, the curves γ_t show that the "width" of the euclidean cylinder $\bar{X}_1$ is the same at every point.

In the hyperbolic case, where X_3 is the strip in $\mathbb{H}^2$ delimited by the geodesics E_1 and E_2 going from -1 to $+1$ and from $-a$ to $+a$, respectively, we glued E_1 to E_2 by the homothety $\varphi_1(z) = az$. For every θ with $-\frac{\pi}{2} < \theta < \frac{\pi}{2}$, we now have a closed curve γ_θ in the quotient space $\bar{X}_3$, which is the image of the euclidean line segment consisting of all $z = re^{i(\frac{\pi}{2}-\theta)}$ with $a \leqslant r \leqslant 1$.

The closed curve γ_0 is geodesic because, for each $\bar{P} \in \gamma_0$, there exists an isometry between a small ball $B_{\bar{d}_{\text{hyp}}}(\bar{P}, \varepsilon)$ and a ball of radius ε in $(\mathbb{H}^2, d_{\text{hyp}})$ sending $\gamma_0 \cap B_{\bar{d}_{\text{hyp}}}(\bar{P}, \varepsilon)$ to a geodesic arc of $\mathbb{H}^2$. The only point $\bar{P}$ where this requires a little checking is when $\bar{P}$ corresponds to the two points i and $ai \in X_3$, in which case the local isometry provided by the proof of Theorem 4.10 is easily seen to satisfy this property. The curves γ_θ with $\theta \neq 0$ are never geodesic.

A few rather immediate computations of hyperbolic lengths show the following:

(1) The curve γ_θ has hyperbolic length $\displaystyle\int_1^a \frac{1}{r \cos\theta}\, dr = \log a \sec\theta$.

(2) Every point $z = re^{i(\frac{\pi}{2}-\theta)}$ of γ_θ is at distance
$$\int_0^{|\theta|} \frac{r}{r\cos t}\, dt = \log(\sec\theta + \tan|\theta|) \text{ from the curve } \gamma_0.$$

Compare Exercise 2.5 for the proof of (2).

Therefore, for every $\delta > 0$, the set of points of $\bar{X}_3$ that are at distance δ from γ_0 consists of the two curves $\gamma_{\pm\theta}$ with $\log(\sec\theta + \tan|\theta|) = \delta$. By elementary trigonometry, this is equivalent to the property that $\sec\theta = \cosh\delta$. Therefore, by (1) above, the length of each of these two curves is equal to $\log a \cosh\delta$. In particular, the width of the cylinder grows exponentially with the distance δ from γ_0, so that γ_0 forms some kind of "narrow waist" for the hyperbolic cylinder $\bar{X}_3$.

The picture on the right-hand side of Figure 5.15 attempts to convey a sense of this exponential growth. This picture is necessarily imperfect. Indeed, as one goes toward one of the ends of the hyperbolic cylinder, its width grows faster than that of any surface of revolution in the 3-dimensional euclidean space $\mathbb{R}^3$.

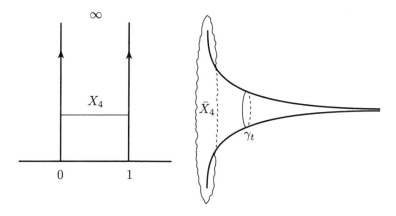

Figure 5.16. Another hyperbolic cylinder

In the hyperbolic plane $\mathbb{H}^2$, there is another type of infinite strip bounded by two complete geodesics E_1 and E_2 which occurs when E_1 and E_2 have one endpoint in common in $\mathbb{R} \cup \{\infty\}$. Applying an isometry of $\mathbb{H}^2$, we can assume without loss of generality that this common point is ∞, namely that E_1 and E_2 are both vertical half-lines. By a horizontal translation followed by a homothety, we can even arrange that E_1 is the vertical half-line with endpoints 0 and ∞, while E_2 goes from 1 to ∞. These two geodesics now delimit the strip $X_4 = \{z \in \mathbb{H}^2; 0 \leqslant \mathrm{Re}(z) \leqslant 1\}$ in $\mathbb{H}^2$.

To glue the edges E_1 and E_2 together, the simplest gluing map $\varphi_1 \colon E_1 \to E_2$ is the horizontal translation defined by $\varphi_1(z) = z + 1$. Let $(\bar{X}_4, \bar{d}_{\mathrm{hyp}})$ be the quotient space obtained by performing this gluing operation on X_4. This quotient space is a hyperbolic surface by Theorems 4.9 and 4.10, and is easily seen to be homeomorphic to the cylinder. We should note that our choice of the gluing map φ_1 is critical here. Indeed, we will see in Section 6.7.1 that other choices lead to hyperbolic cylinders with very different geometric properties.

For every $t \in \mathbb{R}$, let γ_t be the closed curve in the quotient space $\bar{X}_4$ that is the image of the horizontal line segment consisting of those $z \in X_4$ such that $\mathrm{Im}(z) = \mathrm{e}^t$. By definition of the hyperbolic metric, it is immediate that γ_t has hyperbolic length e^{-t}. Also, every point of γ_t is at hyperbolic distance $|t|$ from the central curve γ_0. Therefore, the width of $\bar{X}_4$ grows exponentially toward one end of the cylinder and decreases exponentially toward the other end.

Again, the right-hand side of Figure 5.16 attempts to illustrate this behavior. As in the case of $\bar{X}_3$, this picture is necessarily imperfect for the end with exponential growth. Surprisingly enough, the end with exponential decay can be exactly represented as a surface of revolution in the 3-dimensional space $\mathbb{R}^3$.

More precisely, let X_4^+ denote the upper part of X_4 consisting of those $z \in X_4$ such that $\mathrm{Im}(z) \geqslant \frac{1}{2\pi}$, and let $\bar{X}_4^+$ be its image in $\bar{X}_4$.

In the xy-plane, consider the **tractrix** parametrized by

$$t \longmapsto (t - \tanh t, \operatorname{sech} t), \qquad 0 \leqslant t < \infty,$$

and let the **pseudosphere** S be the surface of revolution in $\mathbb{R}^3$ obtained by revolving the tractrix about the x-axis. This surface is represented in Figure 5.17.

Endow the pseudosphere S with the metric d_S defined by the property that for every P, $Q \in S$, $d_S(P, Q)$ is the infimum of the euclidean lengths $\ell_{\mathrm{euc}}(\gamma)$ of all piecewise differentiable curves γ contained in S and joining P to Q.

As usual, we endow X_4^+ with the metric $d_{X_4^+}$ for which the distance between two points is the infimum of the hyperbolic lengths of all curves in X_4^+ joining these two points. Because hyperbolic geodesics are euclidean semi-circles, it is relatively immediate that

X_4^+ is convex, so that this metric $d_{X_4^+}$ actually coincides with the restriction of the hyperbolic distance. However, it is convenient to keep
a distinct notation because of (minor) subtleties with quotient metrics; compare Lemma 5.8. Let $\bar{d}_{X_4^+}$ be the quotient metric induced
by $d_{X_4^+}$ on the quotient space $\bar{X}_4^+$.

Proposition 5.4. *The metric space $(\bar{X}_4^+, \bar{d}_{X_4^+})$ is isometric to the
surface (S, d_S).*

Proof. The proof will take several steps, in part because of the definition of the quotient metric. It may perhaps be skipped on a first
reading. However, you should at least have a glance atLemma 5.5,
which contains the key geometric idea of the proof.

To
page 114

Let $\rho\colon X_4^+ \to S$ be the map which to $z \in X_4^+$ associates the point
$\rho(z) = (u, v, w)$ with

$$u = \log\left(2\pi\,\mathrm{Im}(z) + \sqrt{4\pi^2\mathrm{Im}(z)^2 - 1}\right) - \frac{\sqrt{4\pi^2\mathrm{Im}(z)^2 - 1}}{2\pi\,\mathrm{Im}(z)},$$

$$v = \frac{1}{2\pi\,\mathrm{Im}(z)}\cos\left(2\pi\,\mathrm{Re}(z)\right),$$

$$w = \frac{1}{2\pi\,\mathrm{Im}(z)}\sin\left(2\pi\,\mathrm{Re}(z)\right).$$

Geometrically, $\rho(z)$ is better understood by setting

$$t = \mathrm{arccosh}(2\pi\,\mathrm{Im}(z)) = \log\left(2\pi\,\mathrm{Im}(z) + \sqrt{4\pi^2\mathrm{Im}(z)^2 - 1}\right).$$

Then $\rho(z) = \left(t - \tanh t, \mathrm{sech}\,t\,\cos\left(2\pi\,\mathrm{Re}(z)\right), \mathrm{sech}\,t\,\sin\left(2\pi\,\mathrm{Re}(z)\right)\right)$ is
obtained by rotating the point of the tractrix corresponding to the
parameter t by an angle of $2\pi\,\mathrm{Re}(z)$ about the x-axis.

In particular, two points z, $z' \in X_4^+$ have the same image under
ρ if and only if $z' - z$ is an integer, namely, if and only if z and z'
are glued together to form a single point of the quotient space $\bar{X}_4^+$.
As a consequence, ρ induces an injective map $\bar{\rho}\colon \bar{X}_4^+ \to S$ defined by
the property that $\bar{\rho}(\bar{P})$ is equal to $\rho(P)$ for every $\bar{P} \in \bar{X}_4^+$ and any
element $P \in \bar{P}$, namely, for any point $P \in X_4^+$ that corresponds to
$\bar{P}$ in the quotient space $\bar{X}_4^+$.

The map ρ is surjective by definition of the pseudosphere S. It
follows that $\bar{\rho}\colon \bar{X}_4^+ \to S$ is surjective and is therefore bijective.

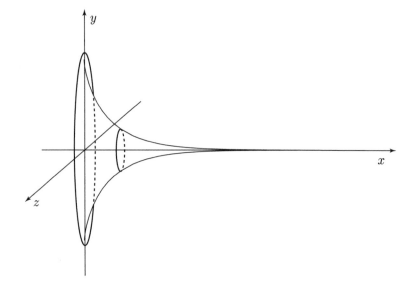

Figure 5.17. The pseudosphere

We will prove that $\bar{\rho}$ is an isometry from $(\bar{X}_4^+, \bar{d}_{X_4^+})$ to (S, d_S). The key step is the following computation.

Lemma 5.5. *The map ρ sends every curve γ in the half-strip X_4^+ to a curve in the surface S whose euclidean length $\ell_{\text{euc}}(\rho(\gamma))$ is equal to the hyperbolic length $\ell_{\text{hyp}}(\gamma)$ of γ.*

Proof. Let us compute the differential map $D_z\rho$. Remember that $D_z\rho$ sends the vector $\vec{v} = (a, b)$ to

$$D_z\rho(\vec{v}) = \left(\frac{\partial u}{\partial x}a + \frac{\partial u}{\partial y}b, \frac{\partial v}{\partial x}a + \frac{\partial v}{\partial y}b, \frac{\partial w}{\partial x}a + \frac{\partial w}{\partial y}b \right)$$

if we write $z = x + iy$.

With $t = \text{arccosh}(2\pi y)$ as before, we find that $\frac{\partial u}{\partial x} = 0$, $\frac{\partial u}{\partial y} = 2\pi \operatorname{sech} t \tanh t$, $\frac{\partial v}{\partial x} = -2\pi \operatorname{sech} t \sin(2\pi x)$, $\frac{\partial v}{\partial x} = 2\pi \operatorname{sech}^2 t \cos(2\pi x)$, $\frac{\partial w}{\partial x} = 2\pi \operatorname{sech} t \cos(2\pi x)$, and $\frac{\partial v}{\partial x} = 2\pi \operatorname{sech}^2 t \sin(2\pi x)$.

After simplifications,

$$\|D_z\rho(\vec{v})\|_{\text{euc}} = 2\pi \operatorname{sech} t \, \|\vec{v}\|_{\text{euc}} = \frac{1}{y} \|\vec{v}\|_{\text{euc}} = \|\vec{v}\|_{\text{hyp}}.$$

If γ is a curve parametrized by $s \mapsto z(s)$, $s_1 \leqslant s \leqslant s_2$, its image $\rho(\gamma)$ is parametrized by $s \mapsto \rho(z(s))$, $s_1 \leqslant s \leqslant s_2$. Therefore,

$$\ell_{\text{euc}}(\varphi(\gamma)) = \int_{s_1}^{s_2} \|(\rho \circ z)'(s)\|_{\text{euc}} \, ds = \int_{s_1}^{s_2} \|D_{z(s)}(z'(s))\|_{\text{euc}} \, ds$$

$$= \int_{s_1}^{s_2} \|z'(s)\|_{\text{hyp}} \, ds = \ell_{\text{hyp}}(\gamma). \qquad \square$$

Lemma 5.6.

$$d_S(\bar{\rho}(\bar{P}), \bar{\rho}(\bar{Q})) \leqslant \bar{d}_{X_4^+}(\bar{P}, \bar{Q})$$

for every $\bar{P}$, $\bar{Q} \in \bar{X}_4^+$.

Proof. Let the points $P = P_1$, $Q_1 \sim P_2$, ..., $Q_{n-1} \sim P_n$, $Q_n = Q$ form a discrete walk w from $\bar{P}$ to $\bar{Q}$. The length of the discrete walk w is $\ell(w) = \sum_{i=1}^n d_{X_4^+}(P_i, Q_i)$.

By convexity of X_4^+, there is a geodesic arc γ_i joining P_i to Q_i whose hyperbolic length is equal to $d_{X_4^+}(P_i, Q_i)$. Lemma 5.5 shows that the image $\gamma_i' = \rho(\gamma_i)$ is a curve joining $\rho(P_i)$ to $\rho(Q_i) = \rho(P_{i+1})$ in S whose euclidean length is equal to the hyperbolic length of γ_i. Chaining together these γ_i' provides a curve γ' joining $\rho(P) = \bar{\rho}(\bar{P})$ to $\rho(Q) = \bar{\rho}(\bar{Q})$ in S. Its euclidean length is

$$\ell_{\text{euc}}(\gamma') = \sum_{i=1}^n \ell_{\text{euc}}(\gamma_i') = \sum_{i=1}^n \ell_{\text{hyp}}(\gamma_i) = \sum_{i=1}^n d_{X_4^+}(P_i, Q_i) = \ell(w).$$

By definition of the metric d_S, this shows that $d_S(\bar{\rho}(\bar{P}), \bar{\rho}(\bar{Q})) \leqslant \ell(w)$. Since this holds for every discrete walk w from $\bar{P}$ to $\bar{Q}$, we conclude that $d_S(\bar{\rho}(\bar{P}), \bar{\rho}(\bar{Q})) \leqslant \bar{d}_{X_4^+}(\bar{P}, \bar{Q})$. $\qquad \square$

Lemma 5.7.

$$\bar{d}_{X_4^+}(\bar{P}, \bar{Q}) \leqslant d_S(\bar{\rho}(\bar{P}), \bar{\rho}(\bar{Q}))$$

for every $\bar{P}$, $\bar{Q} \in \bar{X}_4^+$.

Proof. For a given $\varepsilon > 0$, there exists a piecewise differentiable curve γ going from $\bar{\rho}(\bar{P})$ to $\bar{\rho}(\bar{Q})$ in S whose length is sufficiently close to $d_S(\bar{\rho}(\bar{P}), \bar{\rho}(\bar{Q}))$ that

$$d_S(\bar{\rho}(\bar{P}), \bar{\rho}(\bar{Q})) \leqslant \ell_{\text{euc}}(\gamma) \leqslant d_S(\bar{\rho}(\bar{P}), \bar{\rho}(\bar{Q})) + \varepsilon.$$

We want to decompose γ into pieces coming from X_4^+. For this, we use the following estimate.

In the surface S, consider the tractrix T parametrized by $t \mapsto$ $(t - \tanh t, \operatorname{sech} t, 0)$. If α is a curve in S whose endpoints P' and Q' are both in T, and if β is the portion of T going from P' to Q', we claim that $\ell_{\mathrm{euc}}(\beta) \leqslant \ell_{\mathrm{euc}}(\alpha)$.

To see this, parametrize α by $s \mapsto \big(x(s), y(s), z(s)\big)$, $a \leqslant s \leqslant b$, with

$$\begin{cases} x(s) = t(s) - \tanh t(s) \\ y(s) = \operatorname{sech} t(s) \, \cos \theta(s) \\ z(s) = \operatorname{sech} t(s) \, \sin \theta(s) \end{cases}$$

for some functions $s \mapsto t(s)$ and $s \mapsto \theta(s)$. The curve β has a similar parametrization, where $\theta(s)$ is a constant equal to 0. An immediate computation then yields

$$\begin{aligned} \ell_{\mathrm{euc}}(\alpha) &= \int_a^b \sqrt{x'(s)^2 + y'(s)^2 + z'(s)^2} \, ds \\ &= \int_a^b \sqrt{t'(s)^2 \tanh^2 t(s) + \theta'(s)^2 \operatorname{sech}^2 t(s)} \, ds \\ &\geqslant \int_a^b t'(s) \tanh t(s) \, ds = \ell_{\mathrm{euc}}(\beta). \end{aligned}$$

As a consequence, we can arrange that the intersection of γ with T consists of a single curve contained in T, possibly empty or reduced to a single point. Indeed, if P' and Q' are the first and last points where γ meets T, we can replace the part α of γ going from P' to Q' by the part β of T joining P' to Q' without increasing its length.

We are now ready to conclude. The key observation is that T is also the image under $\bar{\rho}$ of the "suture" of $\bar{X}_4^+$ consisting of those points which correspond to two points of X_4^+ (located on the vertical part of the boundary of the half-strip X_4^+).

If γ is disjoint from T, then γ is the image under ρ of a curve γ' contained in X_4^+ and going from a point P to a point Q. By Lemma 5.5, the hyperbolic length of γ' is equal to the euclidean length of γ. It follows that

$$\bar{d}_{X_4^+}(\bar{P}, \bar{Q}) \leqslant d_{X_4^+}(P, Q) \leqslant \ell_{\mathrm{hyp}}(\gamma') = \ell_{\mathrm{euc}}(\gamma) \leqslant d_S\big(\bar{\rho}(\bar{P}), \bar{\rho}(\bar{Q})\big) + \varepsilon.$$

If γ meets the tractrix T, we can split it into a first piece γ_1 going from $\bar{\rho}(\bar{P})$ to a point of T, a second curve γ_2 contained in T (possibly reduced to a single point), and a third piece γ_3 going from T to the point $\bar\rho(\bar Q)$. Then γ_1, γ_2 and γ_3 are the respective images under ρ of curves γ_1', γ_2' and γ_3' in X_4^+. There are two possible choices for γ_2', one in each of the two vertical sides of X_4^+, and we just pick one of them. We have observed that two points of X_4^+ have the same image under ρ if and only if they are glued together in the quotient space $\bar X_4^+$. Consequently, if P_i and Q_i denote the initial and terminal points of each γ_i, we have that $P \sim P_1$, $Q_1 \sim P_2$, $Q_2 \sim P_3$ and $Q_3 \sim Q$. In particular, the P_i and Q_i form a discrete walk w from $\bar P$ to $\bar Q$ whose length is

$$\ell_{d_{X_4^+}}(w) = \sum_{i=1}^{3} d_{X_4^+}(P_i, Q_i) \leqslant \sum_{i=1}^{3} \ell_{\mathrm{hyp}}(\gamma_i') = \sum_{i=1}^{3} \ell_{\mathrm{euc}}(\gamma_i) = \ell_{\mathrm{euc}}(\gamma)$$

since the euclidean length of γ_i is equal to the hyperbolic length of γ_i' by Lemma 5.5. Therefore,

$$\bar d_{X_4^+}(\bar P, \bar Q) \leqslant d_{X_4^+}(P, Q) \leqslant \ell_{d_{X_4^+}}(w) \leqslant \ell_{\mathrm{euc}}(\gamma) \leqslant d_S\big(\bar\rho(\bar P), \bar\rho(\bar Q)\big) + \varepsilon.$$

We have now proved that $\bar d_{X_4^+}(\bar P, \bar Q) \leqslant d_S\big(\bar\rho(\bar P), \bar\rho(\bar Q)\big) + \varepsilon$ in both cases, and this for every $\varepsilon > 0$. Therefore, $\bar d_{X_4^+}(\bar P, \bar Q) \leqslant d_S\big(\bar\rho(\bar P), \bar\rho(\bar Q)\big)$ as requested. $\qquad\square$

The combination of Lemmas 5.6 and 5.7 shows that $d_S\big(\bar\rho(\bar P), \bar\rho(\bar Q)\big) = \bar d_{X_4^+}(\bar P, \bar Q)$ for every $\bar P$, $\bar Q \in \bar X_4^+$, namely that $\bar\rho$ is an isometry from $(\bar X_4^+, \bar d_{X_4^+})$ to (S, d_S). This completes the proof of Proposition 5.4. $\qquad\square$

We conclude our discussion of Proposition 5.4 by addressing a little subtlety. By convexity of X_4 and X_4^+, $d_{X_4^+}(P, Q) = d_{X_4}(P, Q) = d_{\mathrm{hyp}}(P, Q)$ for every P and $Q \in X_4^+$. However, in the quotient space $\bar X_4^+$, there might conceivably be a difference between the quotient metric $\bar d_{X_4^+}$ and the restriction of the quotient metric $\bar d_{X_4}$ of $\bar X_4$. Indeed, the definition of the quotient metric $\bar d_{X_4}$ involves discrete walks valued in X_4, whereas $\bar d_{X_4^+}$ is defined using discrete walks that are constrained to X_4^+. Compare our discussion of Lemma 5.12 in the next section.

In the specific case under consideration, it turns out that we do not need to worry about this distinction:

Lemma 5.8. *On the subspace $\bar{X}_4^+$ of $\bar{X}_4$, the two metrics $\bar{d}_{X_4^+}$ and $\bar{d}_{X_4}$ coincide.*

Proof. Because X_4^+ is contained in X_4 and because the metrics d_{X_4} and $d_{X_4^+}$ coincide on X_4^+, every discrete walk w valued in X_4^+ is also valued in X_4, and the two lengths $\ell_{d_{X_4^+}}(w)$ and $\ell_{d_{X_4^+}}(w)$ coincide. It follows that $\bar{d}_{X_4}(\bar{P}, \bar{Q}) \leqslant \bar{d}_{X_4^+}(\bar{P}, \bar{Q})$ for every $\bar{P}$, $\bar{Q} \in \bar{X}_4^+$.

Conversely, let $P = P_1$, $Q_1 \sim P_2$, ..., $Q_{n-1} \sim P_n$, $Q_n = Q$ be a discrete walk w valued in X_4 where P and Q both belong to the upper half-strip X_4^+. For each P_i, let P_i' be equal to P_i if P_i is in X_4^+, and otherwise let $P_i' = \mathrm{Re}(P_i) + \frac{1}{2\pi}\mathrm{i}$ be the point of the boundary of X_4^+ that sits right above P_i. Similarly, define Q_i' to be Q_i if $Q \in X_4^+$ and $\mathrm{Re}(Q_i) + \frac{1}{2\pi}\mathrm{i}$ if $Q \notin X_4^+$.

We claim that $d_{\mathrm{hyp}}(P_i', Q_i') \leqslant d_{\mathrm{hyp}}(P_i, Q_i)$. Indeed, if γ is a curve joining P_i to Q_i in X_4, let γ' be obtained from γ by replacing each piece that lies below the line of equation $\mathrm{Im}(z) = \frac{1}{2\pi}$ with the line segment that sits on the line right above that piece. From the definition of hyperbolic length, it is immediate that $\ell_{\mathrm{hyp}}(\gamma') \leqslant \ell_{\mathrm{hyp}}(\gamma)$. Considering all such curves γ and γ', it follows that $d_{\mathrm{hyp}}(P_i', Q_i') \leqslant d_{\mathrm{hyp}}(P_i, Q_i)$.

Since $P_1' = P_1$ and $P_n' = P_n$, we now have a discrete walk $P = P_1'$, $Q_1' \sim P_2'$, ..., $Q_{n-1}' \sim P_n'$, $Q_n' = Q$ which is valued in X_4^+. Since $d_{\mathrm{hyp}}(P_i', Q_i') \leqslant d_{\mathrm{hyp}}(P_i, Q_i)$, this new walk w' has length $\ell_{d_{X_4^+}}(w') \leqslant \ell_{d_{X_4}}(w)$. As a consequence, $\ell_{d_{X_4}}(w) \geqslant \bar{d}_{X_4^+}(\bar{P}, \bar{Q})$.

Considering all such walks w, we conclude that $\bar{d}_{X_4^+}(\bar{P}, \bar{Q}) \leqslant \bar{d}_{X_4}(\bar{P}, \bar{Q})$. Therefore, $\bar{d}_{X_4^+}(\bar{P}, \bar{Q}) = \bar{d}_{X_4}(\bar{P}, \bar{Q})$ for every $\bar{P}$, $\bar{Q} \in \bar{X}_4^+$. $\square$

From page 109

5.5. The once-punctured torus

The **once-punctured torus** is obtained by removing one point from the torus. To explain the terminology, think of what happens to the inner tube of a tire as one drives over a nail. If we describe the torus as a square with opposite edges glued together, we can assume that the point removed is the point corresponding to the four vertices of the square.

This surface of course admits a euclidean metric by restriction of a euclidean metric on the torus. However, we will see in Chapter 6 that such a metric is not complete (see the definition in Section 6.2), and that complete metrics are more desirable. Our goal is to construct a hyperbolic metric on the once-punctured torus, which we will later on prove to be complete.

This example will turn out to be very important. In particular, it will accompany much of our discussion in Chapters 8, 10 and 11.

Consider the hyperbolic polygon X described in Figure 5.18. Namely, X is the region in the hyperbolic plane $\mathbb{H}^2$ bounded by the four complete geodesics E_1, E_2, E_3 and E_4 where E_1 joins -1 to ∞, E_2 joins 0 to 1, E_3 joins 1 to ∞, and E_4 joins 0 to -1.

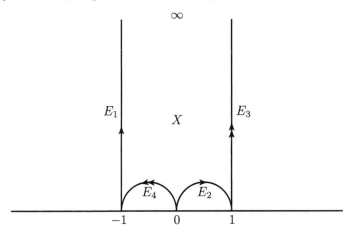

Figure 5.18. A hyperbolic square

As such, X is a "quadrilateral" except that its vertices are in $\mathbb{R} \cup \{\infty\}$, namely, at infinity of $\mathbb{H}^2$. As a subset of $\mathbb{H}^2$, X is therefore a quadrilateral with its four vertices removed.

In a hyperbolic polygon of this type where the vertices are at infinity of $\mathbb{H}^2$ in $\mathbb{R} \cup \{\infty\}$ (and consequently are not really vertices of the polygon in $\mathbb{H}^2$), we say that the vertices are **ideal**. If the vertices of the polygon are all ideal, and if the polygon touches $\mathbb{R} \cup \{\infty\}$ only at these vertices, we say that it is an **ideal polygon**.

With this terminology, X is now an ideal quadrilateral in $\mathbb{H}^2$. It is even an ideal square, in the sense that it has all the symmetries of a square; see Exercise 5.13.

Glue together opposite edges of X, while respecting the orientations indicated in Figure 5.18. Because these geodesics do not have any endpoints in $\mathbb{H}^2$, this is a situation where there are many possible isometric gluings. In particular, the gluing data is not completely determined by the picture. Consequently, we need to be more specific.

To glue the edge E_1 to E_2, we need a hyperbolic isometry φ_1 sending -1 to 0, and ∞ to 1. The simplest one is the linear fractional map

$$\varphi_1(z) = \frac{z+1}{z+2}.$$

Similarly, we can glue E_3 to E_4 by the hyperbolic isometry

$$\varphi_3(z) = \frac{z-1}{-z+2}.$$

As usual, define $\varphi_2 = \varphi_1^{-1}$ and $\varphi_4 = \varphi_3^{-1}$.

Let $(\bar{X}, \bar{d}_X)$ be the quotient metric space obtained from (X, d_X) by performing these edge gluings. Note that X is convex so that the metric d_X is just the restriction of the hyperbolic metric d_{hyp}.

Since X has no vertices in $\mathbb{H}^2$, there is nothing to be checked and Theorem 4.10 shows that $(\bar{X}, \bar{d}_X)$ is a hyperbolic surface. From the description of X as a quadrilateral with its vertices removed, we see that $\bar{X}$ is (homeomorphic to) a once-punctured torus.

We want to better understand the metric $\bar{d}_X$ near the puncture.

For $a > 1$, consider the horizontal line L_a of equation $\mathrm{Im}(z) = a$.

By Proposition 2.18, linear fractional maps send circles to circles in $\widehat{\mathbb{C}} = \mathbb{C} \cup \{\infty\}$. Consequently, φ_1 sends $L_a \cup \{\infty\}$ to a circle C_1 passing through $\varphi_1(\infty) = 1$. Since φ_1 also sends the half-plane $\mathbb{H}^2$ to itself, this circle must be tangent to the real line at 1. The circle C_1 also contains the image $\varphi_1(-2+ai) = 1 + \frac{1}{a}i$ of the point $-2+ai \in L_a$.

It follows that C_1 is the euclidean circle of radius $\frac{1}{2a}$ centered at $1+\frac{1}{2a}i$, and $\varphi_1(L_a) = C_1 - \{1\}$.

Similarly, $\varphi_3(L_a)$ is contained in a circle C_{-1} tangent to the real line at $\varphi_3(\infty) = -1$ and containing $\varphi_3(2+ai) = -1+\frac{1}{a}i$. Namely, C_{-1} is the circle of radius $\frac{1}{2a}$ centered at $-1 + \frac{1}{2a}i$, and $\varphi_3(L_a) = C_{-1} - \{-1\}$.

The map φ_1 also sends -1 to 0. Consequently, it sends the circle C_{-1} to a circle C_0 tangent to the real line at $\varphi_1(-1) = 0$ and passing through $\varphi_1 \circ \varphi_3(3+ai) = \frac{1}{a}i$ (since $\varphi_3(3+ai) \in \varphi_3(L_a) \subset C_{-1}$). Namely, C_0 is the circle of radius $\frac{1}{2a}$ centered at $\frac{1}{2a}i$.

Finally, φ_3 sends C_1 to a circle C_0' tangent to the real line at $\varphi_3(1) = 0$ and passing through $\varphi_3 \circ \varphi_1(-3+ai) = \frac{1}{a}i$. It follows that this circle $C_0' = \varphi_3(C_1)$ is exactly equal to the circle $C_0 = \varphi_1(C_{-1})$.

Let U_∞ be the set of points of X that are on or above the line L_a, and let U_0, U_1 and U_{-1} consist of the points of X that are on or inside the circles C_0, C_1 and C_{-1}, respectively. In addition, because $a > 1$, the U_i are disjoint. See Figure 5.19.

What the above discussion shows is that when P belongs to some of these U_i and is glued to some $P' \in X$, then P' must be in some other U_j. See Figure 5.19.

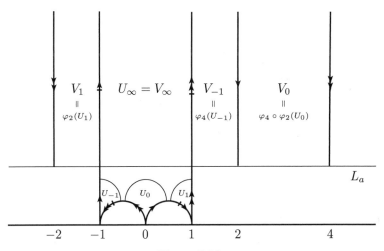

Figure 5.19

Let U denote the union $U_\infty \cup U_0 \cup U_1 \cup U_{-1}$ in X, and let $\bar{U}$ be its image in the quotient $\bar{X}$. Namely, $\bar{U}$ is obtained from U by gluing its eight sides though the restrictions of the maps φ_1, $\varphi_2 = \varphi_1^{-1}$, φ_3 and $\varphi_4 = \varphi_3^{-1}$.

Consider on U the metric d_U defined by the property that $d_U(P,Q)$ is the infimum of the hyperbolic length of all curves joining P to Q in U. In particular, $d_U(P,Q) = \infty$ when P and Q are in different U_i, since they cannot be joined by any curve that is completely contained in U.

Each U_i is convex. This is fairly clear for the vertical half-strip U_∞, since every hyperbolic geodesic is a circle arc centered on the x-axis. The property is slightly less obvious for the other U_i but follows from the fact, proved below, that U_i is isometric to a vertical half-strip V_i. A consequence of this convexity is that $d_U(P,Q) = d_{\mathrm{hyp}}(P,Q)$ when P and Q are in the same U_i.

The metric d_U induces a quotient semi-metric $\bar{d}_U$ on the quotient space $\bar{U}$.

We begin by showing that the metric space $(\bar{U}, \bar{d}_U)$ is isometric to a space that we have already encountered. Let S_a be the surface of revolution obtained by revolving about the x-axis the portion of the pseudosphere parametrized by

$$t \mapsto (t - \tanh t, \operatorname{sech} t), \quad \operatorname{arccosh} \tfrac{a\pi}{3} \leqslant t < \infty.$$

Equivalently, S_a consists of those points (x, y, z) of the pseudosphere S such that $x \geqslant \log\left(\frac{a\pi}{3} + \sqrt{\frac{a^2\pi^2}{9} - 1}\right) - \sqrt{1 - \frac{9}{a^2\pi^2}}$.

As in Proposition 5.4, endow S_a with the metric d_{S_a} defined by the property that $d_{S_a}(P,Q)$ is equal to the infimum of the euclidean lengths of all curves joining P to Q in S_a.

Proposition 5.9. *The metric space $(\bar{U}, \bar{d}_U)$ is isometric to the space (S_a, d_{S_a}).*

Proof. The proof is very similar to that of Proposition 5.4, except that we now have to glue four vertical half-strips instead of a single one.

To page 121

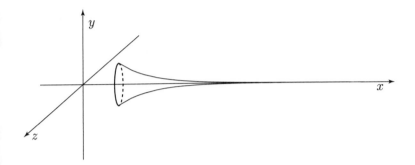

Figure 5.20. The portion S_a of the pseudosphere

In the hyperbolic plane $\mathbb{H}^2$, consider the subsets $V_\infty = U_\infty$, $V_{-1} = \varphi_2(U_{-1}) = \varphi_1^{-1}(U_{-1})$, $V_1 = \varphi_4(U_1) = \varphi_3^{-1}(U_1)$ and $V_0 = \varphi_4 \circ \varphi_2(U_0) = \varphi_3^{-1} \circ \varphi_1^{-1}(U_0)$.

Remember that $V_\infty = U_\infty$ is the vertical half-strip delimited by the horizontal line $L_a = \{z; \mathrm{Im}(z) = a\}$ and by the vertical lines of equation $\mathrm{Re}(z) = \pm 1$.

We already observed that φ_1 sends the line L_a to the circle C_1. It also sends the points ∞, -2 and -1 to 1, ∞ and 0, respectively. Since U_1 is delimited by C_1 and by the geodesics joining 1 to ∞ and 1 to 0, it follows that $V_1 = \varphi_1^{-1}(U_1)$ is delimited by L_a and by the two geodesics joining -1 to ∞ and -2 to ∞. These hyperbolic geodesics are also the vertical lines of equations $\mathrm{Re}(z) = -2$ and $\mathrm{Re}(z) = -1$, so that V_1 is a vertical half-strip.

Similarly, since $\varphi_2(L_a) = C_{-1}$, $V_{-1} = \varphi_2^{-1}(U_{-1})$ is the vertical half-strip delimited by L_a and by the two vertical lines of equations $\mathrm{Re}(z) = 1$ and $\mathrm{Re}(z) = 2$.

Finally, because $C_0 = \varphi_1(C_{-1}) = \varphi_1 \circ \varphi_3(L_a)$, the same type of arguments show that $V_0 = \varphi_3^{-1} \circ \varphi_1^{-1}(U_0)$ is delimited by L_a and by the vertical half-lines of equations $\mathrm{Re}(z) = 2$ and $\mathrm{Re}(z) = 4$.

This situation is illustrated in Figure 5.19.

Let V be the vertical half-strip union of V_1, V_∞, V_{-1} and V_0. Let $\bar{V}$ be the quotient space obtained from V by gluing its left-hand side $\{z \in V; \mathrm{Re}(z) = -2\}$ to its right-hand side $\{z \in V; \mathrm{Re}(z) = 4\}$ by the horizontal translation $\varphi\colon z \mapsto z + 6$. Since V is convex, endow it with the restriction d_V of the hyperbolic metric d_{hyp}, and let $\bar{d}_V$ be the quotient metric induced by d_V on $\bar{V}$.

We now split the argument into two steps.

Lemma 5.10. *The quotient space* $(\bar{V}, \bar{d}_V)$ *is isometric to the subset* (S_a, d_{S_a}) *of the pseudosphere.*

Proof. This is essentially Proposition 5.4.

Indeed, let $V' = \{z \in \mathbb{H}^2; 0 \leqslant \mathrm{Re}(z) \leqslant 1, \mathrm{Im}(z) \geqslant \frac{a}{6}\}$. Let $\bar{V}'$ be the quotient space obtained from V' by gluing its left-hand side $\{z \in V; \mathrm{Re}(z) = 0\}$ to its right-hand side $\{z \in V; \mathrm{Re}(z) = 1\}$ by the horizontal translation $\varphi'\colon z \mapsto z + 1$. Endow V' with the restriction $d_{V'}$ of the hyperbolic metric d_{hyp}, and endow $\bar{V}'$ with the induced quotient metric $\bar{d}_{V'}$.

The hyperbolic isometry $\psi\colon z \mapsto \frac{1}{6}z + \frac{1}{3}$ sends V to V', and it sends the gluing map φ of V to the gluing map φ' of V'. Consequently, $\bar{\psi}$ induces an isometry $\bar{\psi}$ from $(\bar{V}, \bar{d}_V)$ to $(\bar{V}', \bar{d}_{V'})$.

Note that V' is a subset of the half-strip X_4^+ considered in Proposition 5.4, so that $\bar{V} \subset \bar{X}$. The surface S_a is exactly the image of $\bar{V}'$ under the isometry $\bar{\rho}$ constructed in the proof of Proposition 5.4. Therefore, $\bar{\rho}$ restricts to an isometry from $(\bar{V}', \bar{d}_{V'})$ to (S_a, d_{S_a}).

The composition $\bar{\rho} \circ \bar{\psi}$ consequently provides an isometry from $(\bar{V}, \bar{d}_V)$ to (S_a, d_{S_a}).   □

Lemma 5.11. *The quotient spaces* $(\bar{U}, \bar{d}_U)$ *and* $(\bar{V}, \bar{d}_V)$ *are isometric.*

Proof. Let $\psi\colon U \to V$ be the map defined by the property that ψ coincides with $\varphi_2 = \varphi_1^{-1}$ on U_1, with $\varphi_4 = \varphi_3^{-1}$ on U_{-1}, with $\varphi_4 \circ \varphi_2 = \varphi_3^{-1} \circ \varphi_1^{-1}$ on U_0, and with the identity map on U_∞.

A case-by-case inspection of the eight sides of U that are glued together shows that P, $Q \in U$ are glued together in $\bar{U}$ if and only if $\psi(P)$ and $\psi(Q)$ are glued together in $\bar{V}$. For instance, if P is in the intersection of U_1 with the vertical line of equation $\mathrm{Re}(z) = 1$, it is glued to the point $Q = \varphi_3(P)$ contained in the intersection of U_0 with the geodesic joining 0 to -1. Then $\psi(P) = \varphi_2(P)$ and

$\psi(Q) = \varphi_4 \circ \varphi_2(Q)$ differ by the map $\varphi_4 \circ \varphi_2 \circ \varphi_3 \circ \varphi_2^{-1}$. A computation shows that $\varphi_4 \circ \varphi_2 \circ \varphi_3 \circ \varphi_2^{-1}(z) = z + 6$, so that $\psi(P)$ and $\psi(Q)$ are indeed glued together in $\bar{V}$. The other cases are similar, and actually easier since no gluing in $\bar{V}$ is needed.

It follows that ψ induces a map $\bar{\psi} \colon \bar{U} \to \bar{V}$, defined by the property that $\bar{\psi}(\bar{P})$ is equal to the point of $\bar{V}$ corresponding to $\psi(P) \in V$, for an arbitrary point $P \in U$ corresponding to $\bar{P} \in \bar{U}$. Let us show that $\bar{\psi}$ is an isometry.

If $P = P_1$, $Q_1 \sim P_2$, ..., $Q_{n-1} \sim P_n$, $Q_n = Q$ is a discrete walk w from $\bar{P}$ to $\bar{Q}$ in $\bar{U}$, then $\psi(P) = \psi(P_1)$, $\psi(Q_1) \sim \psi(P_2)$, ..., $\psi(Q_{n-1}) \sim \psi(P_n)$, $\psi(Q_n) = \psi(Q)$ is a discrete walk from $\bar{\psi}(\bar{P})$ to $\bar{\psi}(\bar{Q})$ in $\bar{V}$, which has the same length as w. It follows that $\bar{d}_V(\bar{\psi}(\bar{P}), \bar{\psi}(\bar{Q})) \leqslant \bar{d}_U(\bar{P}, \bar{Q})$ for every $\bar{P}, \bar{Q} \in \bar{U}$.

Conversely, let $\psi(P) = P_1'$, $Q_1' \sim P_2'$, ..., $Q_{n-1}' \sim P_n'$, $Q_n' = \psi(Q)$ be a discrete walk w' from $\bar{\psi}(\bar{P})$ to $\bar{\psi}(\bar{Q})$ in $\bar{V}$. Consider the decomposition of V into the four vertical half-strips $\psi(U_1)$, $\psi(U_\infty)$, $\psi(U_{-1})$, $\psi(U_0)$. If P_i' and Q_i' are not in the same half-strip $\psi(U_j)$, draw the geodesic g joining P_i' to Q_i', consider the points R_1, ..., R_k (with $k \leqslant 3$) where g meets the vertical lines $\mathrm{Re}(z) = -1, 1, 2$ separating these half-strips, and replace the part $Q_{i-1}' \sim P_i'$, $Q_i' \sim P_{i+1}'$ of the walk w' by $Q_{i-1}' \sim P_i'$, $R_1 = R_1$, ..., $R_k = R_k$, $Q_i' \sim P_{i+1}'$. By performing finitely many such modifications we can arrange, without changing the length of w', that any two consecutive P_i', Q_i' belong to the same half-strip $\psi(U_j)$. As a consequence, there exists $P_i, Q_i \in U_j$ such that $\psi(P_i) = P_i'$, $\psi(Q_i) = Q_i'$, and $d_U(P_i, Q_i) = d_{\mathrm{hyp}}(P_i, Q_i) = d_{\mathrm{hyp}}(P_i', Q_i')$. Then $P = P_1$, $Q_1 \sim P_2$, ..., $Q_{n-1} \sim P_n$, $Q_n = Q$ is a discrete walk from $\bar{P}$ to $\bar{Q}$ in $\bar{U}$, whose length is equal to the length of w'. This proves that $\bar{d}_U(\bar{P}, \bar{Q}) \leqslant \bar{d}_V(\bar{\psi}(\bar{P}), \bar{\psi}(\bar{Q}))$ for every $\bar{P}, \bar{Q} \in \bar{U}$.

This completes the proof that $\bar{\psi} \colon (\bar{U}, \bar{d}_U) \to (\bar{V}, \bar{d}_V)$ is an isometry. $\square$

The combination of Lemmas 5.10 and 5.11 completes the proof of Proposition 5.9. $\square$ From page 118 ⇧

Without any assumption on a, the metric $\bar{d}_U$ on $\bar{U}$ may be different from the restriction of the metric $\bar{d}_X$ of $\bar{X}$. For instance, when

a is very close to 1, the images $\bar{P}$ and $\bar{Q} \in \bar{U}$ corresponding to $P = ai$ and $Q = \frac{1}{a}i$ are very close with respect to the metric $\bar{d}_X$, since $\bar{d}_X(\bar{P}, \bar{Q}) \leqslant d_X(P, Q) = \log a^2$, but they are quite far from each other with respect to the metric $\bar{d}_U$ since there clearly is a ball $B_{d_{\mathrm{hyp}}}(P, \varepsilon)$ which contains no point that is glued to another one, so that any discrete walk from $\bar{P}$ to $\bar{Q}$ in $\bar{U}$ has length $\geqslant \varepsilon$ in $\bar{U}$.

Lemma 5.12. *If a is chosen large enough that $a \log a > \frac{3}{2}$, the metrics $\bar{d}_X$ and $\bar{d}_U$ coincide on $\bar{U}$.*

To
page 125

Proof. Because every discrete walk valued in U is also valued in X, and because $d_X(P_i, Q_i) \leqslant d_U(P_i, Q_i)$ for every P_i, $Q_i \in U$, it is immediate that $\bar{d}_U(\bar{P}, \bar{Q}) \leqslant \bar{d}_X(\bar{P}, \bar{Q})$ for every $\bar{P}$, $\bar{Q} \in \bar{U}$.

To prove the reverse inequality, pick another number $a' > 1$ sufficiently close to 1 that $\log a' < \log a - \frac{3}{2a}$, and $\varepsilon > 0$ small enough that $\varepsilon < \log \frac{a}{a'} - \frac{3}{2a}$. Let U' and $\bar{U}'$ be associated to a' in the same way as U and $\bar{U}$ were associated to a. Because $a' < a$, the portion $S_{a'}$ of the pseudosphere S contains S_a.

Let $\bar{P}$, $\bar{Q} \in \bar{U}$, and let $P = P_1$, $Q_1 \sim P_2$, $\ldots$, $Q_{n-1} \sim P_n$, $Q_n = Q$ be a discrete walk w from $\bar{P}$ to $\bar{Q}$ in $\bar{X}$, whose d_X length is such that $\ell_{d_X}(w) \leqslant \bar{d}_X(\bar{P}, \bar{Q}) + \varepsilon$. Without loss of generality, we can assume that whenever the hyperbolic geodesic γ_i joining P_i to Q_i meets one of the circles C_0, C_1, C_{-1} and L_a delimiting U in X, it does so only at its endpoints P_i, Q_i; indeed, if this property does not hold, we can just add to the discrete walk w the intersection points of γ_i with these circles, which will not change the d_X-length of w. Adding a few more points if necessary, we can arrange that the same property holds for the circles C_0', C_1', C_{-1}' and $L_{a'}$ similarly associated to a'.

We will show that because of our choice of a' and ε, the discrete walk w stays in U'. Namely, all the P_i and Q_i as well as the geodesic arcs γ_i joining them, are contained in U'.

Indeed, suppose that the property does not hold. Let i_1 be the smallest index for which $\gamma_{i_1} \not\subset U$ and let i_2 be the largest index for which $\gamma_{i_2} \not\subset U$. By the condition that we imposed on the geodesics γ_i, the points P_{i_1} and Q_{i_2} are both contained in the union of the circles C_0, C_1, C_{-1} and L_a. Similarly, let i_1' be the smallest index for which $\gamma_{i_1'} \not\subset U'$ and let i_2' be the largest index for which $\gamma_{i_2'} \not\subset U'$, so that

$P_{i'_1}$ and $Q_{i'_2}$ belong to the union of the circles C'_0, C'_1, C'_{-1} and $L_{a'}$. Note that $i_1 < i'_1 < i'_2 < i_2$.

In the quotient space $\bar{U}'$, the geodesics γ_i with $i_1 \leqslant i \leqslant i'_1$ project to a continuous curve γ joining $\bar{P}_{i_1}$ to $\bar{Q}_{i'_1}$ in $\bar{U}'$. Consider the isometry $\varphi \colon (\bar{U}', \bar{d}_{U'}) \to (S_{a'}, d_{S_{a'}})$ provided by Proposition 5.9. Then $\varphi(\gamma)$ is a piecewise differentiable curve joining $\varphi(\bar{P}_{i_1})$ to $\varphi(\bar{Q}_{i'_1})$ in $S_{a'}$. Note that because P_{i_1} and $Q_{i'_1}$ are in the boundary of U and U', respectively, $\varphi(\bar{P}_{i_1})$ and $\varphi(\bar{Q}_{i'_1})$ respectively belong to the circles ∂S_a and $\partial S_{a'}$ delimiting S_a and $S_{a'}$ in the pseudosphere S. We then use an estimate which already appeared in the proof of Lemma 5.7. Parametrize the curve $\varphi(\gamma)$ by

$$s \mapsto (t(s) - \tanh t(s), \operatorname{sech} t(s) \cos \theta(s), \operatorname{sech} t(s) \sin \theta(s)), \quad 0 \leqslant s \leqslant 1,$$

for some functions $s \mapsto t(s)$ and $s \mapsto \theta(s)$, so that $\varphi(\bar{P}_{i_1}) \in \partial S_a$ and $\varphi(\bar{Q}_{i'_1}) \in \partial S_{a'}$ correspond to $s = 1$ and $s = 0$, respectively. In particular, $t(0) = \operatorname{arccosh} \frac{a'\pi}{3}$ and $t(1) = \operatorname{arccosh} \frac{a\pi}{3}$ by definition of S_a and $S_{a'}$. Then,

$$\sum_{i=i_1}^{i'_1} d_X(P_i, Q_i) = \sum_{i=i_1}^{i'_1} \ell_{\text{hyp}}(\gamma_i) = \ell_{\text{euc}}(\gamma)$$

$$= \int_0^1 \sqrt{x'(s)^2 + y'(s)^2 + z'(s)^2}\, ds$$

$$= \int_0^1 \sqrt{t'(s)^2 \tanh^2 t(s) + \theta'(s)^2 \operatorname{sech}^2 t(s)}\, ds$$

$$\geqslant \int_0^1 t'(s) \tanh t(s)\, ds$$

$$= \log \cosh t(1) - \log \cosh t(0) = \log \frac{a}{a'}.$$

Similarly, $\displaystyle\sum_{i=i'_2}^{i_2} d_X(P_i, Q_i) \geqslant \log \frac{a}{a'}$. As a consequence,

$$\bar{d}_X(\bar{P}, \bar{Q}) \geqslant \ell_{d_X}(w) - \varepsilon = \sum_{i=1}^n d_X(P_i, Q_i) - \varepsilon$$

$$\geqslant \sum_{i=1}^{i_1-1} d_X(P_i, Q_i) + 2\log\frac{a}{a'} + \sum_{i=i_2+1}^n d_X(P_i, Q_i) - \varepsilon.$$

On the other hand, the boundary circle ∂S_a has length $\frac{6}{a}$. Consequently, it is possible to join the two points $\varphi(\bar{P}_{i_1})$ and $\varphi(\bar{Q}_{i_2})$ by a curve of euclidean length $\leqslant \frac{3}{a}$ in S_a so that

$$\bar{d}_U(\bar{P}_{i_1}, \bar{Q}_{i_2}) = \bar{d}_{S_a}\big(\varphi(\bar{P}_{i_1}), \varphi(\bar{Q}_{i_2})\big) \leqslant \frac{3}{a}.$$

It follows that there is a discrete walk $P_{i_1} = P'_1, Q'_1 \sim P'_2, \ldots, P'_{n'-1} \sim P'_{n'}, Q'_{n'} = Q_{i_2}$ of length $\leqslant \frac{3}{a} + \varepsilon$. Chaining this discrete walk with the beginning and the end of w, we obtain a discrete walk $P = P_1$, $Q_1 \sim P_2, \ldots Q_{i_1-1} \sim P_{i_1}, P_{i_1} = P'_1, Q'_1 \sim P'_2, \ldots, P'_{n'-1} \sim P'_{n'}, Q'_{n'} = Q_{i_2}, Q_{i_2} \sim P_{i_2+1}, \ldots, Q_{n-1} \sim P_n, Q_n = Q$ from $\bar{P}$ to $\bar{Q}$ in $\bar{X}$. Therefore,

$$\bar{d}_X(\bar{P}, \bar{Q}) \leqslant \sum_{i=1}^{i_1-1} d_X(P_i, Q_i) + \frac{3}{a} + \varepsilon + \sum_{i=i_2+1}^{n} d_X(P_i, Q_i)$$

Combining this with our earlier estimate for $\bar{d}_X(\bar{P}, \bar{Q})$, we conclude that $2\log\frac{a}{a'} - \varepsilon \leqslant \frac{3}{a} + \varepsilon$, which is impossible by our choices of a, a' and ε.

This contradiction shows that our initial hypothesis was false. Namely, the geodesics γ_i are all in U'. In particular, the P_i and Q_i form a discrete walk from $\bar{P}$ to $\bar{Q}$ in $\bar{U}'$, so that

$$\bar{d}_{U'}(\bar{P}, \bar{Q}) \leqslant \sum_{i=1}^{n} d_{U'}(P_i, Q_i) = \sum_{i=1}^{n} \ell_{\mathrm{hyp}}(\gamma_i) = \ell_{d_X}(w) \leqslant \bar{d}_X(\bar{P}, \bar{Q}) + \varepsilon.$$

Since this holds for every ε that is small enough, we conclude that $\bar{d}_{U'}(\bar{P}, \bar{Q}) \leqslant \bar{d}_X(\bar{P}, \bar{Q})$.

Finally, the two metrics $\bar{d}_U$ and $\bar{d}_{U'}$ coincide on $\bar{U}$. This can be seen by applying the proof of Lemma 5.8 to the space $(\bar{V}, \bar{d}_V)$ of Lemma 5.11, and to $(\bar{V}', \bar{d}_{V'})$ similarly associated to a'. One can also show that the metrics d_{S_a} and $d_{S_{a'}}$ coincide on S_a by a simple estimate of euclidean lengths of curves.

Therefore, $\bar{d}_U(\bar{P}, \bar{Q}) \leqslant \bar{d}_{U'}(\bar{P}, \bar{Q}) \leqslant \bar{d}_X(\bar{P}, \bar{Q})$ for every $\bar{P}$, $\bar{Q} \in \bar{U}$. Since we had already proved the reverse inequality, this shows that the metrics $\bar{d}_U$ and $\bar{d}_X$ coincide on $\bar{U}$. $\square$

In Lemma 5.12, the condition that $a\log a > \frac{3}{2}$ is not quite sharp. With a better hyperbolic distance estimate to improve the inequality

$\bar{d}_U(\bar{P}_{i_1}, \bar{Q}_{i_2}) \leqslant \frac{3}{a}$, one can show that the conclusions of Lemma 5.12 still hold for $a \geqslant 2$. This second estimate is sharp, in the sense that Lemma 5.12 fails for $a < 2$.

From page 122

5.6. Triangular pillowcases

We conclude with an example where the angle condition of Theorems 4.4, 4.10 or 4.13 fails, so that we obtain surfaces with cone singularities, as in Exercise 4.8.

Proposition 5.13. *Let α, β and γ be three numbers in the interval $(0, \pi)$. Then:*

(1) *if $\alpha + \beta + \gamma = \pi$, there exists a triangle T of area 1 in the euclidean plane $(\mathbb{R}^2, d_{\text{euc}})$ whose angles are equal to α, β and γ;*

(2) *if $\alpha + \beta + \gamma < \pi$, there exists a triangle T in the hyperbolic plane $(\mathbb{H}^2, d_{\text{hyp}})$ whose angles are equal to α, β and γ;*

(3) *if $\pi < \alpha + \beta + \gamma < \pi + 2\min\{\alpha, \beta, \gamma\}$, there exists a triangle T in the sphere $(\mathbb{S}^2, d_{\text{sph}})$ whose angles are equal to α, β and γ.*

In addition, in each case, the triangle T is unique up to isometry of $(\mathbb{R}^2, d_{\text{euc}})$, $(\mathbb{H}^2, d_{\text{hyp}})$ of $(\mathbb{S}^2, d_{\text{sph}})$, respectively.

Proof. The euclidean case is well known. See Exercises 5.15 and 3.5 for a proof in the hyperbolic and spherical cases. $\qquad\square$

See also Exercises 2.15 and 3.6 for a proof that the conditions of Proposition 5.13 are necessary.

Given α, β and $\gamma \in (0, \pi)$, let T be the euclidean, hyperbolic or spherical triangle provided by Proposition 5.13.

Choose an isometry φ of $(\mathbb{R}^2, d_{\text{euc}})$, $(\mathbb{H}^2, d_{\text{hyp}})$ or $(\mathbb{S}^2, d_{\text{sph}})$, according to the case, such that $\varphi(T)$ is disjoint from T. The existence of φ is immediate in the euclidean and hyperbolic case. For the spherical case, we observe that the proof of Proposition 5.13(3) in Exercise 3.5 shows that T is always contained in the interior of a hemisphere. Then, we can use for φ the reflection across the great circle delimiting this hemisphere.

Let X be the union of T and $\varphi(T)$. We can consider X as a nonconnected polygon, whose edges are the three sides E_1, E_3, E_5 of T and the corresponding three sides $E_2 = \varphi(E_1)$, $E_4 = \varphi(E_3)$ and $E_6 = \varphi(E_5)$ of $\varphi(T)$. We can then glue these edges by the gluing maps $\varphi_1 \colon E_1 \to E_2$, $\varphi_3 \colon E_3 \to E_4$, $\varphi_5 \colon E_5 \to E_6$ defined by the restrictions of φ to the edges indicated.

By Theorems 4.3, 4.9 or 4.12, this gluing data provides a quotient metric space $(\bar{X}, \bar{d}_X)$. This metric space is locally isometric to $(\mathbb{R}^2, d_{\mathrm{euc}})$, $(\mathbb{H}^2, d_{\mathrm{hyp}})$ or $(\mathbb{S}^2, d_{\mathrm{sph}})$ everywhere, except at the three points that are the images of the vertices under the quotient map. The metric has cone singularities at these three points, with respective cone angles 2α, 2β, $2\gamma < 2\pi$, in the sense of Exercise 4.8.

Note that $\bar{X}$ is obtained by gluing two triangles T and $\varphi(T)$ along their edges. In the euclidean case, this is the familiar construction of a pillowcase obtained by sewing together two triangular pieces of cloth. Figure 5.21 attempts to describe the geometry of $\bar{X}$ in all cases.

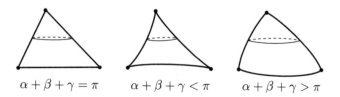

$$\alpha + \beta + \gamma = \pi \qquad \alpha + \beta + \gamma < \pi \qquad \alpha + \beta + \gamma > \pi$$

Figure 5.21. Triangular pillowcases

Exercises for Chapter 5

Exercise 5.1. We want to rigorously prove that the metric spaces $(\bar{X}_1, \bar{d}_{X_1})$ and $(\bar{X}_2, \bar{d}_{X_2})$ of Section 5.1.1 are homeomorphic. Without loss of generality, we can assume that the lower left corners of the rectangle X_1 and the parallelogram X_2 are both equal to the origin $(0, 0) \in \mathbb{R}^2$.

    a. Show that there is a unique linear map $\mathbb{R}^2 \to \mathbb{R}^2$ that sends the bottom edge of X_1 to the bottom edge of X_2, and the left edge of X_1 to the left edge of X_2. Show that this linear map restricts to a homeomorphism $\varphi \colon X_1 \to X_2$.

    b. Show that two points P and $Q \in X_1$ are glued together to form a single point of $\bar{X}_1$ if and only if their images $\varphi(P)$ and $\varphi(Q)$ are glued together in X_2. Conclude that $\varphi\colon X_1 \to X_2$ induces a map $\bar{\varphi}\colon \bar{X}_1 \to \bar{X}_2$, defined by the property that $\bar{\varphi}(\bar{P}) = \overline{\varphi(P)}$ for every $\bar{P} \in \bar{X}_1$.

c. Show that $\bar{\varphi}\colon \bar{X}_1 \to \bar{X}_2$ is a homeomorphism.

Exercise 5.2. Let $\bar{X}$ be the torus obtained by gluing opposite sides of a euclidean parallelogram X. Let $\varphi\colon X \to X$ be the rotation of angle π around the center of the parallelogram, namely, around the point where the two diagonals meet. Show that there exists a unique isometry $\bar{\varphi}\colon \bar{X} \to \bar{X}$ of the quotient metric space $(\bar{X}, \bar{d}_X)$ such that $\bar{\varphi}(\bar{P}) = \overline{\varphi(P)}$ for every $\bar{P} \in \bar{X}$.

Exercise 5.3. Let $(\bar{X}, \bar{d}_X)$ be the torus obtained by gluing opposite edges of a euclidean rectangle $X = [a, b] \times [c, d] \subset \mathbb{R}^2$ by translations, where X is endowed with the euclidean metric $d_X = d_{\text{euc}}$. Recall that a geodesic of $(\bar{X}, \bar{d}_X)$ is a curve γ in $\bar{X}$ such that for every $\bar{P} \in \gamma$ and every $\bar{Q} \in \gamma$ sufficiently close to $\bar{P}$, there is a piece of γ joining $\bar{P}$ to $\bar{Q}$ which is the shortest curve going from $\bar{P}$ to $\bar{Q}$; here the length of a curve in $(\bar{X}, \bar{d}_X)$ is defined as in Exercise 1.11.

a. Let γ be a curve in $\bar{X}$ and let $\pi\colon X \to \bar{X}$ denote the quotient map. Show that if γ is geodesic in $(\bar{X}, \bar{d}_X)$, then its preimage $p^{-1}(\gamma)$ consists of parallel line segments in the rectangle X. You may need to use the main result of Exercise 1.11, which says that the length of a curve in the metric space $(\mathbb{R}^2, d_{\text{euc}})$ coincides with its usual euclidean length ℓ_{euc}.

b. Consider the case where X is the square $X_1 = [0, 1] \times [0, 1]$. Show that every closed geodesic curve in $\bar{X}_1$ has length $\geqslant 1$. Possible hint: Consider the projection of the preimage $p^{-1}(\gamma) \subset X_1 \subset \mathbb{R}^2$ to each of the coordinate axes.

c. Consider the case where X is the square $X_1 = [0, \frac{1}{2}] \times [0, 2]$. Show that $(\bar{X}, \bar{d}_X)$ contains a closed geodesic of length $\frac{1}{2}$.

d. Conclude that the euclidean tori $(\bar{X}_1, \bar{d}_{X_1})$ and $(\bar{X}_2, \bar{d}_{X_2})$ are not isometric.

Exercise 5.4. Let $(\bar{X}, \bar{d}_X)$ be the torus obtained by gluing together opposite sides of the square $X = [0, 1] \times [0, 1]$. Let $\pi\colon X \to \bar{X}$ denote the quotient map.

a. Given $a \in [0, 1]$, let $\varphi_a\colon X \to X$ be the (discontinuous) function defined by the property that $\varphi_a(x, y) = (x + a, y)$ if $0 \leqslant x \leqslant 1 - a$ and $\varphi_a(x, y) = (x + a - 1, y)$ if $1 - a < x \leqslant 1$. Show that φ_a induces a map $\bar{\varphi}_a\colon \bar{X} \to \bar{X}$, uniquely determined by the property that $\bar{\varphi}_a \circ \pi = \pi \circ \varphi_a$.

b. Show that if w is a discrete walk $P = P_1$, $Q_1 \sim P_2$, $\ldots$, $Q_{n-1} \sim P_n$, $Q_n = Q$ is a discrete walk from $\bar{P}$ to $\bar{Q}$ in $\bar{X}$, there exists a discrete walk w' from $\bar{\varphi}_a(\bar{P})$ to $\bar{\varphi}_a(\bar{Q})$ which has the same d_X-length $\ell_{d_X}(w') = \ell_{d_X}(w)$ as w. Hint: When P_i and Q_i are on opposite sides of the line

$x = 1 - a$, add to w the point where the line segment $[P_i Q_i]$ meets this line.

c. Use part b to show that $\bar{d}_X\big(\bar{\varphi}_a(\bar{P}), \bar{\varphi}_a(\bar{Q})\big) \leqslant \bar{d}_X(\bar{P}, \bar{Q})$ for every $\bar{P}$ $\bar{Q} \in \bar{X}$. Then show that $\bar{\varphi}$ is an isometry of $(\bar{X}, \bar{d}_X)$.

d. Use the above construction to show that the metric space $(\bar{X}, \bar{d}_X)$ is homogeneous, namely that, for every $\bar{P}$, $\bar{Q} \in \bar{X}$, there exists an isometry of $(\bar{X}, \bar{d}_X)$ which sends $\bar{P}$ to $\bar{Q}$.

e. More generally, let Y be a parallelogram in $\mathbb{R}^2$ and let $(\bar{Y}, \bar{d}_Y)$ be the quotient metric space obtained by gluing opposite sides of Y by translations as in Section 5.1.1. Show that $(\bar{Y}, \bar{d}_Y)$ is homogeneous.

Exercise 5.5. Let $(\bar{X}, \bar{d}_X)$ be the Klein bottle obtained by gluing together the sides of the square $X = [0,1] \times [0,1]$ as in Section 5.1. Let $\pi \colon X \to \bar{X}$ denote the quotient map.

a. Recall that a geodesic of $(\bar{X}, \bar{d}_X)$ is a curve γ in $\bar{X}$ such that for every $\bar{P} \in \gamma$ and every $\bar{Q} \in \gamma$ sufficiently close to $\bar{P}$, there is a piece of γ joining $\bar{P}$ to $\bar{Q}$ which is the shortest curve going from $\bar{P}$ to $\bar{Q}$, where the length of a curve in $(\bar{X}, \bar{d}_X)$ is defined as in Exercise 1.11. Show that the image $\gamma_1 = \pi\big([0,1] \times \{0\}\big)$ is a closed geodesic of length 1, and that there exists at least one closed geodesic of length 2 which is disjoint from γ_1. Show that $\gamma_2 = \pi\big([0,1] \times \{\frac{1}{2}\}\big)$ satisfies the same two properties, namely, it is a closed geodesic of length 1 which is disjoint from a closed geodesic of length 2.

b. Show that γ_1 and γ_2 are the only two closed geodesics in $\bar{X}$ satisfying the properties of part a. Hint: For a closed geodesic γ in $\bar{X}$ note, as in Exercise 5.3, that the preimage $\pi^{-1}(\gamma)$ must consist of parallel line segments in the square X.

c. Conclude that the Klein bottle $\bar{X}$ is not homogeneous.

Exercise 5.6. In the euclidean plane, let X_1 be the square with vertices $(0,0)$, $(1,0)$, $(0,1)$, $(1,1)$; let X_2 be the parallelogram with vertices $(0,0)$, $(1,1)$, $(0,1)$ and $(1,2)$; and let X_3 be the parallelogram with vertices $(0,0)$, $(1,1)$, $(1,2)$ and $(2,3)$. Let $\bar{X}_1$, $\bar{X}_2$ and $\bar{X}_3$ be the euclidean tori obtained by gluing opposite sides of these parallelograms. Show that the euclidean tori $\bar{X}_1$, $\bar{X}_2$ and $\bar{X}_3$ are all isometric. Hint: Show that each of these euclidean tori is obtained by gluing the sides of two suitably chosen triangles. (You may need to use the result of Exercise 4.4 to justify the fact that the order of the gluings does not matter.)

Exercise 5.7. Given a euclidean hexagon X, index its vertices as P_1, P_2, ..., P_6 in this order as one goes around the hexagon. Suppose that opposite edges have the same length, and that the sum of the angles of X at its odd vertices P_1, P_3 and P_5 is equal to 2π. Use elementary euclidean geometry to show that opposite edges of X are parallel.

Exercise 5.8. A surface is *nonorientable* if it contains a subset homeomorphic to the Möbius strip. For instance, the Klein bottle is a nonorientable euclidean surface. Construct a nonorientable hyperbolic surface.

Exercise 5.9. The *surface of genus* g is the immediate generalization of the case $g = 2$ and consists of a sphere with g handles added. Construct a hyperbolic surface of genus g for $g = 3$, 4 and then for any $g \geqslant 2$. Possible hint: Glue opposite sides of a hyperbolic polygon with suitably chosen angles; Proposition 5.13 may be convenient to construct this polygon.

Exercise 5.10 (The projective plane). Let X be the spherical polygon of Section 5.3, with the gluing data indicated there, and let $(\bar{X}, \bar{d}_X)$ be the corresponding quotient space. Let $\mathbb{RP}^2$ be the set of lines passing through the origin O in $\mathbb{R}^3$. For any two L, $L' \in \mathbb{RP}^2$, let $\theta(L, L') \in [0, \frac{\pi}{2}]$ be the angle between these two lines in $\mathbb{R}^3$.

  a. Show that θ defines a metric on $\mathbb{RP}^2$.

  b. Consider the map $\varphi \colon X \to \mathbb{RP}^2$ which to $P \in X$ associates the line OP. Show that φ induces a bijection $\bar{\varphi} \colon \bar{X} \to \mathbb{RP}^1$.

  c. Show that this bijection $\bar{\varphi}$ is an isometry between the metric spaces $(\bar{X}, \bar{d}_X)$ and $(\mathbb{RP}^2, \theta)$.

  d. Show that the metric space $(\mathbb{RP}^2, \theta)$ (and consequently $(\bar{X}, \bar{d}_X)$ as well) is homogeneous.

Exercise 5.11. Let $(\bar{X}, \bar{d}_X)$ be the projective plane of Section 5.3 (or Exercise 5.10 above). Show that for every great circle C of $\mathbb{S}^2$, the image of $X \cap C$ under the quotient map $\pi \colon X \to \bar{X}$ is a closed geodesic of length π. Conclude that the spherical surface $(\bar{X}, \bar{d}_X)$ is not isometric to the sphere $(\mathbb{S}^2, d_{\mathrm{sph}})$.

Exercise 5.12. Let $(\bar{X}, \bar{d}_X)$ be the projective plane constructed in Section 5.3, by gluing onto itself the boundary of a hemisphere $X \subset \mathbb{S}^2$. Let P_0 be the center of this hemisphere, namely, the unique point such that $X = B_{d_{\mathrm{sph}}}(P_0, \frac{\pi}{2})$. Show that $\bar{X} - \bar{P}_0$ is homeomorphic to the Möbius strip of Section 5.4.

Exercise 5.13. Show that the "hyperbolic square" of Figure 5.18 really has the symmetries of a square, in the sense that there exists an isometry of $\mathbb{H}^2$ which sends 0 to 1, 1 to ∞, ∞ to -1 and -1 to 0.

Exercise 5.14. Let X be the hyperbolic polygon of Figure 5.18, namely, the infinite square in $\mathbb{H}^2$ with vertices at infinity 0, 1, ∞ and -1. We will glue its edges in a different way from the construction of Section 5.5. Namely, we glue the edge 1∞ to the edge 01 by the map $z \mapsto \dfrac{1}{-z+2}$, and the edge $(-1)\infty$ to the edge $0(-1)$ by $z \mapsto \dfrac{1}{-z-2}$. Let $(\bar{X}, \bar{d}_X)$ be the quotient metric space provided by this gluing construction.

a. Show that $(\bar{X}, \bar{d}_X)$ is locally isometric to $(\mathbb{H}^2, d_{\mathrm{hyp}})$.

b. Show that $\bar{X}$ can be decomposed as $\bar{X} = \bar{X}_0 \cup \bar{U}_1 \cup \bar{U}_2 \cup \bar{U}_3$ where each $(\bar{U}_i, \bar{d}_X)$ is isometric to a subset (S_a, d_{S_a}) of the pseudosphere as in Proposiiton 5.9, and where $\bar{X}_0$ is the image of a bounded subset $X_0 \subset X$ under the quotient map $X \to \bar{X}$. Hint: Adapt the analysis of Section 5.5.

c. Show that $(\bar{X}, \bar{d}_X)$ is homeomorphic to the complement of three points in the sphere $\mathbb{S}^2$.

Exercise 5.15 (Hyperbolic triangles). Let α, β and γ be three positive numbers with $\alpha+\beta+\gamma < \pi$. We want to show that there exists a hyperbolic triangle in $\mathbb{H}^2$ whose angles are equal to α, β and γ, and that this triangle is unique up to isometry of $\mathbb{H}^2$. For this, we just adapt the proof of Lemma 5.2.

Let g be the complete hyperbolic geodesic of $\mathbb{H}^2$ with endpoints 0 and ∞. Let h be the complete hyperbolic geodesic passing through the point i and such that the angle from g to h at i, measured counterclockwise, is equal to $+\beta$. For $y < 1$, let k_y be the complete geodesic passing through iy and such that the counterclockwise angle from g to k_y at iy is equal to $-\gamma$. Compare Figure 5.9.

a. Show that the set of those $y < 1$ for which k_y meets h is an interval $(y_0, 1)$.

b. When y is in the above interval $(y_0, 1)$, let α_y be the counterclockwise angle from h to k_y at their intersection point. Show that

$$\lim_{y \to 1^-} \alpha_y = \pi - \beta - \gamma \quad \text{and} \quad \lim_{y \to y_0^+} \alpha_y = 0.$$

Conclude that there exists a value $y \in (y_0, 1)$ for which $\alpha_y = \alpha$.

c. Show that the above $y \in (y_0, 1)$ with $\alpha_y = \alpha$ is unique.

d. Let T and T' be two hyperbolic triangles with the same angles α, β and γ. Show that there is an isometry φ of the hyperbolic plane $(\mathbb{H}^2, d_{\mathrm{hyp}})$ such that $\varphi(T) = T'$. Hint: For g and h as above, apply suitable hyperbolic isometries to send T and T' to triangles with one edge in g and another edge in h, and use part c.

Exercise 5.16 (The Gauss-Bonnet formula). Let X be a bounded polygon in $\mathbb{R}^2$, $\mathbb{H}^2$ or $\mathbb{S}^2$ consisting of finitely many disjoint convex polygons X_1, X_2, ..., X_m. Let $\bar{X}$ be obtained by gluing together pairs of edges of X. As usual we assume that for every vertex P of X, the angles of X at the vertices that are glued to P add up to 2π, so that $\bar{X}$ is a euclidean, hyperbolic or spherical surface.

The quotient space $\bar{X}$ is now decomposed into m images of the convex polygons X_i, n images of the $2n$ edges of X, and p points images of the vertices of X. The **Euler characteristic** of $\bar{X}$ is the integer

$$\chi(\bar{X}) = m - n + p.$$

A deep result, which we cannot prove here, asserts that the Euler characteristic $\chi(\bar{X})$ is independent of the way the space $\bar{X}$ is obtained by gluing edges of a polygon in the sense that if two such surfaces $\bar{X}$ and $\bar{X}'$ are homeomorphic, they have the same Euler characteristic $\chi(\bar{X}) = \chi(\bar{X}')$. See, for instance, [**Massey**, Chap. IX, §4] or [**Hatcher₂**, §2.2]

a. Compute the Euler characteristic of the torus, the Klein bottle, the surface of genus 2, and the projective plane.

b. Show that for each of the convex polygons X_i,

$$\sum_{j=1}^{n_i} \theta_j = (n_i - 2)\pi + K\,\mathrm{Area}(X_i),$$

where: X_i has n_i vertices, and $\theta_1, \theta_2, \ldots, \theta_{n_i}$ are the angles of X_i at these vertices; if $X_i \subset \mathbb{R}^2$ is a euclidean polygon, $\mathrm{Area}(X_i)$ denotes its usual area and $K = 0$; if $X_i \subset \mathbb{H}^2$ is a hyperbolic polygon, $\mathrm{Area}(X_i)$ denotes its hyperbolic area $\mathrm{Area}_{\mathrm{hyp}}(X_i)$ as defined in Exercise 2.14, and $K = -1$; if $X_i \subset \mathbb{S}^2$ is a spherical polygon, $\mathrm{Area}(X_i)$ denotes its surface area in $\mathbb{S}^2$ and $K = +1$. Hint: Use the convexity of X_i to decompose it into $(n_i - 2)$ triangles, and apply the results of Exercises 2.15 and 3.6 to these triangles.

c. Use part b to show that

$$2\pi\chi(\bar{X}) = K\,\mathrm{Area}(X).$$

In particular, if a surface $\bar{X}$ is obtained by gluing together the sides of a euclidean, hyperbolic, or spherical polygon X as above, its Euler characteristic $\chi(\bar{X})$ must be zero, negative or positive, respectively. The equation of part c is a special case of a more general statement known as the **Gauss-Bonnet formula**.

Chapter 6

Tessellations

We are all very familiar with the tiling of a kitchen floor. The floor is divided into polygons (the tiles) with disjoint interiors. When only one type of tile is used, any two such polygons are isometric in the sense that there is an isometry of the euclidean plane sending the first polygons to the second one.

The standard patterns used to tile a kitchen floor can usually be extended to provide a tiling of the whole euclidean plane. In this case, there is a more highfalutin word for tiling, namely "tessellation". In this chapter, we will see how the edge gluings of euclidean or hyperbolic polygons that we considered in Chapters 4 and 5 can be used to obtain interesting tessellations of the euclidean plane, the hyperbolic plane or the sphere.

6.1. Tessellations

We first give a formal definition of tessellations.

A **tessellation** of the euclidean plane, the hyperbolic plane or the sphere is a family of **tiles** X_n, $n \in \mathbb{N}$, such that

(1) each tile X_m is a connected polygon in the euclidean plane, the hyperbolic plane or the sphere;

(2) any two X_m, X_n are isometric;

(3) the X_m cover the whole euclidean plane, hyperbolic plane or sphere, in the sense that their union is equal to this space;

(4) the intersection of any two distinct X_m, X_n consists only of vertices and edges of X_m, which are also vertices and edges of X_n;

(5) (Local Finiteness) for every point P in the plane, there exists an ε such that the ball of radius ε centered at P meets only finitely many tiles X_n.

Figure 6.1 provides an example of a tessellation of the euclidean plane by isometric hexagons.

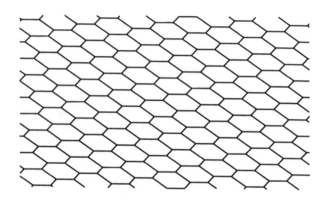

Figure 6.1. A tessellation of the euclidean plane $(\mathbb{R}^2, d_{\mathrm{euc}})$ by hexagons

Exactly like a kitchen can be tiled with tiles with different shapes, one can slightly extend the definition of a tessellation to allow several types of tiles. Namely, condition (2) can be relaxed by asking only that every tile X_n is isometric to one of finitely many model polygons $T_1, T_2, \ldots, T_p$. We will restrict our discussion to the original definition, but will later indicate how to extend the arguments to this more general context.

6.2. Complete metric spaces

In Section 6.3, we will use edge gluings of polygons to construct tessellations of the euclidean plane, the hyperbolic plane or the sphere.

This construction will use a new ingredient, the completeness of a metric space.

In a metric space (X, d), define the **length** of a sequence P_1, P_2, ..., P_n, ... as the infinite sum $\sum_{n=1}^{\infty} d(P_n, P_{n+1})$. This length may be finite or infinite.

The metric space (X, d) is **complete** if every sequence of points P_1, P_2, ..., P_n, ... in X whose length is finite converges to some $P_\infty \in X$.

In another mathematics course you may have encountered a different definition of complete metric spaces involving Cauchy sequences. Exercise 6.1 shows that the two definitions are equivalent. The above formulation has the advantage of being more geometric because of its analogy with the length of a discrete walk. Intuitively, a metric space (X, d) is complete if one cannot escape from it by walking a finite distance. Indeed, a finite length walk is the same as a finite length sequence, and such a sequence $(P_n)_{n \in \mathbb{N}}$ must converge to some point $P_\infty \in X$ if the space is complete.

In Section 6.4, we will establish several general properties which can be used to prove that a metric space is complete. However, we first see how this property is relevant for constructing tessellations of the euclidean and hyperbolic planes.

6.3. From gluing polygon edges to tessellations

Let X be a polygon in the euclidean plane $(\mathbb{R}^2, d_{\text{euc}})$, in the hyperbolic plane $(\mathbb{H}^2, d_{\text{hyp}})$, or in the sphere $(\mathbb{S}^2, d_{\text{sph}})$ with edges E_1, E_2, ..., E_{2p}. As in Chapter 4, group these edges into pairs $\{E_{2k-1}, E_{2k}\}$ and specify isometries $\varphi_{2k-1} \colon E_{2k-1} \to E_{2k}$ and $\varphi_{2k} = \varphi_{2k-1}^{-1} \colon E_{2k} \to E_{2k-1}$. As in Sections 4.4 and 4.5, extend each φ_i to an isometry of the euclidean plane, the hyperbolic plane or the sphere in such a way that $\varphi_i(X)$ is on the side of the edge $\varphi_i(E_i)$ that is opposite to X.

Let Γ consist of all isometries φ of $(\mathbb{R}^2, d_{\text{euc}})$, $(\mathbb{H}^2, d_{\text{hyp}})$ or $(\mathbb{S}^2, d_{\text{sph}})$ that can be written as a composition

$$\varphi = \varphi_{i_l} \circ \varphi_{i_{l-1}} \circ \cdots \circ \varphi_{i_1}$$

of finitely many such gluing maps. By convention this includes the identity map, which can be written as a composition of 0 gluing maps.

We will refer to Γ as the *tiling group* associated to the polygon X and to the edge gluing isometries $\varphi_i \colon E_i \to E_{i\pm 1}$. This is our first encounter with a transformation group, a notion which will be further investigated in Chapter 7.

As usual, endow X with the metric d_X for which $d_X(P,Q)$ is the infimum of the euclidean, hyperbolic or spherical length of all curves joining P to Q in X. Let $(\bar{X}, \bar{d}_X)$ be the quotient metric space of (X, d_X) defined by the above gluing data.

Theorem 6.1 (Tessellation Theorem). *Let X be a euclidean, hyperbolic or spherical connected polygon with gluing data as above. In addition, suppose that for each vertex of X, the angles of X at all vertices that are glued to P add up to $\frac{2\pi}{n}$ for some integer $n > 0$ depending on the vertex; in particular, by Theorems 4.4 or 4.10, the quotient space $(\bar{X}, \bar{d}_X)$ is a euclidean, hyperbolic or spherical surface with cone singularities. Finally, assume that the quotient space $(\bar{X}, \bar{d}_X)$ is complete.*

Then, as φ ranges over all elements of the tiling group Γ, the family of polygons $\varphi(X)$ forms a tessellation of the euclidean plane, the hyperbolic plane or the sphere.

The proof will take a while.

By construction, the tiles $\varphi(X)$ are all isometric. Therefore, the issue is to show that they cover the whole plane and that they have disjoint interiors.

To
page 146 Our strategy will be that of the tile-layer, beginning at X and progressively setting one tile after the other. In this approach, there are two potential snags that need to be ruled out. The first one is that the tiles $\varphi(X)$ might not necessarily cover the whole plane. The second one is that, as we are progressively laying the tiles beginning at X, two distinct tiles might end up overlapping as in the tile-layer's nightmare of Figure 6.2 (well known to weekend do-it-yourselfers, including the author). Ruling out these potential problems will use the completeness of $(\bar{X}, \bar{d})$ in a crucial way; see in particular Lemma 6.4.

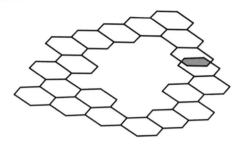

Figure 6.2. The tile-layer's nightmare

Before we proceed, it is convenient to establish some terminology. If the points P and Q of the boundary of X are glued together, then, by definition of the gluing process, $Q = \varphi_{i_l} \circ \varphi_{i_{l-1}} \circ \cdots \circ \varphi_{i_1}(P)$ where the φ_{i_j} are restricted by the condition that $\varphi_{i_{j-1}} \circ \cdots \circ \varphi_{i_1}(P)$ belongs to E_{i_j} for every $j \leqslant l$. In this situation, we observed in the proof of Theorem 4.4 that the tiles $\varphi_{i_1}^{-1} \circ \varphi_{i_2}^{-1} \circ \cdots \circ \varphi_{i_j}^{-1}(X)$ fit nicely side-by-side around the point P. We will say that these tiles are adjacent to X at the point P.

More precisely, a tile $\varphi(X)$ is **adjacent** to X at P if there exists a sequence of gluing maps $\varphi_{i_1}, \varphi_{i_2}, \ldots, \varphi_{i_l}$ such that $\varphi_{i_{j-1}} \circ \cdots \circ \varphi_{i_1}(P)$ belongs to E_{i_j} for every $j \leqslant l$ (including the fact that $P \in E_{i_1}$) and

$$\varphi = \varphi_{i_1}^{-1} \circ \varphi_{i_2}^{-1} \circ \cdots \circ \varphi_{i_l}^{-1}.$$

Note that φ is an element of the tiling group Γ, since each $\varphi_{i_j}^{-1}$ is equal to some gluing map $\varphi_{i_j \pm 1}$. However, the condition that $\varphi_{i_{j-1}} \circ \cdots \circ \varphi_{i_1}(P)$ belongs to E_{i_j} is quite restrictive, so that not every tile $\psi(X)$ with $\psi \in \Gamma$ is adjacent to X.

By convention, we allow the family of gluing maps φ_{i_j} to be empty, in which case φ is the identity map. In particular, the tile X is always adjacent to itself at P

More generally, we will say that the tiles $\varphi(X)$ and $\psi(X)$ are **adjacent** at the point $P \in \varphi(X) \cap \psi(X)$ if $\psi^{-1} \circ \varphi(X)$ is adjacent to X at $\psi^{-1}(P)$, in the above sense.

Lemma 6.2. *There are only finitely many tiles $\varphi(X)$ that are adjacent to X at the point P.*

Proof. If P is in the interior of X, then X is the only tile adjacent to X at P. If P is in an edge E_i but is not a vertex, then E_i is the only edge containing P and $\varphi_i(E_i) = E_{i \pm 1}$ is the only edge containing the unique point $\varphi_i(P)$ that is glued to P; it follows that X and $\varphi_i^{-1}(X)$ are the only tiles adjacent to X at P.

Consequently, we only need to focus on the case where P is a vertex. In this case, the proof will use in a critical way the hypothesis that the angles of X at P and at all the points that are glued to P add up to an angle sum of the form $\frac{2\pi}{n}$, for some integer $n > 0$. When $n = 1$, so that the angle sum is equal to 2π, the result is a property that we have already encountered in the proof of Theorem 4.4. The general argument is just a mild extension of this case.

As in the end of Section 4.3.1 when considering a sequence of gluing maps $\varphi_{i_1}, \varphi_{i_2}, \dots, \varphi_{i_j}, \dots$ such that $P_j = \varphi_{i_{j-1}} \circ \cdots \circ \varphi_{i_1}(P) \in E_{i_j}$ for every $j \leqslant l$, we can always require that the edge E_{i_j} is different from the image $\varphi_{i_{j-1}}(E_{i_{j-1}})$, since otherwise $\varphi_{i_{j-1}}$ and φ_{i_j} are the inverse of each other and cancel out in the composition. As a consequence, the sequence of indices $i_1, i_2, \dots, i_j, \dots$ is uniquely determined once we have chosen an edge E_{i_1} containing $P = P_1$, by induction and because i_j is the unique index for which the edge E_{i_j} contains $P_{i_j} = \varphi_{i_{j-1}} \circ \cdots \circ \varphi_{i_1}(P)$ and is not $\varphi_{i_{j-1}}(E_{i_{j-1}})$.

Consider the sequence of gluing maps $\varphi_{i_1}, \varphi_{i_2}, \dots, \varphi_{i_j}, \dots$ as above, associated to the choice of the edge E_{i_1} containing P. As in the end of Section 4.3.1, the sequence of points P_j eventually returns to $P = P_1$ after visiting all the points of $\bar{P}$, and there exists a number $k \geqslant 1$ such that $P_{j+k} = P_j$ and $i_{j+k} = i_j$ for every j. In particular, the set of points that are glued to P is $\bar{P} = \{P_1, P_2, \dots, P_k\}$.

As in the proof of Theorem 4.4, choose ε small enough that each ball $B_{d_X}(P_j, \varepsilon)$ in X is a disk sector for each $j = 1, 2, \dots, k$. For every $j \geqslant 1$, consider the euclidean, hyperbolic or spherical isometry $\psi_j = \varphi_{i_1}^{-1} \circ \varphi_{i_2}^{-1} \circ \cdots \circ \varphi_{i_{j-1}}^{-1}$ (with the convention that ψ_1 is the identity map). Still as in the proof of Theorem 4.4, the disk sectors $\psi_j\big(B_{d_X}(P_j, \varepsilon)\big)$ fit side-by-side around the vertex $P = P_1$. However, as we return to $P_{k+1} = P_1$, the disk sector $\psi_{k+1}\big(B_{d_X}(P_1, \varepsilon)\big)$ is not equal to $B_{d_X}(P_1, \varepsilon)$ any more and is instead obtained by rotating this disk sector around P_1 by an angle of $\frac{2\pi}{n}$.

If we keep going, $\psi_{2k+1}\big(B_{d_X}(P_1,\varepsilon)\big)$ is obtained by rotating $B_{d_X}(P_1,\varepsilon)$ by an angle of $2\frac{2\pi}{n}$, $\psi_{3k+1}\big(B_{d_X}(P_1,\varepsilon)\big)$ by an angle of $3\frac{2\pi}{n}$, etc..., so that eventually we reach $\psi_{nk+1}\big(B_{d_X}(P_1,\varepsilon)\big) = B_{d_X}(P_1,\varepsilon)$. The isometry ψ_{nk+1} fixes the point P_1 and the two geodesics delimiting the disk sector $B_{d_X}(P_1,\varepsilon)$. It follows that ψ_{nk+1} is the identity. (Use, for instance, Lemma 2.6 to prove this in the hyperbolic case).

As a consequence, $\psi_{j+nk} = \psi_j$ for every $j \geqslant 1$, and there are only finitely many (in fact exactly nk) tiles $\psi_j(X)$.

Similarly if, instead of E_{i_1}, we had selected the other edge $E_{i'_1}$ adjacent to P, the resulting family of gluing maps $\varphi_{i'_1}$, $\varphi_{i'_2}$, $\ldots$, $\varphi_{i'_{j'}}$, $\ldots$ and maps $\psi'_{j'} = \varphi_{i'_1}^{-1} \circ \varphi_{i'_2}^{-1} \circ \cdots \circ \varphi_{i'_{j'-1}}^{-1}$ is such that $\psi'_{j'+nk} = \psi'_{j'}$ for every $j' \geqslant 1$. So this other choice also gives only finitely many tiles $\varphi'_{j'}(X)$ adjacent to X at P.

Since E_{i_1} and $E_{i'_1}$ are the only edges containing the vertex P, this proves that there are only finitely many tiles adjacent to X at P.   □

We can make Lemma 6.2 a little more precise. When P is a vertex of X and with the notation of the above proof, note that $E_{i'_1}$ is the edge containing $P = P_1 = P_{k+1}$ that is different from $E_{i_1} = E_{k+1}$. Therefore, $E_{i'_1}$ is the image of the gluing map $\varphi_{i_k}\colon E_{i_k} \to \varphi_{i_k}(E_{i_k}) = E_{i'_1}$. As a consequence, $\varphi_{i'_1}$ is the inverse of φ_{i_k}. Similarly, $E_{i'_2}$ is the edge containing $\varphi_{i'_1}(P) = \varphi_{i_k}^{-1}(P_{k+1}) = P_k$ that is different from $\varphi_{i'_1}(E_{i_1}) = E_{i_k}$ so that the gluing map $\varphi_{i'_2}$ is equal to the inverse of $\varphi_{i_{k-1}}$. Iterating this argument, we conclude that $\varphi_{i'_{j'}} = \varphi_{i_{k-j'+1}}^{-1}$ for every j' with $1 \leqslant j' \leqslant k$.

Using the property that $i_{j+k} = i_j$ and $i'_{j'+k} = i'_{j'}$ for every j, $j' \geqslant 1$, one concludes that $\varphi_{i'_{j'}} = \varphi_{i_{nk-j'+1}}^{-1}$ for every j' with $1 \leqslant j' \leqslant nk$. As a consequence,

$$\psi'_{j'} = \varphi_{i'_1}^{-1} \circ \varphi_{i'_2}^{-1} \circ \cdots \circ \varphi_{i'_{j'-1}}^{-1}$$

$$= \varphi_{i_{nk}} \circ \varphi_{i_{nk-1}} \circ \cdots \circ \varphi_{i_{nk-j'+2}}$$

$$= \varphi_{i_1}^{-1} \circ \varphi_{i_2}^{-1} \circ \cdots \circ \varphi_{i_{nk-j'+1}}^{-1}$$

$$= \psi_{nk-j'+2}$$

for every j' with $1 \leqslant j' \leqslant nk$, since $\varphi_{i_1}^{-1} \circ \varphi_{i_2}^{-1} \circ \cdots \circ \varphi_{i_{nk}}^{-1} = \psi_{nk+1} = \psi_1$ is the identity map.

The point of this is that we do not need to consider the tiles $\psi'_{j'}(X)$, since they were already included among the tiles $\psi_j(X)$.

In particular, this proves the following.

Complement 6.3. *Under the hypotheses of Theorem 6.1, for every point $P \in X$, there is a small ε such that the tiles $\varphi(X)$ that are adjacent to X at P decompose the disk $B_d(P, \varepsilon) \subset \mathbb{R}^2$, $\mathbb{H}^2$ or $\mathbb{S}^2$ into finitely many euclidean or hyperbolic disk sectors $B_d(P, \varepsilon)$ with disjoint interiors, where $d = d_{\mathrm{euc}}$, d_{hyp} or d_{sph} according to whether X is a euclidean, hyperbolic or spherical polygon.* □

We can rephrase Complement 6.3 by saying that the tiles $\varphi(X)$ that are adjacent to X at P fit nicely side-by-side near P. However, at this point, we have no guarantee that they do not overlap away from P. This is particularly conceivable if X has a funny shape instead of being convex as in most pictures that we have seen so far. We will need the full force of all the hypotheses of Theorem 6.1 to rule out this possibility. Similarly, it will turn out that the adjacent tiles are the only ones meeting X, which is far from obvious at this point.

6.3.1. Hyperbolic tilings. After these preliminary observations, we now begin the proof of Theorem 6.1. Having to systematically write "the euclidean plane, the hyperbolic plane or the sphere" becomes clumsy after a while. Consequently, we will restrict our attention to the less familiar case, for practice, and assume that the plane considered is the hyperbolic plane $(\mathbb{H}^2, d_{\mathrm{hyp}})$. All the arguments will almost automatically extend to the euclidean and spherical context. See Sections 6.3.2 and 6.3.3.

Our first goal is to show that every $P \in \mathbb{H}^2$ is covered by a tile $\varphi(X)$ with $\varphi \in \Gamma$. For this, pick a base point P_0 in the interior of X, consider another point $P \in \mathbb{H}^2$, and let g be the hyperbolic geodesic joining P_0 to P. We will progressively set tiles over the geodesic g.

Look at the first point P_1 where g leaves X. At P_1, g enters one of the finitely many tiles that are adjacent to X at P_1. Let $\psi_1(X)$ be this tile. Then let P_2 be the first point where g leaves $\psi_1(X)$ and enters a tile $\psi_2(X)$. Repeating this process, one inductively defines a sequence of points $P_n \in g$ and of tiles $\psi_n(X)$ such that g leaves $\psi_{n-1}(X)$ at P_n to enter $\psi_n(X)$, and $\psi_n(X)$ is adjacent to $\psi_{n-1}(X)$ at P_n. See Figure 6.3.

Note that the tile $\psi_n(S)$ is not always uniquely determined. This happens when, right after passing P_n, the geodesic g follows an edge separating two of the tiles that are adjacent to $\psi_{n_1}(X)$ at P_n.

This process will terminate when the geodesic g enters a tile $\psi_n(X)$ and never leaves it, namely, when the point P is in the tile $\psi_n(X)$ as requested.

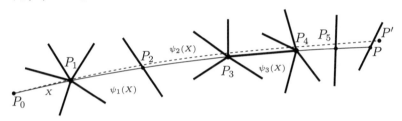

Figure 6.3. Laying tiles over a geodesic

Lemma 6.4. *The above tiling process must terminate after finitely many steps.*

Proof. The hypothesis that the quotient space $(\bar{X}, \bar{d}_X)$ is complete is here crucial.

Suppose that the tiling process continues forever and provides an infinite sequence of points P_n and tiles $\psi_n(X)$.

By construction, P_n is in both the tiles $\psi_n(X)$ and $\psi_{n-1}(X)$. The two points $\psi_n^{-1}(P_n)$ and $\psi_{n-1}^{-1}(P_n) \in X$ are glued together and consequently project to the same point $\bar{P}_n$ in the quotient space $\bar{X}$.

Because the quotient map $X \to \bar{X}$ is distance nonincreasing by Lemma 4.2,

$$\bar{d}_X(\bar{P}_n, \bar{P}_{n+1}) \leqslant d_X(\psi_n^{-1}(P_n), \psi_n^{-1}(P_{n+1})).$$

In addition, the two points P_n and P_{n+1} are joined in $\psi_n(X)$ by a geodesic curve contained in the geodesic g. Therefore,

$$d_X(\psi_n^{-1}(P_n), \psi_n^{-1}(P_{n+1})) = d_{\text{hyp}}(\psi_n^{-1}(P_n), \psi_n^{-1}(P_{n+1})),$$

which is also equal to $d_{\mathrm{hyp}}(P_n, P_{n+1})$ since ψ_n is a hyperbolic isometry. As a consequence, $\bar{d}_X(\bar{P}_n, \bar{P}_{n+1}) \leqslant d_{\mathrm{hyp}}(P_n, P_{n+1})$.

The sequence $(\bar{P}_n)_{n \in \mathbb{N}}$ has length

$$\sum_{n=1}^{\infty} \bar{d}_X(\bar{P}_n, \bar{P}_{n+1}) \leqslant \sum_{n=1}^{\infty} d_{\mathrm{hyp}}(P_n, P_{n+1}) \leqslant \ell_{\mathrm{hyp}}(g) < \infty$$

and therefore converges to some $\bar{P}_\infty \in \bar{X}$ in the complete metric space $(\bar{X}, \bar{d}_X)$.

Let $P_\infty^1, P_\infty^2, \ldots, P_\infty^k \in X$ be the points of $\bar{P}_\infty$, namely, the points of the polygon X that are glued together to form the point $\bar{P}_\infty$ in the quotient space $\bar{X}$. By Lemma 4.5, for $\varepsilon > 0$ sufficiently small, the ball $B_{\bar{d}_X}(\bar{P}_\infty, \varepsilon)$ in $\bar{X}$ is exactly the image of the union of all the balls $B_{d_X}(P_\infty^j, \varepsilon)$ in X. In addition, the balls $B_{d_X}(P_\infty^j, \varepsilon)$ are disjoint for ε sufficiently small.

Fix such an ε. Since $(\bar{P}_n)_{n \in \mathbb{N}}$ converges to $\bar{P}_\infty$, there exists an n_0 such that $\bar{d}_X(\bar{P}_n, \bar{P}_\infty) < \varepsilon/2$ for every $n \geqslant n_0$. By convergence of the sum $\sum_{n=1}^{\infty} d_{\mathrm{hyp}}(P_n, P_{n+1})$, we can also choose n_0 so that $d_{\mathrm{hyp}}(P_n, P_{n+1}) < \varepsilon/2$ for every $n \geqslant n_0$.

Since $\bar{d}(\bar{P}_n, \bar{P}_\infty) < \varepsilon/2$, the point $\psi_n^{-1}(P_n) \in X$ must be in some ball $B_d(P_\infty^{j_n}, \varepsilon/2)$ by another application of Lemma 4.5. Using the property that

$$d_X(\psi_n^{-1}(P_n), \psi_n^{-1}(P_{n+1})) = d_{\mathrm{hyp}}(P_n, P_{n+1}) < \varepsilon/2,$$

we observe that $\psi_n^{-1}(P_{n+1})$ is at distance $< \varepsilon$ from $P_\infty^{j_n}$ in X. The gluing map $\psi_{n+1}^{-1} \circ \psi_n$, which sends $\psi_n^{-1}(P_{n+1}) \in X$ to $\psi_{n+1}^{-1}(P_{n+1}) \in X$, must therefore send $P_\infty^{j_n}$ to some point $P \in \bar{P}_\infty$ and which is at distance $< \varepsilon$ from $\psi_{n+1}^{-1}(P_{n+1})$. By choice of ε, $P_\infty^{j_{n+1}}$ is the only such $P \in \bar{P}_\infty$, so that $\psi_n(P_\infty^{j_n}) = \psi_{n+1}(P_\infty^{j_{n+1}})$. In particular, the tiles $\psi_n(X)$ and $\psi_{n+1}(X)$ are also adjacent at the point $\psi_n(P_\infty^{j_n}) = \psi_{n+1}(P_\infty^{j_{n+1}})$.

If we set $P_\infty = \psi_{n_0}(P_\infty^{j_{n_0}})$, this proves that $\psi_n(P_\infty^{j_n}) = P_\infty$ for every $n \geqslant n_0$ and that the tiles $\psi_n(X)$ are all adjacent to $\psi_{n_0}(X)$ at P_∞. In particular, there are only finitely many such tiles.

By construction, either there is an edge E of the tile $\psi_n(X)$ such that P_n is the unique intersection point of E with the geodesic g, or g coincides with an edge E' of $\psi_n(X)$ for a while and P_n is one of the vertices of $\psi_n(X)$. Since there are only finitely many tiles $\psi_n(X)$ that are adjacent to $\psi_{n_0}(X)$ at P_∞ and since each has only finitely many edges and vertices, we conclude that there are only finitely many points P_n on g.

But this contradicts our original assumption. Therefore, the tiling process of the geodesic g must terminate after finitely many steps. $\square$

An immediate corollary of Lemma 6.4 is that every point $P \in \mathbb{H}^2$ is contained in at least one tile.

More precisely, if $\psi_n(X)$ is the last tile needed to cover the geodesic g joining P_0 to P as in Lemma 6.4, we will say that a tile $\varphi(X)$ is a **canonical tile** for the point P (with respect to the base point P_0) if it is adjacent to $\psi_n(X)$ at P. In particular, if P is contained in the interior of $\psi_n(X)$, then $\psi_n(X)$ is the only canonical tile for P. (This requires a little thought when g follows an edge separating two tiles, in which case there are several possibilities for the tiles X, $\psi_1(X)$, $\psi_2(X)$, ..., $\psi_n(X)$ covering g.)

Lemma 6.5. *For every $P \in \mathbb{H}^2$, there exists an $\varepsilon > 0$ such that for every $P' \in B_{d_{\mathrm{hyp}}}(P, \varepsilon) \subset \mathbb{H}^2$ the canonical tiles for P' are exactly the canonical tiles for P that contain P'.*

Proof. Let us use the notation of the proof of Lemma 6.4. In particular, let $\psi_n(X)$ be the last tile in the covering process, namely, the one containing the endpoint P of the geodesic g.

Let $\mathcal{T}$ be the finite collection of tiles consisting of X, of all the tiles $\psi_i(X)$ and, in addition, of any tile that is adjacent to some $\psi_i(X)$ at P_i.

A case-by-case analysis shows that for every Q in the geodesic g there is a small ball $B_{d_{\mathrm{hyp}}}(P, \varepsilon)$ in $\mathbb{H}^2$ which is contained in the union of the tiles $Y \in \mathcal{T}$. Indeed, there are three types of such points $Q \in g$: those in the interior of some tile $\psi_i(X)$, the points P_i, and the points where g follows an edge of $\psi_i(X)$.

As one slightly moves the point P to a nearby point P', the geodesic g moves to the geodesic g' joining P_0 to P' which stays very close to g. If we investigate what happens near the points P_i, we see that the tiling process of g' will only involve tiles of $\mathcal{T}$, and that the final tile will be adjacent to $\psi_n(X)$ at P. The result easily follows. See Figure 6.3.   □

Lemma 6.6. *Let P and Q be in the interior of the same tile $\varphi(X)$. If $\varphi(X)$ is canonical for P, it is also canonical for Q. In addition, $\varphi(X)$ is the only canonical tile for P and Q.*

Proof. This is a relatively simple consequence of Lemma 6.5.

Because of our assumption that X is connected, there is a curve in $\varphi(X)$, parametrized by $t \mapsto z(t)$, $a \leqslant t \leqslant b$, and joining $P = z(a)$ to $Q = z(b)$. After pushing it in the interior of $\varphi(X)$, we can even assume that this curve is completely contained in this interior. We can then consider

$$t_0 = \sup\{t; \; \varphi(X) \text{ is a canonical tile for } z(t)\}.$$

By definition of t_0, there are points $z(t)$ arbitrarily close to $z(t_0)$ such that the tile $\varphi(X)$ is canonical for $z(t)$. By Lemma 6.5, it follows that $\varphi(X)$ is canonical for $z(t_0)$. Because $z(t_0)$ is in the interior of $\varphi(X)$, this tile is the only canonical tile for $z(t_0)$ by definition of canonicity. Another application of Lemma 6.5 then shows that $\varphi(X)$ is canonical for $z(t)$ for every t sufficiently close to t_0. If t_0 was less than b, this would contradict the definition of t_0 as a supremum. Therefore, $t_0 = b$, and $\varphi(X)$ is canonical for $z(t_0) = z(b) = Q$, as requested.

Since P and Q are in the interior of $\varphi(X)$, no other tile can be canonical for them.   □

Lemma 6.7. *Every tile $\varphi(X)$ is canonical for some P in its interior.*

Proof. Let us first prove that if $\varphi(X)$ is canonical for some P in its interior, then for every gluing map φ_i the tile $\varphi \circ \varphi_i(X)$ is also canonical for some P' in its own interior. Later, we will show how this proves Lemma 6.7.

The tiles $\varphi(X)$ and $\varphi \circ \varphi_i(X)$ meet along the edge $\varphi \circ \varphi_i(E_i)$. Let Q be a point of this edge, not a vertex.

Let P'' be a point of the interior of $\varphi(X)$ that is close to Q. By Lemma 6.6, $\varphi(X)$ is the unique canonical tile for P''. Choosing P'' sufficiently close to Q and applying Lemma 6.5, we conclude that

$\varphi(X)$ is also canonical for Q. Therefore, every tile that is canonical for Q is adjacent to $\varphi(X)$ at Q, namely, it is either $\varphi(X)$ or $\varphi \circ \varphi_i(X)$.

If P' is a point of the interior of $\varphi \circ \varphi_i(X)$ that is sufficiently close to Q, Lemma 6.5 again shows that each canonical tile for P' will be either $\varphi(X)$ or $\varphi \circ \varphi_i(X)$. This canonical tile cannot be $\varphi(X)$ since P' is not in $\varphi(X)$. Therefore, $\varphi \circ \varphi_i(X)$ is canonical for P'.

This concludes the proof that if $\varphi(X)$ is canonical for some P in its interior, then $\varphi \circ \varphi_i(X)$ is canonical for some P' in its interior. By definition of Γ, every $\varphi \in \Gamma$ is obtained from the identity by successive multiplication on the right by gluing maps φ_i. By an induction starting with the fact that X is canonical for P_0, it follows that every tile $\varphi(X)$ is canonical for some $P \in \mathbb{H}^2$. $\qquad\square$

Proof of Theorem 6.1. Putting all these steps together, we can now conclude the proof of Theorem 6.1, namely that the tiles $\varphi(X)$ with $\varphi \in \Gamma$ form a tessellation of the hyperbolic plane $\mathbb{H}^2$. We have to check that:

(1) each tile $\varphi(X)$ is a polygon, and any two tiles $\varphi(X)$ and $\varphi'(X)$ are isometric;

(2) the plane $\mathbb{H}^2$ is equal to the union of the tiles $\varphi(X)$;

(3) any two distinct tiles $\varphi(X)$ and $\varphi'(X)$ have disjoint interiors;

(4) every point P is the center of a ball $B_{d_{\text{hyp}}}(P, \varepsilon)$ which meets only finitely many tiles $\varphi(X)$.

The first condition is trivial.

Lemma 6.4 shows that every point $P \in \mathbb{H}^2$ is contained in a tile $\varphi(X)$. This proves (2).

If the interiors of the tiles $\varphi(X)$ and $\varphi'(X)$ contain the same point P, the combinations of Lemmas 6.6 and 6.7 show that $\varphi(X)$ and $\varphi'(X)$ each are the unique canonical tile for P. Therefore, $\varphi(X) = \varphi'(X)$. This proves (3).

Finally, a point $P \in \mathbb{H}^2$ has only finitely many canonical tiles, and there exists a ball $B_{d_{\text{hyp}}}(P, \varepsilon)$ contained in the union of these canonical tiles. No other tile $\varphi(X)$ can meet this ball; indeed, its interior would otherwise meet the interior of one of the canonical tiles of P, contradicting (3). Therefore, the ball $B_{d_{\mathrm{hyp}}}(P, \varepsilon)$ meets only finitely many tiles, namely, the tiles that are canonical for P.   □

6.3.2. Euclidean tilings. The proof of Theorem 6.1 in the euclidean setup is identical to that for the hyperbolic plane.

6.3.3. Spherical tilings. When X is a polygon in the sphere $\mathbb{S}^2$, the only difference is that the shortest geodesic g from P_0 to P may not be unique. We need to prove that the set of canonical tiles for P is independent of the choice of this geodesic g. Several such shortest geodesics exist only when P is the antipodal point $-P_0$ of P_0. In this case, any two geodesics g and g' connecting P_0 to $P = -P_0$ are great semi-circles and can be moved from one to the other by a rotation about the line OP_0. We then have to check that the set of canonical tiles for P does not change as one rotates from g to g', which is easily

From page 136
done by the arguments used in the proofs of Lemmas 6.5 and 6.6. Once this is proved, the rest of the proof of Theorem 6.1 is identical.

6.3.4. Slight generalizations. We briefly indicate two ways in which the hypotheses of Theorem 6.1 can be somewhat relaxed.

Self-gluing of edges. So far, the edges of the polygon X were paired by gluing maps $\varphi_i\colon E_i \to E_{j_i}$ with $i_j = i \pm 1$ and $\varphi_{j_i} = \varphi_i^{-1}$. However, we never used the fact that these two edges E_i and E_{i_j} were distinct. As a consequence, we can allow the possibility that $j_i = i$. In this case, because we still need that $\varphi_i = \varphi_{j_i} = \varphi_i^{-1}$, the isometry φ_i of E_i must be the identity or a flip reflecting E_i across its midpoint.

The case where φ_i is a flip can actually be reduced to the previous setup, by converting the midpoint of E_i into a new vertex and splitting E_i into two new edges.

Beware that when $\varphi_i \colon E_i \to E_i$ is the identity, the isometric extension φ_i to all of $\mathbb{R}^2$, $\mathbb{H}^2$ or $\mathbb{S}^2$ occurring in the tiling group will be the reflection across the complete geodesic containing E_i because it must send X to the opposite side of E_i.

The statement and proof of Theorem 6.1 immediately extends to this context.

Allowing nonconnected polygons. We can also relax the hypothesis that X is connected and allow it to be the union of finitely many disjoint connected polygons X_1, X_2, ..., X_p. Then, the tiling process will provide a tessellation where each tile is isometric to one of the model tiles X_1, X_2, ..., X_p. This again follows the intuition of practical tile setting.

More precisely, we just need to take into account some additional information, namely, which of these connected polygons X_i contains each edge E_i. We then introduce the **tiling groupoid** Γ consisting of all isometries of the form

$$\varphi = \varphi_{i_l} \circ \varphi_{i_{l-1}} \circ \cdots \circ \varphi_{i_1},$$

where, for each $k = 1$, 2, ..., $l - 1$, the edge E_{i_k} is in the same connected polygon X_i as the edge $\varphi_{i_{k-1}}(E_{i_{k-1}})$ that is glued to $E_{i_{k-1}}$ by $\varphi_{i_{k-1}}$. The need for the barbaric terminology "groupoid" arises from the fact that Γ is not a group in the sense that we will encounter in Chapter 7; indeed, the composition of two elements of Γ is not always in Γ.

The tiles of the tessellation are then all the tiles of the form $\varphi(X_i)$, where $\varphi \in \Gamma$ is decomposed as $\varphi = \varphi_{i_l} \circ \varphi_{i_{l-1}} \circ \cdots \circ \varphi_{i_1}$ as above and where X_i is the connected polygon of X that contains the starting edge E_{i_1}.

The fact that these tiles form a tessellation is proved by exactly the same arguments as Theorem 6.1.

6.4. Completeness and compactness properties

Theorem 6.1, and in particular Lemma 6.4, points to the importance of the completeness property for our geometric endeavors. We will need a few criteria to guarantee that a space is complete. This section is devoted to results of this type.

6.4.1. The euclidean and hyperbolic plane are complete.

Theorem 6.8. *The euclidean plane* $(\mathbb{R}^2, d_{euc})$ *is complete.*

Proof. This is an immediate consequence of the following deep property of real numbers.

Fact 6.9. *The real line* $\mathbb{R}$, *endowed with the usual metric* $d(x,y) = |x-y|$, *is complete.* □

We refer to any of the many textbooks on real analysis for a proof of Fact 6.9, which requires a deep understanding of the definition of real numbers. Historically, real numbers were introduced precisely for this property to hold true. (Of course, it took many centuries to state the property in this way).

To prove that $(\mathbb{R}^2, d_{\mathrm{euc}})$ is complete, consider a sequence of points $P_1, P_2, \ldots, P_n, \ldots$ in $\mathbb{R}^2$ with finite length $\sum_{n=1}^{\infty} d_{\mathrm{euc}}(P_n, P_{n+1}) < \infty$. We want to show that the sequence $(P_n)_{n \in \mathbb{N}}$ converges to some point $P_\infty \in \mathbb{R}^2$.

If $P_n = (x_n, y_n)$, note that

$$d_{\mathrm{euc}}(P_n, P_{n+1}) \geqslant |x_{n+1} - x_n| = d(x_n, x_{n+1})$$

by equation (1.3). As a consequence, the sequence of real numbers $x_1, x_2, \ldots, x_n, \ldots$ has length

$$\sum_{n=1}^{\infty} d(x_n, x_{n+1}) \leqslant \sum_{n=1}^{\infty} d_{\mathrm{euc}}(P_n, P_{n+1}) < \infty.$$

By completeness of $(\mathbb{R}, d)$ (Fact 6.9), it follows that the sequence $(x_n)_{n \in \mathbb{N}}$ converges to some number x_∞ in $\mathbb{R}$.

The same argument shows that the sequence $(y_n)_{n \in \mathbb{N}}$ converges to some number y_∞.

Now, consider the point $P_\infty = (x_\infty, y_\infty)$ in $\mathbb{R}^2$. For every $\varepsilon > 0$, by definition of the convergence of sequences, there exists a number n_1 such that $|x_n - x_\infty| < \frac{\varepsilon}{\sqrt{2}}$ for every $n \geqslant n_1$, and there exists another number n_2 such that $|y_n - y_\infty| < \frac{\varepsilon}{\sqrt{2}}$ for every $n \geqslant n_2$. If n_0 is the larger of n_1 and n_2, we conclude that

$$d_{\mathrm{euc}}(P_n, P_\infty) = \sqrt{(x_n - x_\infty)^2 + (y_n - y_\infty)^2} \leqslant \sqrt{\tfrac{\varepsilon^2}{2} + \tfrac{\varepsilon^2}{2}} = \varepsilon$$

for every $n \geqslant n_0$.

This proves that the sequence $(P_n)_{n \in \mathbb{N}}$ converges to the point P_∞ in $(\mathbb{R}^2, d_{\mathrm{euc}})$, and completes the proof that this metric space is complete. □

We now use the completeness of the euclidean plane to prove that the hyperbolic plane is also complete. This is one more property that these two spaces have in common.

Theorem 6.10. *The hyperbolic plane* $(\mathbb{H}^2, d_{\mathrm{hyp}})$ *is a complete metric space.*

Proof. Let $P_1, P_2, \ldots, P_n, \ldots$ be a sequence of points in $\mathbb{H}^2$ with finite length

$$L = \sum_{n=1}^{\infty} d_{\mathrm{hyp}}(P_n, P_{n+1}) < \infty.$$

We need to show that the sequence $(P_n)_{n \in \mathbb{N}}$ converges to some P_∞ for the hyperbolic metric d_{hyp}.

Let γ_n be the hyperbolic geodesic arc joining P_n to P_{n+1}. Iterated uses of the Triangle Inequality give that for every $P \in \gamma_n$,

$$d_{\mathrm{hyp}}(P, P_1) \leqslant \sum_{k=1}^{n-1} d_{\mathrm{hyp}}(P_k, P_{k+1}) + d(P_n, P)$$

$$\leqslant \sum_{k=1}^{n-1} d_{\mathrm{hyp}}(P_k, P_{k+1}) + d(P_n, P_{n+1}) \leqslant L.$$

The estimate provided by Lemma 2.5 shows that $d_{\mathrm{hyp}}(P, P_1) \geqslant \left| \ln \frac{y}{y_1} \right|$ if $P = (x, y)$ and $P_1 = (x_1, y_1)$. Combining these two inequalities, we conclude that

$$y_1 e^{-L} < y < y_1 e^{L}$$

for every $P = (x, y) \in \gamma_n$. To ease the notation, set $c_1 = y_1 e^{-L}$ and $c_2 = y_1 e^{L}$ so that the y-coordinate of P is between c_1 and c_2 for every $P \in \gamma_n$.

Comparing the formulas for the euclidean and hyperbolic lengths ℓ_{euc} and ℓ_{hyp}, we conclude that

$$d_{\mathrm{hyp}}(P_n, P_{n+1}) = \ell_{\mathrm{hyp}}(\gamma_n) \geqslant \frac{1}{c_2} \ell_{\mathrm{euc}}(\gamma_n) = \frac{1}{c_2} d_{\mathrm{euc}}(P_n, P_{n+1}).$$

In particular, the sum $\sum_{n=1}^{\infty} d_{\mathrm{euc}}(P_n, P_{n+1})$ is finite. Since $(\mathbb{R}^2, d_{\mathrm{euc}})$ is complete (Theorem 6.8), we conclude that P_n converges to some point $P_\infty \in \mathbb{R}^2$ for the metric d_{euc}.

We showed that the y-coordinate y_n of P_n is such that $y_n > c_1$ for every n. It follows that the y-coordinate of the limit P_∞ is also $\geqslant c_1 > 0$, so that P_∞ is in the upper half-space $\mathbb{H}^2$. Comparing the hyperbolic and euclidean lengths of the line segment $[P_n, P_\infty]$ also gives

$$d_{\text{hyp}}(P_n, P_\infty) \leqslant \ell_{\text{hyp}}([P_n, P_\infty]) \leqslant \frac{1}{c_1}\ell_{\text{euc}}([P_n, P_\infty]) = \frac{1}{c_1}d_{\text{euc}}(P_n, P_\infty).$$

Therefore, since the sequence $\left(P_n\right)_{n\in\mathbb{N}}$ converges to P_∞ for the euclidean metric d_{euc}, it also converges to P_∞ for the hyperbolic metric d_{hyp}. $\qquad\square$

6.4.2. Compactness properties. For metric spaces, the notion of completeness is closely related to another property, namely, compactness.

This compactness property is based on subsequences. Given a sequence of points P_1, P_2, ..., P_n, ... in a space X, any increasing sequence of integers $n_1 < n_2 < \cdots < n_k < \cdots$ provides a new sequence P_{n_1}, P_{n_2}, ..., P_{n_k}, ... in X. This new sequence $(P_{n_k})_{k\in\mathbb{N}}$ is a **subsequence** of the original sequence $(P_n)_{n\in\mathbb{N}}$. In other words, one goes from a sequence to a subsequence by forgetting elements of the original sequence, while keeping infinitely many of them so that they still form a sequence.

A metric space (X, d) is **compact** if every sequence $(P_n)_{n\in\mathbb{N}}$ in X admits a converging subsequence, namely, a subsequence $(P_{n_k})_{k\in\mathbb{N}}$ such that

$$\lim_{k\to\infty} P_{n_k} = P_\infty$$

for some $P_\infty \in X$.

The connection between compactness and completeness is provided by the following result.

Proposition 6.11. *Every compact metric space (X, d) is complete.*

Proof. Let P_1, P_2, ..., P_n, ... be a sequence with finite length $\sum_{n=1}^\infty d(P_n, P_{n+1}) < \infty$. We want to show that the sequence $(P_n)_{n\in\mathbb{N}}$ converges.

By compactness, it admits a converging subsequence $(P_{n_k})_{k \in \mathbb{N}}$ such that

$$\lim_{k \to \infty} P_{n_k} = P_\infty.$$

For an arbitrary $\varepsilon > 0$, there exists a number k_0 such that $d(P_{n_k}, P_\infty) < \frac{\varepsilon}{2}$ for every $k \geqslant k_0$ by definition of the limit of a sequence.

Similarly, by convergence of the series $\sum_{i=1}^{\infty} d(P_i, P_{i+1})$, there exists a number n_0 such that $\sum_{i=n}^{\infty} d(P_i, P_{i+1}) < \frac{\varepsilon}{2}$ for every $n \geqslant n_0$. As a consequence, the Triangle Inequality shows that

$$d(P_n, P_{n'}) \leqslant \sum_{i=n}^{n'-1} d(P_i, P_{i+1}) \leqslant \sum_{i=n}^{\infty} d(P_i, P_{i+1}) < \frac{\varepsilon}{2}$$

for every n, $n' \geqslant n_0$ with $n \leqslant n'$.

Then, for every $n \geqslant \max\{n_0, n_{k_0}\}$ after picking any k such that $n_k \geqslant n$, we conclude that

$$d(P_n, P_\infty) \leqslant d(P_n, P_{n_k}) + d(P_{n_k}, P_\infty) < \frac{\varepsilon}{2} + \frac{\varepsilon}{2} = \varepsilon.$$

Therefore, for every $\varepsilon > 0$, we found an $n'_0 = \max\{n_0, n_{k_0}\}$ such that $d(P_n, P_\infty) < \varepsilon$ for every $n \geqslant n'_0$. This proves that the sequence $(P_n)_{n \in \mathbb{N}}$ converges to P_∞. $\qquad\square$

The advantage of compactness over completeness is that it is often easier to check. For instance, the following criterion is particularly useful.

Proposition 6.12. *Let $\varphi : X \to X'$ be a continuous map from a metric space (X, d) to a metric space (X', d'). If, in addition, X is compact, then its image $\varphi(X) \subset X'$ is compact for the restriction of the metric d'.*

Proof. Let P'_1, P'_2, ..., P'_n, ... be a sequence in $\varphi(X) \subset X'$. By definition of the image, each P'_n is the image of some $P_n \in X$ under φ, namely, $P'_n = \varphi(P_n)$.

Since X is compact, there is a subsequence P_{n_1}, P_{n_2}, ..., P_{n_k}, ... of the sequence $(P_n)_{n \in \mathbb{N}}$ which converges to some point $P_\infty \in X$. By continuity of φ, $P'_{n_k} = \varphi(P_{n_k})$ converges to $\varphi(P_\infty) \in \varphi(X)$ as k tends to ∞.

Therefore, for every sequence $(P'_n)_{n \in \mathbb{N}}$ in $\varphi(X)$, we can find a subsequence $(P'_{n_k})_{k \in \mathbb{N}}$ which converges in $\varphi(X)$. This proves that $\varphi(X)$ is compact. $\square$

6.4.3. Compactness properties in the euclidean plane, the hyperbolic plane and the sphere.

A subset Y of a metric space (X, d) is **bounded** if, for an arbitrary point $P_0 \in X$, there exists a number K such that $d(P, P_0) \leqslant K$ for every $P \in Y$. This is equivalent to the property that Y is contained in a large ball $B_d(P_0, K)$. Using the Triangle Inequality, one easily verifies that this property is independent of the choice of the point P_0.

A subset Y of the metric space (X, d) determines three types of points in X. The **interior points** P are those for which there exists an $\varepsilon > 0$ such that the ball $B_d(P, \varepsilon)$ is completely contained in Y. An **exterior point** is a point $P \in X$ such that for some $\varepsilon > 0$, the ball $B_d(P, \varepsilon)$ is disjoint from Y. A **boundary point** is a point $P \in X$ which is neither interior nor exterior, namely, such that every ball $B_d(P, \varepsilon)$ centered at P contains points that are in Y and points that are not in Y.

A subset Y of the metric space (X, d) is **closed** if it contains all its boundary points.

Theorem 6.13. *Every closed bounded subset of the euclidean plane* $(\mathbb{R}^2, d_{\mathrm{euc}})$ *is compact.*

Proof. Again, the key ingredient is a deep property of real numbers, whose proof can be found in most textbooks on real analysis.

Fact 6.14. *In the real line $\mathbb{R}$ endowed with the usual metric $d(x, y) = |x - y|$, every closed interval $[a, b]$ is compact.* $\square$

Let X be a closed bounded subset of $\mathbb{R}^2$. To show that X is compact, consider a sequence $(P_n)_{n \in \mathbb{N}}$ of points $P_n = (x_n, y_n) \in X$.

Because X is bounded, there is a closed interval $[a, b] \in \mathbb{R}$ which contains all the x_n. Therefore, by Fact 6.14, there is a subsequence $(x_{n_k})_{k \in \mathbb{N}}$ which converges to some number $x_\infty \in [a, b]$.

Turning our attention to the y-coordinates, the boundedness of Y again implies that the y_{n_k} are all contained in some interval $[c, d]$. We can therefore extract from the subsequence $(y_{n_k})_{k \in \mathbb{N}}$ a subsequence $(y_{n_{k_i}})_{i \in \mathbb{N}}$ which converges to some $y_\infty \in [c, d]$.

Note that the sequence $(P_{n_{k_i}})_{i \in \mathbb{N}}$ is a subsequence of $(P_n)_{n \in \mathbb{N}}$. We now have

$$\lim_{i \to \infty} x_{n_{k_i}} = x_\infty \quad \text{and} \quad \lim_{i \to \infty} y_{n_{k_i}} = y_\infty.$$

The argument concluding the proof of Theorem 6.8 can again be used to show that the subsequence $(P_{n_{k_i}})_{i \in \mathbb{N}}$ converges to $P_\infty = (x_\infty, y_\infty)$.

The point P_∞ cannot be an exterior point of X since there are points $P_{n_{k_i}}$ that are arbitrarily close to it. Therefore, P_∞ is either an interior point or a boundary point of X. Since X is closed, it follows that P_∞ is in X. Therefore, every sequence $(P_n)_{n \in \mathbb{N}}$ in X admits a subsequence which converges to some point $P_\infty \in X$. Namely, (X, d_{euc}) is compact. $\qquad \square$

Theorem 6.15. *Every closed bounded subset of the hyperbolic plane* $(\mathbb{H}^2, d_{\text{hyp}})$ *is compact.*

Proof. Let X be a bounded subset of the hyperbolic plane $(\mathbb{H}^2, d_{\text{hyp}})$. In particular, the subset X is contained in a large ball $B_{d_{\text{hyp}}}(P_0, K)$.

Let $(P_n)_{n \in \mathbb{N}}$ be a sequence in X, and write $P_n = (x_n, y_n)$. Since $d_{\text{hyp}}(P_n, P_0) < K$, the same argument as in the proof of Theorem 6.10 shows that $c_1 < y_n < c_2$ with $c_1 = y_0 e^{-K}$ and $c_2 = y_0 e^K$, and that

$$d_{\text{euc}}(P_n, P_0) < c_2 \, d_{\text{hyp}}(P_n, P_0) < c_2 K.$$

In particular, the sequence $(P_n)_{n \in \mathbb{N}}$ is contained in a large euclidean ball $B_{d_{\text{euc}}}(P_0, c_2 K)$. Applying Theorem 6.13 to the closed ball with the same radius and the same center, there consequently exists a subsequence $(P_{n_k})_{k \in \mathbb{N}}$ which converges to some point $P_\infty = (x_\infty, y_\infty)$ in $(\mathbb{R}^2, d_{\text{euc}})$.

Since $y_{n_k} > c_1$ for every k, the coordinate y_∞ is greater than or equal to $c_1 > 0$, so that P_∞ belongs to the hyperbolic plane $\mathbb{H}^2$.

Borrowing another argument from the proof of Theorem 6.10,

$$d_{\text{hyp}}(P_n, P_\infty) \leqslant \frac{1}{c_1} d_{\text{euc}}(P_n, P_\infty)$$

for every n. It follows that P_{n_k} also converges to P_∞ for the hyperbolic metric d_{hyp}, as k tends to ∞.

Since X is closed, it again follows that P_∞ is an element of X.

This proves that every sequence in X has a converging subsequence. In other words, (X, d_{hyp}) is compact. □

For subsets of the sphere $\mathbb{S}^2$, the boundedness hypothesis is irrelevant since $d_{\mathrm{sph}}(P, Q) \leqslant \pi$ for every $P, Q \in \mathbb{S}^2$.

Theorem 6.16. *Every closed bounded subset of the sphere $(\mathbb{S}^2, d_{\mathrm{sph}})$ is compact.*

Proof. Let X be a closed subset of $\mathbb{S}^2$, and let $(P_n)_{n \in \mathbb{N}}$ be a sequence in X. By the immediate generalization of Theorem 6.13 to three dimensions, the sphere $\mathbb{S}^2$ is compact for the restriction of the 3-dimensional euclidean metric d_{euc}. Therefore, there exists a subsequence $\left(P_{n_k}\right)_{k \in \mathbb{N}}$ converging to some $P_\infty \in \mathbb{S}^2$ for the euclidean metric d_{euc}.

By an elementary euclidean geometry argument comparing the length of a circle arc to the length of its chord, the euclidean and spherical distances are related by the property that

$$d_{\mathrm{sph}}(P_{n_k}, P_\infty) = 2 \arcsin \frac{d_{\mathrm{euc}}(P_{n_k}, P_\infty)}{2}.$$

It follows that $d_{\mathrm{sph}}(P_{n_k}, P_\infty)$ converges to 0 as k tends to ∞, namely that the subsequence $\left(P_{n_k}\right)_{k \in \mathbb{N}}$ converges to P_∞ in $(\mathbb{S}^2, d_{\mathrm{sph}})$. □

6.4.4. A few convenient properties. For future reference, we mention here several easy properties which will be convenient in the future.

Lemma 6.17. *In a metric space (X, d), let $(P_n)_{n \in \mathbb{N}}$ be a sequence with finite length. If there exists a subsequence $\left(P_{n_k}\right)_{k \in \mathbb{N}}$ which converges to some point P_∞, then the whole sequence converges to P_∞.*

Proof. This property was the main step in the proof of Proposition 6.11. □

Lemma 6.18. *The length of a sequence $(P_n)_{n \in \mathbb{N}}$ is greater than or equal to the length of any of its subsequences $\left(P_{n_k}\right)_{k \in \mathbb{N}}$.*

Proof. By an iterated use of the Triangle Inequality,

$$d(P_{n_k}, P_{n_{k+1}}) \leqslant \sum_{n=n_k}^{n_{k+1}-1} d(P_n, P_{n+1}).$$

It follows that

$$\sum_{k=1}^{\infty} d(P_{n_k}, P_{n_{k+1}}) \leqslant \sum_{n=1}^{\infty} d(P_n, P_{n+1}). \qquad \square$$

Lemma 6.19. *Suppose that the metric space (X, d) splits as the union $X = X_1 \cup X_2$ of two subsets X_1 and X_2. If, for the restrictions of the metric d, the two metric spaces (X_1, d) and (X_2, d) are complete, then (X, d) is complete.*

Proof. Let $(P_n)_{n \in \mathbb{N}}$ be a sequence with finite length

$$\sum_{n=1}^{\infty} d(P_n, P_{n+1}) < \infty.$$

We want to show that this sequence converges to some $P_\infty \in X$.

At least one of X_1 and X_2 must contain P_n for infinitely many $n \in \mathbb{N}$. Without loss of generality, let us assume that this is X_1. Consequently, we have a subsequence $(P_{n_k})_{k \in \mathbb{N}}$ which is completely contained in X_1.

This subsequence has finite length by Lemma 6.18. Since (X_1, d) is complete, this implies that $(P_{n_k})_{k \in \mathbb{N}}$ converges to some limit $P_\infty \in X_1$. By Lemma 6.17, it follows that the whole sequence $(P_n)_{n \in \mathbb{N}}$ converges to $P_\infty \in X$. $\qquad \square$

6.5. Tessellations by bounded polygons

We now apply the results of the previous sections to construct various examples of tessellations of the euclidean plane, the hyperbolic plane or the sphere.

The simplest examples come from bounded polygons. Indeed, because of our convention that a polygon X contains all of its edges, it is always closed in $(\mathbb{R}^2, d_{\mathrm{euc}})$, $(\mathbb{H}^2, d_{\mathrm{hyp}})$ or $(\mathbb{S}^2, d_{\mathrm{sph}})$. If, in addition, the polygon X is bounded, it is therefore compact by Theorems 6.13, 6.15 or 6.16. Recall that a spherical polygon is always bounded.

If $(\bar{X}, \bar{d}_X)$ is the quotient metric space obtained by gluing together the edges of X, recall from Lemma 4.2 that the quotient map $X \to \bar{X}$ is continuous. It follows that $(\bar{X}, \bar{d}_X)$ is compact by Proposition 6.12, and therefore complete by Proposition 6.11. This proves:

Proposition 6.20. *Let X be a bounded polygon in the euclidean space $(\mathbb{R}^2, d_{\mathrm{euc}})$, the hyperbolic plane $(\mathbb{H}^2, d_{\mathrm{hyp}})$ or the sphere $(\mathbb{S}^2, d_{\mathrm{sph}})$. Then, if we glue together the edges of X, the resulting quotient metric space $(\bar{X}, \bar{d}_X)$ is compact, and therefore complete.* $\qquad\square$

In particular, we can now apply Theorem 6.1 to several of the examples that we considered in Chapter 5.

6.5.1. Tessellations of the euclidean plane. Let us look at the euclidean polygon gluings that we examined in Section 5.1. All these polygons were bounded so that we can apply Proposition 6.20 and Theorem 6.1 to create tessellations of the euclidean plane.

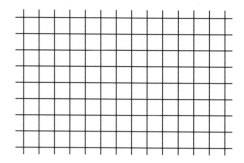

Figure 6.4. A tessellation of the euclidean plane by squares

First, we considered the case where X is a square, and where we glue opposite sides by translations. In this case, the tiling group Γ consists entirely of translations, and the corresponding tessellation of the euclidean plane by squares is illustrated in Figure 6.4.

If we consider the same square, but glue its sides with a twist to obtain a Klein bottle, the tessellation of the euclidean plane associated to this polygon gluing is the same as for the torus. The difference here is that the tiling group Γ is different. In addition to translations, it contains glide reflections, namely, compositions of a reflection with a nontrivial translation parallel to the axis of the reflection.

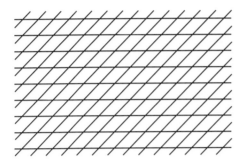

Figure 6.5. A tessellation of the euclidean plane by parallelograms

We also considered the case where X is a parallelogram, and where opposite sides are glued by translations. Since the parallelogram is bounded, this provides a tessellation of the euclidean plane by parallelograms. The tiling group Γ in this case again consists entirely of translations, and a tessellation of this type is illustrated in Figure 6.5.

A different, and perhaps more interesting, tessellation of the plane by parallelograms is associated to the gluing of edges of the parallelogram that give the Klein bottle. See Exercise 6.4.

Finally, we saw that the torus can also be obtained by gluing opposite sides of a hexagon. This leads to the tessellation of the euclidean plane by hexagons that we already encountered in Figure 6.1.

6.5.2. Tessellations of the hyperbolic plane by bounded polygons. Let us now switch to hyperbolic polygons. The hyperbolic octagon of Section 5.2 is bounded in the hyperbolic plane. The hyperbolic surface of genus 2 that we obtain by gluing opposite edges of this octagon consequently gives rise to a tessellation of the hyperbolic plane $\mathbb{H}^2$. This tessellation is represented in Figure 6.6.

The tiles of this tessellation may look very different from each other, but they are actually all isometric by isometries of the hyperbolic plane. Some of these isometries are a little more apparent in

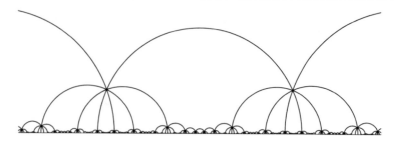

Figure 6.6. A tessellation of the hyperbolic plane by octagons

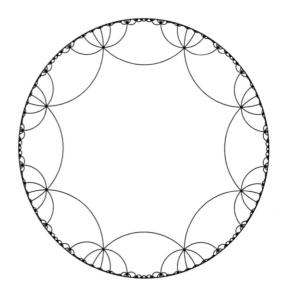

Figure 6.7. A tessellation of the disk model by hyperbolic octagons

Figure 6.7, which represents the image of the same tessellation in the disk model $\mathbb{B}^2$ for the hyperbolic plane. Note the nice rotational symmetry. Compare Figure 5.11.

More hyperbolic rotational symmetries of the tessellation are revealed if one transports it by an isometry of the disk model that sends to the euclidean center of the disk any of the points where eight octagons meet. See Figure 6.8.

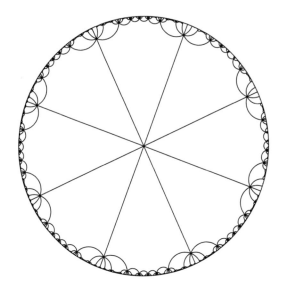

Figure 6.8. Another view of the tessellation of the disk model by hyperbolic octagons

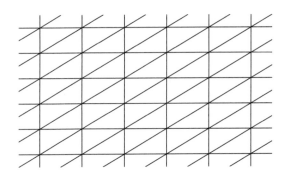

Figure 6.9. A tessellation of the euclidean plane $\mathbb{R}^2$ by triangles of angles $\frac{\pi}{2}$, $\frac{\pi}{3}$, $\frac{\pi}{6}$

6.5.3. Tessellations by triangles.

Theorem 6.21. *For any three integers* a, b, $c \geqslant 2$,

(1) *if* $\frac{\pi}{a} + \frac{\pi}{b} + \frac{\pi}{c} = \pi$, *there exists a tessellation of the euclidean plane* $\mathbb{R}^2$ *by euclidean triangles of angles* $\frac{\pi}{a}$, $\frac{\pi}{b}$, $\frac{\pi}{c}$;

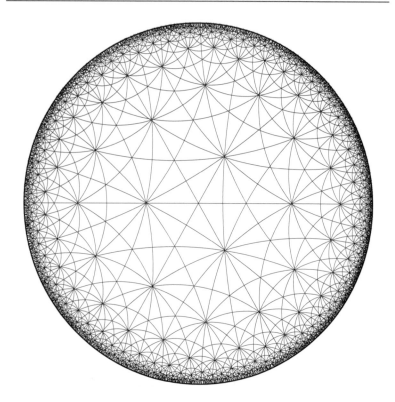

Figure 6.10. A tessellation of the disk model $\mathbb{B}^2$ of the hyperbolic plane by triangles of angles $\frac{\pi}{2}, \frac{\pi}{3}, \frac{\pi}{7}$

(2) *if $\frac{\pi}{a} + \frac{\pi}{b} + \frac{\pi}{c} < \pi$, there exists a tessellation of the hyperbolic plane $\mathbb{H}^2$ by hyperbolic triangles of angles $\frac{\pi}{a}, \frac{\pi}{b}, \frac{\pi}{c}$;*

(3) *if $\frac{\pi}{a} + \frac{\pi}{b} + \frac{\pi}{c} > \pi$, there exists a tessellation of the sphere $\mathbb{S}^2$ by spherical triangles of angles $\frac{\pi}{a}, \frac{\pi}{b}, \frac{\pi}{c}$.*

Proof. Proposition 5.13 provides a triangle X with the angles indicated. As gluing data, let us take for each edge the self-gluing of Section 6.3.4 so that the tiling group Γ is generated by the reflections across the complete geodesics containing each side of the triangle. Since the triangle X is bounded, Theorem 6.1 (generalized

as in Section 6.3.4) and Proposition 6.20 then provide the requested
tessellations. □

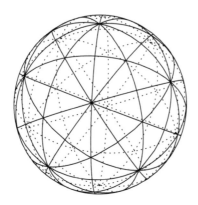

Figure 6.11. A tessellation of the sphere $\mathbb{S}^2$ by triangles of
angles $\frac{\pi}{2}$, $\frac{\pi}{3}$, $\frac{\pi}{5}$

Examples of these tessellations are illustrated in Figures 6.9, 6.10
and 6.11.

Note that the spherical case can only occur for $\{a, b, c\} = \{2, 2, c\}$,
$\{2, 3, 3\}$, $\{2, 3, 4\}$ or $\{2, 3, 5\}$, with $c \geqslant 2$ arbitrary in the first case.
The euclidean case arises only when $\{a, b, c\} = \{2, 3, 6\}$, $\{2, 4, 4\}$ or
$\{3, 3, 3\}$.

6.6. Tessellations by unbounded polygons

The punctured torus discussed in Section 5.5 offers an interesting
challenge because it is less obviously complete.

Let X be the hyperbolic square of Figure 5.18. Consider $U = U_\infty \cup U_1 \cup U_0 \cup U_{-1}$ as in Section 5.5, and let Y be the union of the
complement $X - U$ and of all its boundary points. Let $\bar{U}$ and $\bar{Y}$ be
their respective images in the quotient space $(\bar{X}, \bar{d}_X)$.

The subset Y is closed and bounded in the hyperbolic plane $\mathbb{H}^2$.
If follows that Y is compact by Theorem 6.15, and therefore that $\bar{Y}$
is also compact by Proposition 6.12. As a consequence, $(\bar{Y}, \bar{d}_X)$ is
complete by Proposition 6.11.

In Proposition 5.9 and Lemma 5.12 we showed that $(\bar{U}, \bar{d}_X)$ is isometric to the portion (S_a, d_{S_a}) of the pseudosphere S in $\mathbb{R}^3$. Let us show that (S_a, d_{S_a}) is complete.

Let $P_1, P_2, \ldots, P_n, \ldots$ be a sequence of finite length in (S_a, d_{S_a}). Recall that the distance $d_{S_a}(P, Q)$ is defined as the infimum of the euclidean lengths of all curves joining P to Q and contained in S_a; it follows that $d_{euc}(P, Q) \leqslant d_{S_a}(P, Q)$ for every P, $Q \in S_a$. In particular, the sequence $(P_n)_{n \in \mathbb{N}}$ has finite length in $(\mathbb{R}^3, d_{euc})$, and consequently it converges to some point P_∞ in the complete space $(\mathbb{R}^3, d_{euc})$ (by the immediate generalization of Theorem 6.8 to three dimensions). Since the surface S_a is defined by finitely many equations and weak inequalities involving continuous functions, the limit P_∞ is in S_a, and one easily checks that the sequence $(P_n)_{n \in \mathbb{N}}$ converges to P_∞ in (S_a, d_{S_a}).

This proves that (S_a, d_{S_a}) is complete; therefore, so is the isometric metric space $(\bar{U}, \bar{d})$.

The space $(\bar{X}, \bar{d})$ is the union of the two subsets $\bar{Y}$ and $\bar{U}$, each of which is complete for the metric $\bar{d}$. Lemma 6.19 shows that this implies that $(\bar{X}, \bar{d})$ is complete. We will see another proof that $(\bar{X}, \bar{d})$ is complete in Theorem 6.25.

As before, an application of Theorem 6.1 provides a tessellation of the hyperbolic plane by tiles isometric to the hyperbolic square X. This tessellation is illustrated in Figure 6.12.

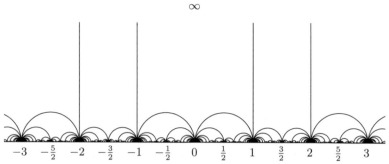

Figure 6.12. A tessellation of the hyperbolic plane coming from the once-punctured torus

6.7. Incomplete hyperbolic surfaces

We now look at a few examples where the quotient space $(\bar{X}, \bar{d}_X)$ is incomplete.

6.7.1. Incomplete hyperbolic cylinders. Let us revisit the example of Figure 5.16. Namely, X is the vertical strip delimited in the upper half-space $\mathbb{H}^2$ by the two vertical half-lines E_1 and E_2 of the respective equations $\mathrm{Re}(z) = 0$ and $\mathrm{Re}(z) = 1$, and we glue the two sides by an isometry $\varphi_1 \colon E_1 \to E_2$. As usual, extend φ_1 to an isometry of H_2 that sends X to the side of E_2 opposite X. An elementary algebraic computation shows that φ_1 is of the form $\varphi_1(z) = az + 1$ for a real number $a > 0$.

Since we only have two edges, the tiling group Γ consists of all possible compositions of φ_1 and $\varphi_2 = \varphi_1^{-1}$. As a consequence, every element of the tiling group is of the form $\varphi_1^n = \varphi_1 \circ \varphi_1 \circ \cdots \circ \varphi_1$ or $\varphi_1^{-n} = \varphi_2^n = \varphi_1^{-1} \circ \varphi_1^{-1} \circ \cdots \circ \varphi_1^{-1}$, where $n \geqslant 0$ is the number of maps occurring in the composition.

In Section 5.4.2, we considered the case where $a = 1$. In this case, an immediate computation shows that all elements of the tiling group Γ are horizontal translations of the form $\varphi_1^n(z) = z + n$ with $n \in \mathbb{Z}$. In particular, each tile $\varphi(X)$ is a vertical strip of the form $\{z \in \mathbb{H}^2; n \leqslant \mathrm{Im}(z) \leqslant n + 1\}$, and it is immediate that these tiles form a tessellation of the hyperbolic plane. As in the case of the punctured torus that we just examined in Section 6.6, the quotient space $(\bar{X}, \bar{d}_X)$ is complete.

The situation is completely different when $a \neq 1$. In this case, the tiles $\varphi_1^n(z)$ are still vertical strips, delimited by the vertical half-lines $\mathrm{Im}(z) = a_n$ and $\mathrm{Im}(z) = a_{n+1}$, where $a_n = \varphi_1^n(0)$. However, by induction on n, we see that for every $n \geqslant 0$

$$\varphi_1^n(z) = a^n z + a^{n-1} + a^{n-2} + \cdots + a + 1 = a^n z + \frac{1 - a^n}{1 - a},$$

while

$$\varphi_1^{-n}(z) = a^{-n} z - a^{-n} - a^{-(n-1)} - \cdots - a^{-2} - a^{-1}$$
$$= a^{-n} z - a^{-1} \frac{1 - a^{-n}}{1 - a^{-1}} = a^{-n} z + \frac{1 - a^{-n}}{1 - a}.$$

As a consequence, $a_n = \varphi_1^n(0) = \dfrac{1 - a^n}{1 - a}$ for every $n \in \mathbb{Z}$.

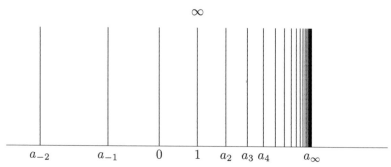

Figure 6.13. A partial tessellation coming from an incomplete hyperbolic cylinder

Consider the case where $a < 1$. Then a_n converges to $a_\infty = \frac{1}{1-a}$ as n tends to $+\infty$ and, more importantly, $a_n < a_\infty$ for every $n \in \mathbb{Z}$. This implies that for every element φ_1^n of the tiling group, the corresponding tile $\varphi_1^n(X)$ stays to the left of the vertical half-line of equation $\mathrm{Re}(z) = a_\infty$. In particular, the tiles $\varphi_1^n(X)$ cannot cover the whole hyperbolic plane $\mathbb{H}^2$. The case where $a = 0.75$ is represented in Figure 6.13.

A symmetric phenomenon occurs when $a > 1$, in which case the tiles $\varphi_1^n(X)$ are all to the right of the vertical half-line of equation $\mathrm{Im}(z) = a_\infty$, where $a_\infty = \frac{1}{1-a} = \lim_{n \to -\infty} a_n$. Again, this prevents the tiles from covering the whole plane $\mathbb{H}^2$.

If we combine these observations with Theorem 6.1, we conclude that the quotient metric space $(\bar{X}, \bar{d}_X)$ is not complete if $a \neq 1$. The accumulation of the tiles along the line $\mathrm{Im}(z) = a_\infty$ explains why the proof of Lemma 6.4 fails in this case.

This abstract argument showing that $(\bar{X}, \bar{d}_X)$ is incomplete is fine mathematically, but it will be more gratifying to exhibit a finite-length sequence $(\bar{P}_n)_{n \in \mathbb{N}}$ in $(\bar{X}, \bar{d}_X)$ that does not converge. We restrict our attention to the case where $a < 1$. Set $P_n = \mathrm{i}a^{-n}$ for every integer $n \geqslant 1$. Note that P_n is glued to the point $Q_n = 1 + \mathrm{i}a^{-n+1}$, and that the points P_n and Q_{n+1} can be joined by a horizontal line segment whose hyperbolic length is equal to a^n. Therefore, the length of the sequence $(\bar{P}_n)_{n\in\mathbb{N}}$ is

$$\sum_{n=1}^{\infty} \bar{d}_X(\bar{P}_n, \bar{P}_{n+1}) = \sum_{n=1}^{\infty} \bar{d}_X(\bar{P}_n, \bar{Q}_{n+1}) \leqslant \sum_{n=1}^{\infty} a^n < \infty,$$

where the geometric series converges since $a < 1$.

We claim that this finite length sequence $(\bar{P}_n)_{n\in\mathbb{N}}$ cannot converge to any point $\bar{P}_\infty$ in $(\bar{X}, \bar{d}_X)$. Indeed, the proof of Theorem 4.10 shows that there is a small ball $B_{\bar{d}_X}(\bar{P}_\infty, \varepsilon)$ such that the set of points $P \in E_1 \cup E_2$ projecting to some $\bar{P} \in B_{\bar{d}_X}(\bar{P}_\infty, \varepsilon)$ under the quotient map is either empty if $\bar{P}_\infty$ corresponds to a point P_∞ in the interior of X or two geodesic arcs of hyperbolic length 2ε when $\bar{P}_\infty$ corresponds to a point $P_\infty \in E_1$ glued to another point $Q_\infty \in E_2$. In the first case, $B_{\bar{d}_X}(\bar{P}_\infty, \varepsilon)$ does not contain any $\bar{P}_n$. In the second case, each of the two geodesic arcs can contain at most finitely many points P_n and Q_n, since P_n and Q_n converge to ∞ as n tends to ∞. In both cases, this shows that $B_{\bar{d}_X}(\bar{P}_\infty, \varepsilon)$ contains no $\bar{P}_n$ with n large enough, so that $(\bar{P}_n)_{n\in\mathbb{N}}$ cannot converge to any $\bar{P}_\infty \in \bar{X}$.

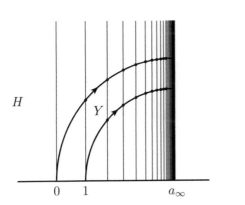
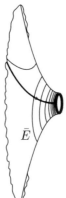

Figure 6.14. The incomplete cylinder of Figure 6.13

It is interesting to understand a little better the geometry of $(\bar{X}, \bar{d}_X)$. Again, we focus our attention on the case where $a < 1$. The case where $a > 1$ is similar.

In this case where $a < 1$, the tiles $\varphi_1^n(X)$, with $n \in \mathbb{Z}$, tessellate the open quadrant H consisting of those $z \in \mathbb{H}^2$ with $\operatorname{Re}(z) < a_\infty =$

$\frac{1}{1-a}$. On H, consider the partition $\bar{H}$ consisting of the subsets of H of the form $\{\varphi_1^n(P); n \in \mathbb{Z}\}$ with $P \in H$. Since φ_1 is the gluing map used to glue the two edges of X together, there is a natural map $\varphi\colon \bar{X} \to \bar{H}$, uniquely determined by the property that it associates $\bar{P} \in \bar{H}$ to $\bar{P} \in \bar{X}$ for every $P \in X$. Because the $\varphi_1^n(X)$ tessellate H, one easily sees that H is bijective. Anticipating results from Chapter 7, we will prove in Theorem 7.12 that ψ is actually an isometry from $(\bar{X}, \bar{d}_X)$ to $(\bar{H}, \bar{d}_H)$.

It turns out that there is another natural way to tessellate the quadrant H. Let Y be the quarter annulus delimited in H by the two euclidean quarter circles C_1 and C_2 centered at $a_\infty = \frac{1}{1-a}$ and passing through the points 0 and 1, respectively. Note that $\varphi_1(z) = az + 1$ can also be written as $\varphi_1(z) = a(z - a_\infty) + a_\infty$, so that φ_1 sends the larger quarter circle to the smaller one. It easily follows that the $\varphi_1^n(Y)$ also tessellate H. By the same argument as above, $(\bar{H}, \bar{d}_H)$ is therefore isometric to the quotient space $(\bar{Y}, \bar{d}_Y)$ obtained from (Y, d_Y) by gluing together its sides C_1 and C_2 by φ_1.

As a consequence, the two metric spaces $(\bar{X}, \bar{d}_X)$ and $(\bar{Y}, \bar{d}_Y)$ are isometric.

It turns out that we have already encountered the quotient space $(\bar{Y}, \bar{d}_Y)$ in Section 5.4.2. More precisely, it corresponds to one half of the hyperbolic cylinder $\bar{X}_3$ of Figure 5.15 because Y is one half of the hyperbolic strip X_3. Comparing the left-hand side of Figure 5.15 to the left-hand side of Figure 6.14, we conclude that $(\bar{Y}, \bar{d}_Y)$ is isometric to any one of the two halves of the hyperbolic annulus $\bar{X}_3$ delimited by the waist γ_0. The correspondence is rigorously described by the hyperbolic isometry $\varphi(z) = a_\infty z + a_\infty$, and is illustrated by the right-hand sides of Figures 5.15 and 6.14. In particular, the open annulus $(\bar{Y}, \bar{d}_Y)$ has one end that flares out with exponential growth, while the other one gets very close to the closed curve γ_0.

It is interesting to consider the image $\bar{E}$ of the edges E_1 and E_2 of X under the isometric correspondence between $(\bar{X}, \bar{d}_X)$ and $(\bar{Y}, \bar{d}_Y)$. By careful inspection of Figure 6.14, one easily sees that this image $\bar{E}$ consists of an infinite curve which, in one direction, goes to infinity in the end of $\bar{Y}$ with exponential growth whereas, in the other direction, it spirals around the closed curve γ_0. This is illustrated on the right-hand side of Figure 6.14. A more detailed description of this construction can also be found in [**Weeks$_4$**, §2].

6.7.2. Incomplete punctured tori. We can similarly deform the hyperbolic once-punctured torus of Section 5.5. We keep the same polygon X, namely, in this case the hyperbolic square delimited by the complete geodesics E_1, E_2, E_3, E_4, where E_1 goes from -1 to ∞, E_2 from 0 to 1, E_3 from 1 to ∞ and E_4 from 0 to -1.

To glue E_1 to E_2, we need a hyperbolic isometry φ_1 sending -1 to 0, ∞ to 1, and X to the side of E_2 opposite X. These isometries are the linear fractional maps of the form

$$\varphi_1(z) = \frac{z+1}{z+a}$$

with $a > 1$. Similarly, the E_3 is glued to E_4 by a map of the form

$$\varphi_3(z) = \frac{z-1}{-z+b}$$

with $b > 1$. In Section 5.5, we considered the case where $a = b = 2$, but we now consider the general case.

As in Section 5.5, chop off little pieces U_{-1}, U_0, U_1 and U_∞ near the four corners -1, 0, 1 and ∞ of the square X, in such a way that the intersections $U_i \cap E_j$ are glued together by the gluing maps φ_j. However, in this general case, we cannot always arrange that the U_i are delimited by euclidean circles tangent to $\mathbb{R}$ as in Section 5.5. To construct the U_i, first select P_1 and $Q_1 \in E_1$ with $\mathrm{Im}(P_1) > \mathrm{Im}(Q_1)$, P_3 and $Q_3 \in E_3$ with $\mathrm{Im}(P_3) > \mathrm{Im}(Q_3)$, and set $P_2 = \varphi_1(P_1)$, $Q_2 = \varphi_1(Q_1)$, $P_4 = \varphi_3(P_3)$ and $Q_4 = \varphi_3(Q_3)$. Then pick disjoint curves γ_{-1}, γ_0, γ_1, γ_∞ in X connecting Q_1 to P_4, Q_4 to Q_2, P_2 to Q_3, and P_3 to P_1, respectively. Finally, define U_i as the portion of X that is delimited by γ_i and is adjacent to the corner i.

As in Section 5.5, define $V_\infty = U_\infty$, $V_1 = \varphi_2(U_1)$, $V_{-1} = \varphi_4(U_{-1})$ and $V_0 = \varphi_4 \circ \varphi_2(U_0)$, and let $U = U_{-1} \cup U_0 \cup U_1 \cup U_\infty$ and $V = V_{-1} \cup V_0 \cup V_1 \cup V_\infty$. As before, V and the subsets V_i are vertical half-strips delimited on the sides by vertical lines, and from below by curves that are the images of the curves γ_i. See Figure 6.15, and compare Figure 5.19 in Section 5.5.

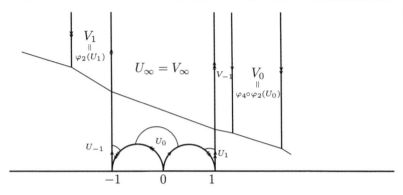

Figure 6.15. An incomplete punctured torus

The left side of V is glued to its right side by the map

$$\varphi_4 \circ \varphi_2 \circ \varphi_3 \circ \varphi_1(z) = \left(\frac{b-1}{a-1}\right)^2 z + \frac{(ab-1)(a+b-2)}{(a-1)^2}$$

which, in general, is not a horizontal translation any more (unless $a = b$). So we are essentially in the same situation as for the incomplete cylinder of Section 6.7.1, and, when $a \neq b$, the tiles $(\varphi_4 \circ \varphi_2 \circ \varphi_3 \circ \varphi_1)^n(X)$ accumulate along the vertical half-line of equation $\mathrm{Im}(z) = a_\infty$, where $a_\infty = \frac{ab-1}{b-a}$ is the only real number such that $\varphi_4 \circ \varphi_2 \circ \varphi_3 \circ \varphi_1(a_\infty) = a_\infty$.

Looking a little more closely into the tiling process, we can easily convince ourselves that the tiles $\varphi(X)$ do not overlap. The simplest way to see this is to observe that the tiling construction provides a natural correspondence between the tiles constructed here and the tiles coming from the complete punctured torus of Section 6.6. In particular, if an edge of one tiling separates two tiles of the same tiling, then in the other tiling the corresponding edge separates the corresponding tiles. Since there is no overlap between the tiles of the example of Section 6.6, we conclude that there cannot be any overlap between the tiles in the example of this section either.

This implies that all the tiles stay on the same the side of the line $\mathrm{Im}(z) = a_\infty$. More precisely, the tiles $\varphi(X)$ are all disjoint from the hyperbolic half-plane[1] $H = \{z \in \mathbb{H}^2; \text{Im}(z) \geqslant a_\infty\}$ if $a < b$, and $H = \{z \in \mathbb{H}^2; \text{Im}(z) \leqslant a_\infty\}$ if $a > b$.

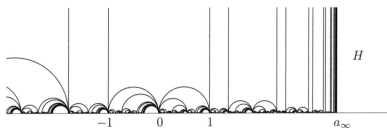

$-1 \qquad 0 \qquad 1 \qquad\qquad\qquad a_\infty$

Figure 6.16. The tessellation associated to the incomplete punctured torus of Figure 6.15

The tessellation (in a case where $a < b$) is illustrated in Figure 6.16.

Note that the tiles $\varphi(X)$ are also disjoint from every hyperbolic half-plane of the form $\psi(H)$, where ψ is an element of the tiling group Γ. Indeed, if $\varphi(X)$ met $\psi(H)$, then $\psi^{-1} \circ \varphi(X)$ would meet H; however, it immediately follows from the definition of the tiling group that $\psi^{-1} \circ \varphi$ is also an element of Γ, contradicting the fact that all tiles are disjoint from H. Therefore, the tiles are disjoint from all the hyperbolic half-planes $\psi(H)$ as ψ ranges over all the elements of the tiling group Γ. Some of these half-planes are visible in Figure 6.16 where they appear as white euclidean half-disks.

6.8. Poincaré's polygon theorem

Let $(\bar{X}, \bar{d}_X)$ be the quotient space obtained by gluing together the edges of a polygon (X, d_X) in the hyperbolic plane. We will provide a relatively simple criterion to check whether or not the quotient metric space $(\bar{X}, \bar{d}_X)$ is complete.

The fundamental ideas involved have already appeared in our analysis of the complete example of Section 6.6 and in the incomplete examples of Section 6.7. The completeness criterion essentially says that there is no new phenomenon. For this reason, this section may be skipped on a first reading.

To page 181

---

[1]Not to be confused with $\mathbb{H}^2$, which is a euclidean half-plane but is the whole hyperbolic plane.

We already observed in Proposition 6.20 that when X is bounded the quotient space $(\bar{X}, \bar{d}_X)$ is compact and therefore complete. Thus, the only relevant case here is the one where X is unbounded and, in particular, when some of its edges are infinite.

Edges E of the polygon X can be of three types. A bounded edge goes from one vertex of the polygon to another one. A *singly infinite* edge goes from a vertex to a point of $\mathbb{R} \cup \{\infty\}$, which we will call a *vertex at infinity* or an *ideal vertex*. A *doubly infinite* edge joins two points of $\mathbb{R} \cup \{\infty\}$, which consequently are two ideal vertices of the polygon.

To simplify the exposition, we will restrict attention to the case where a vertex at infinity is the endpoint of at most two edges of X. Note that we had already imposed a similar, and equally natural, condition for the vertices of X that are in $\mathbb{H}^2$.

When we glue the edges E_i and $E_{i\pm 1}$, the gluing map $\varphi_i\colon E_i \to E_{i\pm 1}$ sends each ideal vertex $\xi \in \mathbb{R} \cup \{\infty\}$ adjacent to E_i to an ideal vertex adjacent to $E_{i\pm 1}$. In this way, we can extend the gluing of edges to gluings between the ideal vertices of X, as in the case of nonideal vertices. An element of the corresponding quotient space will be an *ideal vertex* of the quotient space $\bar{X}$ of X under the gluing operation. Note that exactly as ideal vertices of the polygon X are not elements of X, ideal vertices of the quotient space $\bar{X}$ are not elements of $\bar{X}$.

Let $\bar{\xi} = \{\xi_1, \xi_2, \ldots, \xi_k\}$ be an ideal vertex of $\bar{X}$. Namely, $\bar{\xi}$ is the set of ideal vertices of X that are glued to a given ideal vertex ξ. The gluing maps identifying the ξ_j can be organized in a nice manner, as in the case of nonideal vertices in Section 4.3.1, but with a minor twist because an ideal vertex is allowed to be adjacent to only one edge of X. We state this as a lemma for future reference.

Lemma 6.22. *The indexing of the ideal vertices in $\bar{\xi} = \{\xi_1, \xi_2, \ldots, \xi_k\}$ can be chosen so that there exists gluing maps $\varphi_{i_j}\colon E_{i_j} \to E_{i'_{j+1}} = E_{i_j \pm 1}$, with $j = 1, 2, \ldots, k-1$, such that*

> (1) *ξ_j is an endpoint of E_{i_j}, ξ_{j+1} is an endpoint of $E_{i'_{j+1}}$, and the gluing map φ_{i_j} sends ξ_j to ξ_{j+1};*

(2) *for every* j *with* $1 < j < k - 1$, *the two edges* E_{i_j} *and* $E_{i'_j}$ *adjacent to* ξ_j *are distinct, so that* φ_{i_j} *is different from* $\varphi_{i_{j-1}}^{-1}$;

(3) *exactly one of the following holds:*

   (i) *either there exists a gluing map* $\varphi_{i_k} \colon E_{i_k} \to E_{i'_1}$ *such that* E_{i_k} *is an edge adjacent to* ξ_k *and different from the range* $E_{i'_k}$ *of* $\varphi_{i_{k-1}}$, *such that* $E_{i'_1}$ *is an edge adjacent to* ξ_1 *and different from the domain* E_{i_1} *of* φ_{i_1}, *and such that* φ_{i_k} *sends* ξ_k *to* ξ_1; *or*

   (ii) *each of* ξ_1 *and* ξ_k *is adjacent to a unique edge of* X, *namely,* E_{i_1} *and* $E_{i'_k}$, *respectively.*

Proof. As in Section 4.3.1, start with any element $\xi_1 = \xi \in \bar{\xi}$. Let E_{i_1} be one of the $\leqslant 2$ edges containing ξ_1. Set $\xi_2 = \varphi_{i_1}(\xi_1)$. If $\varphi_{i_1}(E_{i_1})$ is the only edge that is adjacent to ξ_2, stop here. Otherwise, apply the same process to the other edge E_{i_2} leading to ξ_2, define $\xi_3 = \varphi_{i_2}(\xi_2)$, and iterate this construction.

If this process goes on forever, we eventually reach an index k such that $\xi_{k+1} = \xi_j$ for some $j \leqslant k$. If k is the smallest such index, one easily checks that $\xi_{k+1} = \xi_1$ and that we are in the situation of alternative 3(i).

Otherwise, we reach an index k such that ξ_k is adjacent to only one edge, namely, $\varphi_{i_{k-1}}(E_{i_{k-1}}) = E_{i'_k}$. In this case, we restart the process beginning at ξ_1, but going backward. Let $E_{i'_1}$ be the edge that is adjacent to ξ_1 and is different from E_{i_1} (if any). Let E_{i_0} be the edge that is glued to $E_{i'_1}$ by the gluing map $\varphi_{i_0} \colon E_{i_0} \to E_{i'_1}$, and set $\xi_0 = \varphi_{i_0}^{-1}(\xi_1)$. If E_{i_0} is the only edge that is adjacent to ξ_0, stop here. Otherwise, let $E_{i'_0}$ be the other edge adjacent to ξ_0, consider the edge $E_{i_{-1}}$ that is glued to E_{i_0} by the gluing map $\varphi_{i_{-1}} \colon E_{i_{-1}} \to E_{i'_0}$, set $\xi_{-1} = \varphi_{i_{-1}}^{-1}(\xi_0)$, and iterate the construction. In this case, the process must eventually conclude by the discussion of the previous case, and we reach an index $-l \leqslant 0$ such that ξ_{-l} is adjacent to only one edge $E_{i_{-l}}$ of X. Shifting all indices by $l - 1$ to make them positive, we are now in the situation of alternative 3(ii).

In both cases, conditions (1) and (2) are satisfied by construction. $\qquad\square$

In the case of alternative 3(i), we will say that we have an *edge cycle* around the ideal vertex $\bar\xi = \{\xi_1, \xi_2, \ldots, \xi_k\}$ of $\bar X$. In this case, it is convenient to consider the indices j modulo k, namely, in such a way that j is identified with $j + k$. This convention has the advantage that each gluing map φ_{i_j} then goes from E_{i_j} to $E_{i'_{j+1}}$. Note that the composition $\varphi_{i_k} \circ \varphi_{i_{k-1}} \circ \cdots \circ \varphi_{i_2} \circ \varphi_{i_1}$ sends ξ_1 to itself. Our completeness criterion will be expressed in terms of this composition of gluing maps, considered for all edge cycles around ideal vertices.

A *horocircle* or *horocycle* centered at the point $\xi \in \mathbb{R} \cup \{\infty\}$ is a curve $C - \{\xi\}$ in $\mathbb{H}^2$ where C is a euclidean circle tangent to $\mathbb{R} \cup \{\infty\}$ at ξ (and lying "above" $\mathbb{R} \cup \{\infty\}$). When $\xi = \infty$, this means that $C - \{\xi\}$ is a horizontal line. Note that any isometry of $\mathbb{H}^2$ must send any horocycle centered at ξ to a horocycle centered at $\varphi(\xi)$, since φ sends a euclidean circle to a euclidean circle. See Exercises 6.10–6.12 for an interpretation of a horocircle as a hyperbolic circle of infinite radius centered at ξ.

An isometry φ of the hyperbolic plane $\mathbb{H}^2$ is *horocyclic* at ξ if it respects some horocircle centered at ξ. When $\xi = \infty$, this just means that φ is a horizontal translation $z \mapsto z + b$ (possibly the identity map) or a reflection $z \mapsto -\bar z + b$ across a vertical line.

In particular, a hyperbolic isometry which is horocyclic at ξ respects every horocircle centered at ξ. When $\xi = \infty$, this immediately follows from the above observation. In the general case, it suffices to apply this special case to $\psi \circ \varphi \circ \psi^{-1}$, where ψ is any hyperbolic isometry sending ξ to ∞, for instance $z \mapsto 1/(z - \xi)$.

As usual, extend each gluing map $\varphi_i \colon E_i \to E_{i+1}$ to a hyperbolic isometry $\varphi_i \colon \mathbb{H}^2 \to \mathbb{H}^2$ that near E_i, sends X to the side of E_{i+1} opposite X.

Proposition 6.23. *The following two conditions are equivalent:*

(1) *(Horocircle Condition) At each ideal vertex ξ of X, one can choose a horocircle C_ξ centered at ξ such that whenever the gluing map $\varphi_i \colon E_i \to E_{i\pm1}$ sends ξ to another ideal vertex ξ', its extension $\varphi_i \colon \mathbb{H}^2 \to \mathbb{H}^2$ sends C_ξ to $C_{\xi'}$;*

(2) *(Edge Cycle Condition) for every edge cycle around an ideal vertex $\bar\xi = \{\xi_1, \xi_2, \ldots, \xi_k\}$ of $\bar X$ with gluing maps $\varphi_{i_j} \colon E_{i_j} \to$*

$E_{i'_{j+1}} = E_{i_j \pm 1}$ *sending* ξ_j *to* ξ_{j+1} *for* $j = 1, 2, \ldots, k$ *as in alternative* 3(i) *of Lemma* 6.22 *(counting indices modulo* k *so that* $j + k$ *is considered the same as* j), *the corresponding composition* $\varphi_{i_k} \circ \varphi_{i_{k-1}} \circ \cdots \circ \varphi_{i_2} \circ \varphi_{i_1}$ *is horocyclic at* ξ_1.

Proof. Suppose that the first condition (1) holds. Then, for an edge cycle around an ideal vertex consisting of gluing maps $\varphi_{i_j} \colon E_{i_j} \to E_{i'_{j+1}} = E_{i_j \pm 1}$, with $j = 1, 2, \ldots, k$, the composition $\varphi_{i_k} \circ \varphi_{i_{k-1}} \circ \cdots \circ \varphi_{i_2} \circ \varphi_{i_1}$ sends the ideal vertex ξ_1 to itself, and consequently sends the horocircle C_{ξ_1} to itself. It follows that $\varphi_{i_k} \circ \varphi_{i_{k-1}} \circ \cdots \circ \varphi_{i_2} \circ \varphi_{i_1}$ is horocyclic. This proves that the second condition (2) holds.

Conversely, suppose that the second condition (2) holds. For each ideal vertex $\bar{\xi} = \{\xi_1, \xi_2, \ldots, \xi_k\}$ of $\bar{X}$, organize the gluing data as in Lemma 6.22. Pick an arbitrary horocircle C_{ξ_1} centered at ξ_1, and define

$$C_{\xi_j} = \varphi_{i_{j-1}} \circ \varphi_{i_{j-2}} \circ \cdots \circ \varphi_{i_1}(C_{\xi_1})$$

for $j \leqslant k$. Note that $\varphi_{i_{j-1}} \circ \varphi_{i_{j-2}} \circ \cdots \circ \varphi_{i_1}$ sends ξ_1 to ξ_j, so that C_{ξ_j} is really a horocircle at ξ_j.

If we are in the case of alternative 3(ii) of Lemma 6.22, there is nothing to prove. Otherwise, we have to make sure that the gluing map φ_{i_k} sends the horocircle C_{ξ_k} to C_{ξ_1}. This is where the hypothesis that $\varphi_{i_k} \circ \varphi_{i_{k-1}} \circ \cdots \circ \varphi_{i_2} \circ \varphi_{i_1}$ is horocylic is needed, since it shows that

$$\varphi_{i_k}(C_{\xi_k}) = \varphi_{i_k} \circ \varphi_{i_{k-1}} \circ \cdots \circ \varphi_{i_2} \circ \varphi_{i_1}(C_{\xi_1}) = C_{\xi_1}.$$

Performing this construction for every ideal vertex $\bar{\xi} = \{\xi_1, \xi_2, \ldots, \xi_k\}$ of $\bar{X}$ provides the conclusions of condition (1). □

Complement 6.24. *If the conditions of Proposition* 6.23 *hold, we can choose the horocircles* C_ξ *in condition* (1) *so that they are arbitrarily small. (When* $\xi = \infty$, *this means that* C_ξ *is a horizontal line which is arbitrarily high.)*

Proof. In the above proof that condition (2) implies condition (1), we have a degree of freedom in the choice of the initial horocircle C_{ξ_1} for each ideal vertex $\bar{\xi} = \{\xi_1, \xi_2, \ldots, \xi_k\}$ of $\bar{X}$. In particular, this horocircle can be chosen small enough that all other horocircles C_{ξ_j} are also arbitrarily small. □

In practice, the Edge Cycle Condition (2) of Proposition 6.23 is often easier to check than the Horocircle Condition (1). However, the Horocircle Condition (1) will be better suited for the proof of our main Theorem 6.25 below.

Exercise 6.14 shows that in the specific case of the composition map $\varphi_{i_k} \circ \varphi_{i_{k-1}} \circ \cdots \circ \varphi_{i_2} \circ \varphi_{i_1}$ of gluing maps that occurs in the Edge Cycle Condition (2) of Proposition 6.23, this isometry is horocyclic if and only if it is parabolic in the sense defined in Exercise 2.7. However, we will not need this property.

Theorem 6.25 (Poincaré's Polygon Theorem). *Let $(\bar{X}, \bar{d}_X)$ be the quotient space obtained by gluing together the edges of a polygon (X, d_X) in $\mathbb{H}^2$, using edge gluing maps $\varphi_i \colon E_i \to E_{i\pm 1}$. The quotient space $(\bar{X}, \bar{d}_X)$ is complete if and only if the (equivalent) conditions of Proposition 6.23 are satisfied.*

We will split the proof of equivalence into two parts.

Proof of the "only if" part of Theorem 6.25. We need to show that the conditions of Proposition 6.23 hold if $(\bar{X}, \bar{d}_X)$ is complete. We will actually prove the contrapositive of this statement. More precisely, we will assume that the Edge Cycle Condition (2) of Proposition 6.23 does not hold, and then show that $(\bar{X}, \bar{d}_X)$ is not complete.

By our assumption, there is an edge cycle $\bar{\xi} = \{\xi_1, \xi_2, \ldots, \xi_k\}$, with data as in alternative 3(i) of Lemma 6.22, such that the composition

$$\varphi = \varphi_{i_k} \circ \varphi_{i_{k-1}} \circ \cdots \circ \varphi_{i_2} \circ \varphi_{i_1}$$

is not horocyclic.

Modifying X and all gluing data by a hyperbolic isometry sending ξ_1 to ∞, we can arrange that $\xi_1 = \infty$ without loss of generality. Then φ sends ∞ to ∞, and is consequently of the form $\varphi(z) = az + b$ or $-a\bar{z} + b$ with $a > 0$, $b \in \mathbb{R}$. The fact that φ is not horocyclic is equivalent to the property that $a \neq 1$.

The argument is essentially the one that we already encountered in Sections 6.7.1 and 6.7.2. First consider the case where $a > 1$.

Near each ideal vertex ξ_j of X, chop off an infinite half-strip U_j delimited by a curve joining the two edges E_{i_j} and $E_{i'_j}$ that are adjacent to ξ_i. Then U_1, $\varphi_{i_1}^{-1}(U_2)$, $\varphi_{i_1}^{-1} \circ \varphi_{i_2}^{-1}(U_3)$, $\ldots$, $\varphi_{i_1}^{-1} \circ \varphi_{i_2}^{-1} \circ$ $\cdots \circ \varphi_{i_{k-1}}^{-1}(U_k)$ are vertical half-strips which are adjacent to each other, and their union forms a vertical half-strip V. In addition, φ^{-1} sends the vertical side of V that is contained in E_{i_1} to the other vertical side of V.

Pick a point $P_1 \in E_{i_1}$ so that the horizontal line L_1 passing through P_1 lies above the curves delimiting V from below. Let Q_2 be the point where this line L_1 hits the other vertical side of V. We already observed that this other vertical side is contained in $\varphi^{-1}(E_{i_1})$, so that we can consider the point $P_2 = \varphi(Q_2) \in E_{i_1}$. Repeating the process, we can thus define a sequence of points $P_n \in E_{i_1}$ such that P_n is the image under φ of the point Q_n of the other vertical side of V that is at the same euclidean height as P_{n-1}.

We now consider the sequence $(\bar{P}_n)_{n \in \mathbb{N}}$ in the quotient space $\bar{X}$.

Let y_1 be the y-coordinate of P_1 (and Q_2). By construction, the y-coordinates of P_n and Q_{n+1} are both equal to $a^n y_1$, and P_n and Q_{n+1} have the same x-coordinate as P_1 and Q_2, respectively. Considering the hyperbolic length of the horizontal line segment s_n going from P_n to Q_{n+1}, we conclude that

$$\bar{d}_X(\bar{P}_n, \bar{P}_{n+1}) = \bar{d}_X(\bar{P}_n, \bar{Q}_{n+1}) \leqslant d_X(P_n, Q_{n+1})$$
$$\leqslant \ell_{\mathrm{hyp}}(s_n) = a^{-n} y_1^{-1} \ell_{\mathrm{euc}}(s_n) = a^{-n} y_1^{-1} \ell_{\mathrm{euc}}(s_1).$$

As a consequence, the total length

$$\sum_{n=1}^{\infty} \bar{d}_X(\bar{P}_n, \bar{P}_{n+1}) \leqslant y_1^{-1} \ell_{\mathrm{euc}}(s_1) \sum_{n=1}^{\infty} a^{-n} < \infty$$

of the sequence $(\bar{P}_n)_{n \in \mathbb{N}}$ is finite, since $a > 1$. However, by the argument that we used in Section 6.7.1, this sequence cannot converge to any point of the quotient space $\bar{X}$. In particular, the metric space $(\bar{X}, \bar{d}_X)$ is not complete.

The argument is very similar when $a < 1$. This time, we will choose the points P_n on the vertical side of V that is not contained in E_{i_1}. Namely, we start with P_1 in that side, consider the point $Q_2 \in E_{i_1}$ that is at the same euclidean height as P_1, set $P_2 = \varphi^{-1}(Q_2)$, and iterate the process. This again gives a sequence $(\bar{P}_n)_{n \in \mathbb{N}}$ in $\bar{X}$

that has finite length (this time because $a < 1$) but cannot converge in $\bar{X}$. Therefore, $(\bar{X}, \bar{d}_X)$ is not complete in this case as well. $\quad\square$

Proof of the "if" part of Theorem 6.25. Let us assume that the Horocircle Condition (1) of Proposition 6.23 holds. We will prove that the quotient space $(\bar{X}, \bar{d}_X)$ is complete.

For this, consider in $\bar{X}$ a sequence $(\bar{P}_n)_{n \in \mathbb{N}}$ with finite length. We want to show that this sequence converges. The problem is that of course, there is no reason for the corresponding sequence $(P_n)_{n \in \mathbb{N}}$ to have finite length in X.

We will split the proof into four distinct cases, only two of which involve significant geometric ideas. The first case is relatively easy to deal with.

CASE 1: *There exists a hyperbolic ball $B_{d_{\mathrm{hyp}}}(P_0, r)$ that contains infinitely many P_n.*

The P_n that are contained in $B_{d_{\mathrm{hyp}}}(P_0, r)$ form a subsequence $\left(P_{n_k}\right)_{k \in \mathbb{N}}$ valued in that ball. Since the closed hyperbolic ball B of center P_0 and radius r is compact by Theorem 6.15, this subsequence itself admits a subsequence $\left(P_{n_{k_l}}\right)_{l \in \mathbb{N}}$ which converges to some point $P_\infty \in B$ for the hyperbolic metric d_{hyp}; the point P_∞ must be in X since all polygons are closed. It easily follows that $P_{n_{k_l}}$ also converges to $P_\infty \in X$ for the path metric d_X, as l tends to ∞. (Hint: For every $P \in X$, there exists a small ball $B_{d_{\mathrm{hyp}}}(P, \varepsilon)$ in $\mathbb{H}^2$ such that $d_X(P, Q) = d_{\mathrm{hyp}}(P, Q)$ for every $Q \in X \cap B_{d_{\mathrm{hyp}}}(P, \varepsilon)$.)

Since the quotient map $X \mapsto \bar{X}$ is continuous by Lemma 4.2, the (sub)subsequence $\left(P_{n_{k_l}}\right)_{l \in \mathbb{N}}$ converges to $\bar{P}_\infty$ in $(\bar{X}, \bar{d}_X)$.

Now the sequence $(\bar{P}_n)_{n \in \mathbb{N}}$ has finite length and admits a subsequence converging to $\bar{P}_\infty \in \bar{X}$. The whole sequence consequently converges to $\bar{P}_\infty$ in $(\bar{X}, \bar{d}_X)$ by Lemma 6.17.

We now consider the case where the P_n stay away from the edges of the polygon X.

CASE 2: *There exists an $\varepsilon > 0$ such that every P_n is at hyperbolic distance $> \varepsilon$ from every edge of X.*

In particular, no two distinct points of the ball $B_{d_X}(P_n, \varepsilon)$ are glued together.

Since the total length $\sum_{n=1}^{\infty} \bar{d}_X(\bar{P}_n, \bar{P}_{n+1})$ of the sequence $(\bar{P}_n)_{n \in \mathbb{N}}$ is finite, $\bar{d}_X(\bar{P}_n, \bar{P}_{n+1})$ tends to 0 as n tends to ∞, and in particular there exists an n_0 such that $\bar{d}_X(\bar{P}_n, \bar{P}_{n+1}) < \varepsilon$ for every $n \geqslant n_0$.

Consider a discrete walk w from $\bar{P}_n$ to $\bar{P}_{n+1}$ in $\bar{X}$, whose length $\ell_{d_X}(w)$ is close enough to the infimum $\bar{d}_X(\bar{P}_n, \bar{P}_{n+1})$ that $\ell_{d_X}(w) < \varepsilon$. Let $P_n = Q_1$, $R_1 \sim Q_2$, ..., $R_{k-1} \sim Q_k$, $R_k = P_{n+1}$ be the steps of this discrete walk. By induction and using the Triangle Inequality, the points Q_i stay in the ball $B_{d_X}(P_n, \varepsilon)$ and are equal to the points R_{i+1}, since there is no nontrivial gluing in this ball. Therefore, by the Triangle Inequality,

$$d_X(P_n, P_{n+1}) \leqslant \sum_{i=1}^{k} d_X(Q_k, R_k) = \ell_{d_X}(w).$$

Since this holds for every discrete walk w whose length is sufficiently close to $\bar{d}_X(\bar{P}_n, \bar{P}_{n+1})$, it follows that $d_X(P_n, P_{n+1}) \leqslant \bar{d}_X(\bar{P}_n, \bar{P}_{n+1})$. As the reverse inequality always holds by Lemma 4.2, this inequality is actually an equality, and $d_X(P_n, P_{n+1}) = \bar{d}_X(\bar{P}_n, \bar{P}_{n+1})$ for every $n \geqslant n_0$.

In particular, the sequence $(P_n)_{n \in \mathbb{N}}$ has finite length in X. Since $(\mathbb{H}^2, d_{\text{hyp}})$ is complete, the finite length sequence $(P_n)_{n \in \mathbb{N}}$ converges to some point $P_\infty \in \mathbb{H}^2$. This point P_∞ must be in the interior of X since the P_n stay at distance $> \varepsilon$ from its boundary. Since the quotient map $X \to \bar{X}$ is continuous (Lemma 4.2), it follows that the sequence $(\bar{P}_n)_{n \in \mathbb{N}}$ converges to the point $\bar{P}_\infty$ in the quotient space $(\bar{X}, \bar{d}_X)$, which concludes the proof in this case.

Next, we consider the case that is completely opposite to Case 2. This is the crucial case, which will strongly use our hypothesis that the Horocircle Condition (1) of Proposition 6.23 holds.

CASE 3: *All the P_n are contained in edges of X.*

The Horocircle Condition (1) of Proposition 6.23 asserts that there exists a horocircle C_ξ centered at each ideal vertex ξ such that whenever the gluing map $\varphi_i \colon E_i \to E_{i+1}$ sends ξ to another ideal vertex ξ', its extension $\varphi_i \colon \mathbb{H}^2 \to \mathbb{H}^2$ sends C_ξ to $C_{\xi'}$.

Let B_ξ be the "inside" of C_ξ in $\mathbb{H}^2$. Namely, when $\xi \neq \infty$, B_ξ is the closed euclidean disk bounded by C_ξ minus the point ξ. When $\xi = \infty$, B_ξ is the euclidean half-plane bounded by C_ξ from below. Such a B_ξ is called a **horodisk** centered at ξ.

By Complement 6.24, we can assume without loss of generality that the horocircles are small enough that they are disjoint from each other, and that the only edges of X that are met by B_ξ are those which are adjacent to the ideal vertex ξ.

Let $B = \bigcup_\xi B_\xi$ denote the union of all the B_ξ, as ξ ranges over all ideal vertices of X. There is a convenient function $h \colon B \to \mathbb{R}$ defined as follows. For $P \in B_\xi$, let Q be the point of C_ξ that is closest to P; then $h(P) = d_{\mathrm{hyp}}(P, Q)$. The point Q can be easily constructed from the property that it is the intersection of C_ξ with the complete geodesic g passing through ξ and P. This property is easily checked when $\xi = \infty$, for instance using Lemma 2.5; for the general case, just transport everything by an isometry of $\mathbb{H}^2$ sending ξ to ∞.

The function h is called the **Busemann function**. See Exercises 6.10, 6.11 and 6.12 for another geometric interpretation of h. This function has the following two important properties:

(1) if P and $Q \in B$ are glued together, then $h(P) = h(Q)$;

(2) for any two P, Q in the same horodisk B_ξ, $d_{\mathrm{hyp}}(P, Q) \geqslant |h(P) - h(Q)|$.

The first property is an immediate consequence of the fact that the gluing maps are isometries and send each C_ξ to some $C_{\xi'}$. The second property is a consequence of Lemma 2.5 in the case where $\xi = \infty$. The general case follows from this special case by modifying the geometric setup through a hyperbolic isometry sending ξ to ∞.

After bringing up all this machinery, we now return to our sequence of points P_n in the edges of X.

Fix an arbitrary $\varepsilon > 0$, whose precise value will not be important. By the same argument as in Case 2, there is a number n_1 such that $\bar{d}_X(\bar{P}_n, \bar{P}_{n+1}) < \varepsilon$ for every $n \geqslant n_1$.

In addition, we can assume that we are not in the situation of Case 1, since otherwise we are done. This guarantees that there exists an $n_0 \geqslant n_1$ such that for every $n \geqslant n_0$, the point P_n is sufficiently far away on the edge E containing it that it belongs to the horodisk B_{ξ_n} associated to some ideal vertex ξ_n, and that it is even at distance $> \varepsilon$ from the point where the edge E meets the horocircle C_{ξ_n} delimiting B_{ξ_n}.

For $n \geqslant n_0$, consider a discrete walk w from $\bar{P}_n$ to $\bar{P}_{n+1}$ in $\bar{X}$, whose length $\ell_{d_X}(w)$ is close enough to the infimum $\bar{d}_X(\bar{P}_n, \bar{P}_{n+1})$ that $\ell_{d_X}(w) < \varepsilon$ and whose steps are $P_n = Q_1$, $R_1 \sim Q_2$, ..., $R_{k-1} \sim Q_k$, $R_k = P_{n+1}$. We will prove by induction that all the Q_i and R_{i-1} belong to $B = \bigcup_\xi B_\xi$, and that

$$(6.1) \quad h(P_n) - \sum_{j=1}^{i-1} d_X(Q_j, R_j) \leqslant h(Q_i) \leqslant h(P_n) + \sum_{j=1}^{i-1} d_X(Q_j, R_j).$$

Of course, the Q_i and R_{i-1} may jump from one horodisk B_ξ to another.

We start the induction with the case where $i = 1$, in which case the property trivially holds since $Q_1 = P_n$.

Suppose that the induction hypothesis holds for i. If Q_i is in the horodisk B_ξ, the distance from Q_i to the horocycle C_ξ bounding B_ξ is $h(Q_i)$. By the induction hypothesis,

$$h(Q_i) - d_X(Q_i, R_i) \geqslant h(P_n) - \sum_{j=1}^{i} d_X(Q_j, R_j)$$
$$\geqslant h(P_n) - \ell_{d_X}(w) > h(P_n) - \varepsilon > 0$$

since $\ell_{d_X}(w) \leqslant \varepsilon$ and since P_n is at distance $> \varepsilon$ from the horocycle bounding the horodisk B_{ξ_n} that contains it. In particular, the distance from $Q_i \in B_\xi$ to R_i is less than the distance from Q_i to the boundary horocircle C_ξ bounding B_ξ. It follows that R_i is also in B_ξ.

Since Q_{i+1} is glued to R_i, and since the gluing maps sendpoints of B to points of B, we conclude that Q_{i+1} is also in B.

Also, combining (6.1) with the two fundamental properties of h,

$$h(Q_{i+1}) = h(R_i) \leqslant h(Q_i) + d_X(Q_i, R_i) \leqslant h(P_n) + \sum_{j=1}^{i} d_X(Q_j, R_j)$$

and

$$h(Q_{i+1}) = h(R_i) \geqslant h(Q_i) - d_X(Q_i, R_i) \geqslant h(P_n) - \sum_{j=1}^{i} d_X(Q_j, R_j).$$

This proves that (6.1) holds for $i + 1$.

This completes our proof by induction that all Q_i and R_{i-1} with $1 \leqslant i \leqslant k$ belong to B and satisfy (6.1).

One more step in the same proof gives that

$$h(P_{n+1}) = h(R_k) \leqslant h(Q_k) + d_X(Q_k, R_k) \leqslant h(P_n) + \sum_{j=1}^{k} d_X(Q_k, R_k)$$

and

$$h(P_{n+1}) = h(R_k) \geqslant h(Q_k) - d_X(Q_k, R_k) \geqslant h(P_n) - \sum_{j=1}^{k} d_X(Q_k, R_k),$$

so that

$$|h(P_{n+1}) - h(P_n)| \leqslant \sum_{j=1}^{k} d_X(Q_k, R_k) = \ell_{d_X}(w).$$

Since this holds for every discrete walk w from $\bar{P}_n$ to $\bar{P}_{n+1}$ whose length is sufficiently close to the infimum $\bar{d}_X(\bar{P}_n, \bar{P}_{n+1})$, we conclude that

$$|h(P_{n+1}) - h(P_n)| \leqslant \bar{d}_X(\bar{P}_n, \bar{P}_{n+1}).$$

A consequence of this inequality is that the length of the sequence $\big(h(P_n)\big)_{n \in \mathbb{N}}$ in $\mathbb{R}$ is bounded by the length of the sequence $\big(\bar{P}_n\big)_{n \in \mathbb{N}}$ in $\bar{X}$, and in particular is finite. Since $(\mathbb{R}, d_{\mathrm{euc}})$ is complete (Fact 6.9), if follows that the sequence $\big(h(P_n)\big)_{n \in \mathbb{N}}$ converges in $\mathbb{R}$. As a consequence, it is bounded by some number M.

However, this shows that for $n \geqslant n_0$, the P_n stay at hyperbolic distance $\leqslant M$ from the finitely many points where the edges of X meet the horocircles C_ξ. In particular, these P_n are all contained in a very large hyperbolic ball. So we are in the situation of Case 1 after all, which proves that the sequence $(\bar{P}_n)_{n \in \mathbb{N}}$ converges in $\bar{X}$.

We are now ready to conclude.

CASE 4: *General case.*

We will use the three previous cases to extract from $(\bar{P}_n)_{n\in\mathbb{N}}$ a converging subsequence and then apply Lemma 6.17.

If we are in the situation of Case 2, namely, if the P_n stay away from the edges of X, we are done. Consequently, we can assume that this is not the case.

We then construct by induction a subsequence $(P_{n_k})_{k\in\mathbb{N}}$ such that for every k, there exists a point Q_{n_k} in an edge of X such that $d_X(P_{n_k}, Q_{n_k}) < 2^{-k}$. (We could replace 2^{-k} by the kth term of any convergent series.) Indeed, supposing the first k terms of the subsequence have been constructed, the fact that the hypothesis of Case 2 does not hold implies that there exists an index $n_{k+1} > n_k$ such that $P_{n_{k+1}}$ is at distance $< 2^{-(k+1)}$ from the edges of X. This means that there exists a point $Q_{n_{k+1}}$ in the boundary of X which is at distance $< 2^{-(k+1)}$ from $P_{n_{k+1}}$. This inductive process clearly provides the subsequence $(P_{n_k})_{k\in\mathbb{N}}$ and the sequence $(Q_{n_k})_{k\in\mathbb{N}}$ requested.

We claim that the sequence $(Q_{n_k})_{k\in\mathbb{N}}$ has finite length. Indeed,

$$d_X(Q_{n_k}, Q_{n_{k+1}}) \leqslant d(Q_{n_k}, P_{n_k}) + d_X(P_{n_k}, P_{n_{k+1}}) + d_X(P_{n_{k+1}}, Q_{n_{k+1}})$$
$$\leqslant 2^{-k} + d_X(P_{n_k}, P_{n_{k+1}}) + 2^{-(k+1)}$$

by the Triangle Inequality, so that

$$\sum_{k=1}^{\infty} d_X(Q_{n_k}, Q_{n_{k+1}}) \leqslant \sum_{k=1}^{\infty} 2^{-k} + \sum_{k=1}^{\infty} d_X(P_{n_k}, P_{n_{k+1}}) + \sum_{k=1}^{\infty} 2^{-(k+1)}$$

is finite since the length of $(P_{n_k})_{k\in\mathbb{N}}$ is finite.

Therefore, we can apply Case 3 to the sequence $(Q_{n_k})_{k\in\mathbb{N}}$. Our analysis of this case shows that this sequence converges. Since $d_X(P_{n_k}, Q_{n_k}) < 2^{-k}$ tends to 0 as k tends to ∞, it follows that the sequence $(P_{n_k})_{k\in\mathbb{N}}$ converges to the same limit.

Thus, our original finite length sequence $(\bar{P}_n)_{n\in\mathbb{N}}$ has a converging subsequence $(P_{n_k})_{k\in\mathbb{N}}$. It therefore converges by Lemma 6.17. From page 169

This concludes the proof of the "if" part of Theorem 6.25, and therefore completes the proof of this statement. □

Exercises for Chapter 6

Exercise 6.1. A *Cauchy sequence* in a metric space (X, d) is a sequence of points $P_1, P_2, \ldots, P_n, \ldots$ in X such that for every $\varepsilon > 0$, there exists a number n_0 such that $d(P_n, P_{n'}) < \varepsilon$ for every $n, n' \geqslant n_0$.

    a. Show that a sequence $(P_n)_{n \in \mathbb{N}}$ that has finite length is Cauchy.

    b. Conversely, let $(P_n)_{n \in \mathbb{N}}$ be a Cauchy sequence. Show, by induction on k, that it contains a subsequence $\left(P_{n_k}\right)_{k \in \mathbb{N}}$ such that $d\left(P_{n_k}, P_{n_{k+1}}\right) \leqslant 2^{-k}$ for every k. Show that this subsequence $\left(P_{n_k}\right)_{k \in \mathbb{N}}$ has finite length.

    c. Show that if a Cauchy sequence admits a converging subsequence, then the whole sequence is convergent.

    d. Combine the previous steps to show that a metric space (X, d) is complete in the sense of Section 6.2 if and only if every Cauchy sequence in (X, d) is convergent.

Exercise 6.2. Let (X, d) and (X', d') be two metric spaces for which there exists a homeomorphism $\varphi \colon X \to X'$. Prove, or disprove by a counterexample, each of the following two statements:

    a. If (X, d) is compact, then (X', d') is also compact.

    b. If (X, d) is complete, then (X', d') is also complete.

Exercise 6.3. Show that a tessellation of the sphere $\mathbb{S}^2$ has only finitely many tiles. Hint: The sphere is compact.

Exercise 6.4. In the euclidean plane, let X be the parallelogram with vertices $(0, 0)$, $(2, 1)$, $(2, 2)$, $(0, 1)$. Glue the left edge of X to the right edge by the reflection-translation $(x, y) \mapsto (x + 2, -y + 2)$, and the bottom edge to the top edge by the translation $(x, y) \mapsto (x, y + 1)$; we saw in Section 5.1 that the corresponding quotient space $\bar{X}$ is homeomorphic to the Klein bottle. Draw the tessellation associated by Theorem 6.1 to this gluing data.

Exercise 6.5. Construct a tessellation of the euclidean or hyperbolic plane by tiles which are not convex. Extra credit: Find one which is particularly pretty (solution not unique).

Exercise 6.6. For three integers a, b, $c \geqslant 2$ with $\frac{1}{a} + \frac{1}{b} + \frac{1}{c} > 1$, consider a tessellation of the sphere $\mathbb{S}^2$ by spherical triangles of angles $\frac{\pi}{a}, \frac{\pi}{b}, \frac{\pi}{c}$ as in Theorem 6.21. Show that this tessellation consists of $\frac{4abc}{ab+bc+ac-abc}$ triangles. Hint: Use the area formula of Exercise 3.6.

Exercise 6.7.

    a. Show that for every euclidean triangle $T \subset \mathbb{R}^2$, there exists a tessellation of the euclidean plane $\mathbb{R}^2$ whose tiles are all isometric to T.

b. Give an example of a spherical triangle $T \subset \mathbb{S}^2$ for which there exists no tessellation of the sphere $\mathbb{S}^2$ by triangles isometric to T. Hint: Use the area formula of Exercise 3.6 or the hint for part c below.

c. Give an example of a hyperbolic triangle $T \subset \mathbb{H}^2$ for which there exists no tessellation of the hyperbolic plane $\mathbb{H}^2$ by triangles isometric to T. Hint: Consider the angles of the tiles of the tessellation around one vertex.

Exercise 6.8. Show that for every integer p, $q \geqslant 2$ with $\frac{1}{p} + \frac{1}{q} < \frac{1}{2}$, there exists a tessellation of the hyperbolic plane $\mathbb{H}^2$ by p-gons such that exactly q of these p-gons meet at each vertex. Possible hint: Use appropriate triangles.

Exercise 6.9. Let X be a polygon in the euclidean plane $\mathbb{R}^2$, and let $(\bar{X}, \bar{d}_X)$ be a quotient space obtained by isometrically gluing its edges together. Show that $(\bar{X}, \bar{d}_X)$ is always complete. Hint: Adapt the proof of Poincaré's Polygon Theorem 6.25, noting that Case 3 of this proof becomes much simpler in the euclidean case.

Exercise 6.10. Fix a point $P_0 \in \mathbb{H}^2$ in the hyperbolic plane and a point $\xi \in \mathbb{R} \cup \{\infty\}$ at infinity of $\mathbb{H}^2$. Show that as $Q \in \mathbb{H}^2$ tends to ξ in $\mathbb{R}^2 \cup \{\infty\}$, the limit

$$h(P) = \lim_{Q \to \xi} d_{\mathrm{hyp}}(Q, P) - d_{\mathrm{hyp}}(Q, P_0)$$

exists for every $P \in \mathbb{H}^2$. Possible hint: Use the explicit formula for the hyperbolic metric provided by Exercise 2.2.

Exercise 6.11. In Exercise 6.10, consider the case where $\xi = \infty$. If $P_0 = (x_0, y_0)$ and $P = (x, y)$, show that $h(P) = \log \frac{y_0}{y}$.

Exercise 6.12. In Exercise 6.10, show that the set of $P \in \mathbb{H}^2$ with $h(P) = 0$ is exactly the horocircle centered at ξ and passing through P_0. Possible hint: Use Exercise 6.11 and a hyperbolic isometry sending ξ to ∞.

Exercise 6.13. In Figure 6.15, V_0 is delimited from below by a straight line segment with negative slope, whereas $U_0 = \varphi_2^{-1} \circ \varphi_1^{-1}(V_0)$ is delimited from above by a curve C which appears to be a euclidean circle arc.

a. Show that C is indeed a circle arc contained in a circle passing through the point 0.

b. This circle appears to be tangent to the x-axis at 0. Is this actually the case? Explain.

Exercise 6.14. Consider an edge cycle as in condition (2) of Proposition 6.23, and use the same notation as in that statement.

a. Show that $\varphi_{i_k} \circ \varphi_{i_{k-1}} \circ \cdots \circ \varphi_{i_2} \circ \varphi_{i_1}$ cannot be an antilinear fractional map and is different from the identity. Possible hint: Look at the way the tiles $\varphi_{i_1}^{-1} \circ \varphi_{i_2}^{-1} \circ \cdots \circ \varphi_{i_{k-1}}^{-1} \circ \varphi_{i_k}^{-1}(X)$ sit side-by-side near ξ_1.

b. Show that $\varphi_{i_k} \circ \varphi_{i_{k-1}} \circ \cdots \circ \varphi_{i_2} \circ \varphi_{i_1}$ is horocyclic if and only if it is parabolic in the sense of Exercise 2.7.

Exercise 6.15. All the tessellations constructed in this chapter are invariant under a transformation group Γ, as defined in Chapter 7. Do an Internet search for examples of tessellations which are invariant under no nontrivial transformation group. Suggested key word: Penrose.

Chapter 7

Group actions and fundamental domains

The tiling groups we encountered in Chapter 6 are examples of isometric group actions. The current chapter is devoted to definitions and basic properties of group actions.

7.1. Transformation groups

A *transformation group* on a set X is a family Γ of bijections $\gamma \colon X \to X$ such that

(1) if γ and γ' are in Γ, their composition $\gamma \circ \gamma'$ is also in Γ;

(2) the group Γ contains the identity map Id_X;

(3) if γ is in Γ, its inverse map γ^{-1} is also in Γ.

Recall that the *identity map* $\mathrm{Id}_X \colon X \to X$ is defined by the property that $\mathrm{Id}_X(x) = x$ for every $x \in X$. Also, the *inverse* of a bijection $\varphi \colon X \to Y$ is the map $\varphi^{-1} \colon Y \to X$ such that $\varphi^{-1}(y)$ is the number x such that $\varphi(x) = y$. This is equivalent to the properties that $\varphi \circ \varphi^{-1} = \mathrm{Id}_Y$ or $\varphi^{-1} \circ \varphi = \mathrm{Id}_X$.

A transformation group Γ naturally gives rise to a map $\Gamma \times X \to X$ which to (γ, x) associates $\gamma(x)$. This map is traditionally defined as the *group action* of Γ over X.

We are particularly interested in the situation where X is a metric space (X, d) and where all the elements of Γ are isometries of (X, d). In this case we say that Γ is a **group of isometries** of (X, d) or that Γ **acts by isometries** on (X, d) or that we have an **isometric action** of Γ on (X, d).

It is time for a few examples.

Example 7.1. The **trivial group** $\Gamma = \{\mathrm{Id}_X\}$, consisting of the single bijection Id_X, satisfies all three conditions since $\mathrm{Id}_X \circ \mathrm{Id}_X = \mathrm{Id}_X$ and $\mathrm{Id}_X^{-1} = \mathrm{Id}_X$. It is the smallest possible transformation group for X, and clearly acts by isometries.

Example 7.2. The set Γ of all isometries $\varphi \colon X \to X$ of (X, d) is a transformation group, obviously acting by isometries. By definition, this is the **isometry group** of (X, d). Clearly, it is the largest group of isometries acting on (X, d).

Example 7.3. In the hyperbolic plane $(\mathbb{H}^2, d_{\mathrm{hyp}})$, the set of linear fractional maps $z \mapsto \frac{az+b}{cz+d}$ with integer coefficients a, b, c, $d \in \mathbb{Z}$ and with $ad - bc = 1$ forms a group of isometries. Indeed, a computation shows that the composition of two such linear fractionals also has integer coefficients. The identity map can be written as $z \mapsto \frac{z+0}{0z+1}$, and the inverse of $z \mapsto \frac{az+b}{cz+d}$ is $z \mapsto \frac{dz-b}{-cz+a}$.

The set of antilinear fractional maps with integer coefficients is *not* a transformation group. Indeed, the composition of two antilinear fractionals is linear fractional, and not antilinear fractional. Also, the identity is not antilinear fractional.

Example 7.4. Let φ_1, φ_2, ..., φ_n be a family of bijections of a set X. Let Γ be the set of all bijections φ of X which can be written as a composition of finitely many φ_i and their inverses. Namely, Γ is the set of all φ of the form $\varphi = \varphi_{i_1}^{\pm 1} \circ \varphi_{i_2}^{\pm 1} \circ \cdots \circ \varphi_{i_k}^{\pm 1}$ where the indices $i_j \in \{1, 2, \ldots, n\}$ are not necessarily distinct. Then Γ is a transformation group. It is actually the smallest transformation group containing φ_1, φ_2, ..., φ_n. By definition, the transformation group Γ is then **generated** by φ_1, φ_2, ..., φ_n.

As a fundamental example, all the tiling groups constructed in Chapter 6 are transformation groups of the euclidean plane, the hyperbolic plane or the sphere, acting by isometries.

7.2. Group actions and quotient spaces

Let Γ be a group of isometries of the metric space (X, d). We will use the action of Γ to create a new type of quotient space $\bar{X}$.

For every P, let $\bar{P}$ consist of those $Q \in X$ which are of the form $Q = \gamma(P)$ for some $\gamma \in \Gamma$. This subset $\bar{P} \subset X$ is the **orbit** of the point P under the action of Γ. The orbit is also denoted as $\bar{P} = \Gamma(P)$.

Lemma 7.5. *As P ranges over all points of X, the orbits $\bar{P}$ form a partition of X.*

Proof. The proof uses all three conditions in the definition of transformation groups at the beginning of Section 7.1.

Since Id_X must be in Γ, the point $P = \mathrm{Id}_X(P)$ is an element of $\bar{P}$. Therefore, the sets $\bar{P}$ cover all of X.

It remains to show that distinct $\bar{P}$ and $\bar{Q}$ must be disjoint. For this, suppose that $\bar{P}$ and $\bar{Q}$ share a common point R. We want to show that $\bar{P} = \bar{Q}$. By definition, there exists $\alpha, \beta \in \Gamma$ such that $R = \alpha(P)$ and $R = \beta(Q)$. As a consequence, $Q = \beta^{-1} \circ \alpha(P)$.

For every $S \in \bar{Q}$, there exists an element $\gamma \in \Gamma$ such that $S = \gamma(Q)$. Therefore, $S = \gamma \circ \beta^{-1} \circ \alpha(P)$. The conditions in the definition of a transformation group imply that $\gamma \circ \beta^{-1} \circ \alpha$ is an element of Γ. Therefore, S is an element of the orbit $\bar{P}$. Since this holds for every $S \in \bar{Q}$, we conclude that $\bar{Q}$ is contained in $\bar{P}$.

Conversely, $\bar{P}$ is contained in $\bar{Q}$ by symmetry so that $\bar{P} = \bar{Q}$ whenever $\bar{P}$ and $\bar{Q}$ intersect. $\square$

Let $\bar{X}$ be the partition defined by the orbits $\bar{P}$. Lemma 4.1 shows that the metric d induces a quotient semi-metric $\bar{d}$ on $\bar{X}$.

When the semi-metric $\bar{d}$ is a metric, the metric space $(\bar{X}, \bar{d})$ is the **quotient space** of X under the action of Γ, or the **orbit space** of the action of Γ on X. It is also denoted by $\bar{X} = X/\Gamma$.

In this situation, the semi-metric $\bar{d}$ is much simpler than in the general setting of Chapter 4.

Proposition 7.6. *Let the group Γ act by isometries on the metric space (X, d), and consider the quotient space $(\bar{X}, \bar{d})$. Then, for every*

$\bar{P}, \bar{Q} \in \bar{X}$,

$$\bar{d}(\bar{P}, \bar{Q}) = \inf\{d(P', Q'); P' \in \bar{P}, Q' \in \bar{Q}\}$$
$$= \inf\{d(P, \gamma(Q)); \gamma \in \Gamma\}.$$

In other words, the (semi-)distance from $\bar{P}$ to $\bar{Q}$ is equal to the infimum of the distances from points in the orbit of P to points in the orbit of Q, which is also equal to the infimum of the distances from the point P itself to points in the orbit of Q.

Proof. Define

$$\bar{d}'(\bar{P}, \bar{Q}) = \inf\{d(P', Q'); P' \in \bar{P}, Q' \in \bar{Q}\}.$$

We want to show that $\bar{d} = \bar{d}'$.

Let us first show that $\bar{d}'$ satisfies the Triangle Inequality, namely that $\bar{d}'(\bar{P}, \bar{R}) \leqslant \bar{d}'(\bar{P}, \bar{Q}) + \bar{d}'(\bar{Q}, \bar{R})$ for every $\bar{P}, \bar{Q}, \bar{R} \in \bar{X}$.

By definition of the infimum there exists, for every $\varepsilon > 0$, points $P' \in \bar{P}$ and $Q' \in \bar{Q}$ such that

$$d(P', Q') \leqslant \bar{d}'(\bar{P}, \bar{Q}) + \varepsilon.$$

Similarly, there exists $Q'' \in \bar{Q}$ and $R'' \in \bar{R}$ such that

$$d(Q'', R'') \leqslant \bar{d}'(\bar{Q}, \bar{R}) + \varepsilon.$$

The fact that both Q' and Q'' are in the orbit $\bar{Q}$ means that there exists $\gamma', \gamma'' \in \Gamma$ such that $Q' = \gamma'(Q)$ and $Q'' = \gamma''(Q)$. As a consequence, $Q'' = \gamma(Q')$ for $\gamma = \gamma'' \circ \gamma'^{-1} \in \Gamma$. Consider the point $R' = \gamma^{-1}(R'')$, which is also in the orbit $\bar{R}$ of R. Then,

$$\bar{d}'(\bar{P}, \bar{R}) \leqslant d(P', R') \leqslant d(P', Q') + d(Q', R')$$
$$\leqslant d(P', Q') + d(\gamma(Q'), \gamma(R'))$$
$$\leqslant d(P', Q') + d(Q'', R'')$$
$$\leqslant \bar{d}'(\bar{P}, \bar{Q}) + \bar{d}'(\bar{Q}, \bar{R}) + 2\varepsilon$$

using the fact that γ is an isometry of (X, d).

Since this folds for every $\varepsilon > 0$, it follows that $\bar{d}'(\bar{P}, \bar{R}) \leqslant \bar{d}'(\bar{P}, \bar{Q}) + \bar{d}'(\bar{Q}, \bar{R})$. Namely, $\bar{d}'$ satisfies the Triangle Inequality.

Also, note that the definition of $\bar{d}'$ immediately implies that $\bar{d}'(\bar{P}, \bar{Q}) \leqslant d(P, Q)$ for every $P, Q \in X$.

After these two preliminary observations, we are now ready to complete the proof. By definition, $\bar{d}(\bar{P}, \bar{Q})$ is equal to the infimum of the lengths $\ell_d(w)$ of all discrete walks w from P to Q, namely, walks of the form $P = P_1, Q_1 \sim P_2, \ldots, Q_{n-1} \sim P_n, Q_n = Q$. Then,

$$\ell_d(w) = \sum_{i=1}^n d(P_i, Q_i) \geqslant \sum_{i=1}^n \bar{d}'(\bar{P}_i, \bar{Q}_i) \geqslant \bar{d}'(\bar{P}_1, \bar{Q}_n) = \bar{d}'(\bar{P}, \bar{Q})$$

by repeated use of the Triangle Inequality for $\bar{d}'$. Since this holds for every discrete walk w, we conclude that $\bar{d}(\bar{P}, \bar{Q}) \geqslant \bar{d}'(\bar{P}, \bar{Q})$.

Conversely, if $P' \in \bar{P}$ and $Q' \in \bar{Q}$, $\bar{d}(\bar{P}, \bar{Q}) = \bar{d}(\bar{P}', \bar{Q}') \leqslant d(P', Q')$ by Lemma 4.2. Since $\bar{d}'(\bar{P}, \bar{Q})$ is defined as the infimum of those distances $d(P', Q')$, we conclude that $\bar{d}(\bar{P}, \bar{Q}) \leqslant \bar{d}'(\bar{P}, \bar{Q})$.

The combination of the two inequalities shows that

$$\bar{d}(\bar{P}, \bar{Q}) = \bar{d}'(\bar{P}, \bar{Q}) = \inf\{d(P', Q'); P' \in \bar{P}, Q' \in \bar{Q}\}.$$

It remains to show that $\bar{d}'(\bar{P}, \bar{Q})$ is equal to

$$\bar{d}''(\bar{P}, \bar{Q}) = \inf\{d(P, Q'); Q' \in \bar{Q}\} = \inf\{d(P, \gamma(Q)); \gamma \in \Gamma\}.$$

This is immediate, once we realize the following equalities between subsets of $\mathbb{R}$

$$\begin{aligned}\{d(P', Q'); P' \in \bar{P}, Q' \in \bar{Q}\} &= \{d(\alpha(P), \beta(Q)); \alpha, \beta \in \Gamma\} \\ &= \{d(P, \alpha^{-1} \circ \beta(Q)); \alpha, \beta \in \Gamma\} \\ &= \{d(P, \gamma(Q)); \gamma \in \Gamma\}\end{aligned}$$

using the fact that every $\alpha \in \Gamma$ is an isometry of (X, d). $\qquad \square$

The isometric action of Γ on the metric space (X, d) is **discontinuous** if, for every $P \in X$, there exists a ball $B_d(P, \varepsilon)$ centered at P such that there are only finitely many $\gamma \in \Gamma$ with $\gamma(P) \in B_d(P, \varepsilon)$.

Theorem 7.7. *If the transformation group Γ acts by isometries and discontinuously on the metric space (X, d), the semi-metric $\bar{d}$ induced on the quotient space $\bar{X} = X/\Gamma$ is a metric. In particular, the quotient space $(\bar{X}, \bar{d})$ is a metric space.*

Proof. We have to show that $\bar{d}(\bar{P}, \bar{Q}) \neq 0$ whenever $\bar{P} \neq \bar{Q}$.

Because the action is by isometries, Proposition 7.6 asserts that $\bar{d}(\bar{P}, \bar{Q})$ is equal to $\inf\{d(P, Q'); Q' \in \bar{Q}\}$. Since the action is discontinuous, there exists an $\varepsilon > 0$ for which there are only finitely many $\gamma \in \Gamma$ such that $\gamma(P) \in B_d(P, \varepsilon)$.

This implies that the ball $B_d(P, \frac{\varepsilon}{2})$ can contain only finitely many points Q' of the orbit $\bar{Q}$ of Q. Indeed, if $\gamma(Q)$ and $\gamma'(Q) \in \bar{Q}$ are both in the ball $B_d(P, \frac{\varepsilon}{2})$, then

$$d\big(P, \gamma' \circ \gamma^{-1}(P)\big) = d\big(\gamma'^{-1}(P), \gamma^{-1}(P)\big)$$
$$\leqslant d\big(\gamma'^{-1}(P), Q\big) + d\big(Q, \gamma^{-1}(P)\big)$$
$$\leqslant d\big(P, \gamma'(Q)\big) + d\big(\gamma(Q), P\big) < \tfrac{\varepsilon}{2} + \tfrac{\varepsilon}{2} = \varepsilon,$$

using the fact that γ and γ' are both isometries. By discontinuity of the action, $\gamma' \circ \gamma^{-1}$ can take only finitely many values in Γ, so there can be only finitely many such γ. This proves that the intersection of the orbit $\bar{Q}$ with the ball $B_d(P, \frac{\varepsilon}{2})$ is a finite set $\{Q_1, Q_2, \ldots, Q_n\}$, possibly empty.

If the intersection is nonempty, note that no Q_i can be equal to P since $\bar{P} \neq \bar{Q}$. Therefore,

$$\bar{d}(\bar{P}, \bar{Q}) = \min\{d(P, Q_1), d(P, Q_2), \ldots, d(P, Q_n)\} > 0$$

since the numbers $d(P, Q_i)$ are all different from 0 and since there are only finitely many of them.

If $B_d(P, \frac{\varepsilon}{2})$ contains no $Q' \in \bar{Q}$, then $\bar{d}(\bar{P}, \bar{Q}) \geqslant \frac{\varepsilon}{2} > 0$. Therefore, $\bar{d}(\bar{P}, \bar{Q}) \neq 0$ in both cases. ☐

The **stabilizer** of the point $P \in X$ for the action of Γ on X is $\Gamma_P = \{\gamma \in \Gamma; \gamma(P) = P\}$. The stabilizer Γ_P is easily seen to be a transformation group acting on X.

If the action of Γ on the metric space (X, d) is by isometries, note that every element of the stabilizer Γ_P sends each ball $B_d(P, \varepsilon)$ to itself. In other words, Γ_P acts on $B_d(P, \varepsilon)$ and we can consider the quotient space $B_d(P, \varepsilon)/\Gamma_P$ with the quotient metric induced by d.

Theorem 7.8. *If the transformation group Γ acts by isometries and discontinuously on the metric space (X, d), then, for every $\bar{P}$ in the quotient metric space $(\bar{X}, \bar{d})$, there exists a ball $B_{\bar{d}}(\bar{P}, \varepsilon)$ which is isometric to the quotient space $B_d(P, \varepsilon)/\Gamma_P$.*

Proof. By hypothesis, there exists a ball $B_d(P, \varepsilon_1)$ which contains $\gamma(P)$ for only finitely many $\gamma \in \Gamma$. As a consequence, there exists an $\varepsilon > 0$ such that $d(P, \gamma(P)) \geqslant 5\varepsilon$ for every $\gamma(P) \in \bar{P}$ different from P.

Let $\widehat{B}_d(P, \varepsilon)$ denote the quotient space $B_d(P, \varepsilon)/\Gamma_P$, endowed with the quotient metric $\widehat{d}$ induced by the restriction of d to $B_d(P, \varepsilon)$. Also, let $\widehat{P}$ denote the point of $\widehat{B}_d(P, \varepsilon)$ corresponding to $P \in B_d(P, \varepsilon)$.

Since Γ_P is contained in Γ, any orbit under the action of Γ_P is contained in a unique orbit of Γ. It follows that there is a well-defined map $\varphi \colon \widehat{B}_d(P, \varepsilon) \to B_{\bar{d}}(\bar{P}, \varepsilon)$ defined by the property that $\varphi(\widehat{Q}) = \bar{Q}$ for every $Q \in B_d(P, \varepsilon)$. Note that the point $\bar{Q} \in \bar{X}$ is indeed in the ball $B_{\bar{d}}(\bar{P}, \varepsilon)$ since the quotient map $\bar{X} \to X$ is distance nonincreasing by Lemma 4.2. We will show that φ is an isometry.

Let $Q, Q' \in B_d(X, \varepsilon)$. Proposition 7.6 shows that

$$\bar{d}(\varphi(\widehat{Q}), \varphi(\widehat{Q}')) = \bar{d}(\bar{Q}, \bar{Q}') = \inf\{d(Q, \gamma(Q')), \gamma \in \Gamma\} \leqslant d(Q, Q').$$

If $d(Q, \gamma(Q'))$ is sufficiently close to the infimum that $d(Q, \gamma(Q')) \leqslant \bar{d}(\bar{Q}, \bar{Q}') + \varepsilon$, a repeated use of the Triangle Inequality shows that

$$\begin{aligned} d(P, \gamma(P)) &\leqslant d(P, Q) + d(Q, \gamma(Q')) + d(\gamma(Q'), \gamma(P)) \\ &\leqslant d(P, Q) + d(Q, Q') + \varepsilon + d(Q', P) \\ &\leqslant d(P, Q) + d(Q, P) + d(P, Q') + \varepsilon + d(Q', P) < 5\varepsilon. \end{aligned}$$

By choice of ε, this implies that $\gamma(P) = P$. What this proves is that if $d(Q, \gamma(Q'))$ is very close to the infimum $\bar{d}(\bar{Q}, \bar{Q}')$, then γ is in $\Gamma_P \subset \Gamma$. As a consequence, the two infimums

$$\bar{d}(\bar{Q}, \bar{Q}') = \inf\{d(Q, \gamma(Q')); \gamma \in \Gamma\}$$

and

$$\widehat{d}(\widehat{Q}, \widehat{Q}') = \inf\{d(Q, \gamma(Q')); \gamma \in \Gamma_P\}$$

are equal. This proves that $\bar{d}(\varphi(\widehat{Q}), \varphi(\widehat{Q}')) = \bar{d}(\bar{Q}, \bar{Q}') = \widehat{d}(\widehat{Q}, \widehat{Q}')$ for every $\widehat{Q}, \widehat{Q}' \in \widehat{B}_d(P, \varepsilon)$.

In particular, φ is injective.

If $\bar{Q}$ is a point of $B_{\bar{d}}(\bar{P}, \varepsilon)$, then $\inf\{d(P', Q'); P' \in \bar{P}, Q' \in \bar{Q}\} = \bar{d}(\bar{P}, \bar{Q}) < \varepsilon$, so that there exists a $Q' \in \bar{Q}$ such that $d(P, Q') < \varepsilon$. As a consequence, $\bar{Q} = \varphi(\widehat{Q}')$ is in the image of φ. This proves that $\varphi \colon \widehat{B}_d(P, \varepsilon) \to B_{\bar{d}}(\bar{P}, \varepsilon)$ is surjective, and concludes the proof that φ is an isometry. $\square$

The action of Γ on X is **free** if, for every $\gamma \in \Gamma - \{\mathrm{Id}_X\}$, $\gamma(P) \neq P$ for every $P \in X$. In other words, the action is free if the stabilizer Γ_P of every point $P \in X$ is the trivial group $\{\mathrm{Id}_X\}$.

Corollary 7.9. *If the transformation group Γ acts by isometries, discontinuously, and freely on the metric space (X, d), then the quotient metric space $(\bar{X}, \bar{d})$ is locally isometric to (X, d).*

Proof. If $\Gamma_P = \{\mathrm{Id}_X\}$, then $\widehat{B}_d(P, \varepsilon)$ is equal to $B_d(P, \varepsilon)$, and the quotient metric $\widehat{d}$ coincides with d (compare Exercise 4.3). Therefore, Theorem 7.8 shows that every $\bar{P} \in \bar{X}$ is the center of a ball $\bar{B}_{\bar{d}}(P, \varepsilon)$ which is isometric to a ball $B_d(P, \varepsilon)$ in X. $\qquad\square$

When (X, d) is the hyperbolic plane $(\mathbb{H}^2, d_{\mathrm{hyp}})$, Exercises 7.12 and 7.14 discuss the possible types for the quotient space $B_d(P, \varepsilon)/\Gamma_P$ when the stabilizer is nontrivial. It is isometric, either to a hyperbolic cone with cone angle $\frac{2\pi}{n}$, or to a hyperbolic disk sector of angle $\frac{\pi}{n}$, for some integer $n \geqslant 1$. By Theorem 7.8, this consequently describes the local geometry of the quotient of the hyperbolic plane by a discontinuous isometric group action.

The same results (and proofs) hold without modifications for quotients of the euclidean plane $(\mathbb{R}^2, d_{\mathrm{euc}})$ or of the sphere $(\mathbb{S}^2, d_{\mathrm{sph}})$ under a discontinuous isometric group action.

7.3. Fundamental domains

Let the group Γ act by isometries on the euclidean plane $(\mathbb{R}^2, d_{\mathrm{euc}})$, the hyperbolic plane $(\mathbb{H}^2, d_{\mathrm{hyp}})$ or the sphere $(\mathbb{S}^2, d_{\mathrm{sph}})$. For simplicity, write $(X, d) = (\mathbb{R}^2, d_{\mathrm{euc}})$, $(\mathbb{H}^2, d_{\mathrm{hyp}})$ or $(\mathbb{S}^2, d_{\mathrm{sph}})$, according to the case considered.

A **fundamental domain** for the action of Γ on X is a connected polygon $\Delta \subset X$ such that as γ ranges over all elements of Γ, the polygons $\gamma(\Delta)$ are all distinct and form a tessellation of X.

The tessellations that we constructed in Chapter 6 provide many examples of fundamental domains. For instance, Figures 6.4 and 6.5 provide two different fundamental domains for the same group Γ acting isometrically on the euclidean plane $(\mathbb{R}^2, d_{\mathrm{euc}})$, and consisting of all integral translations $(x, y) \mapsto (x + m, y + n)$ with $m, n \in \mathbb{Z}$; see Exercise 7.7.

We will see that fundamental domains can be very useful for proving that an action is discontinuous and for determining the geometry of the quotient space X/Γ.

7.3.1. Fundamental domains and discontinuity.

Proposition 7.10. *If the isometric action of* Γ *on* $(X, d) = (\mathbb{R}^2, d_{\mathrm{euc}})$, $(\mathbb{H}^2, d_{\mathrm{hyp}})$ *or* $(\mathbb{S}^2, d_{\mathrm{sph}})$ *admits a fundamental domain* Δ, *the action of* Γ *is discontinuous.*

Proof. Consider $P \in X$. By the Local Finiteness Condition in the definition of tessellations, there exists an $\varepsilon > 0$ and a finite subset $\{\gamma_1, \gamma_2, \ldots, \gamma_n\} \subset \Gamma$ such that the γ_i are the only $\gamma \in \Gamma$ for which $\gamma(\Delta)$ meets the ball $B_d(P, \varepsilon)$. Without loss of generality, P is in the polygon $\gamma_1(\Delta)$.

If $\gamma \in \Gamma$ is such that $\gamma(P) \in B_d(P, \varepsilon)$, then $\gamma \circ \gamma_1(\Delta)$ contains $\gamma(P)$, and consequently meets the ball $B_d(P, \varepsilon)$. Therefore, $\gamma \circ \gamma_1 = \gamma_i$ and $\gamma = \gamma_i \circ \gamma_1^{-1}$ for some $i = 1, 2, \ldots, n$. It follows that there are only finitely many such $\gamma \in \Gamma$. $\qquad\square$

In particular, by Proposition 7.6, there is a well-defined quotient metric space $(\bar{X}, \bar{d}) = (X/\Gamma, \bar{d})$ when Γ admits a fundamental domain Δ. The next section describes this quotient space in terms of edge gluings of the polygon Δ.

7.3.2. Fundamental domains and quotient spaces.
Let Δ be a fundamental domain for the isometric action of Γ on $(X, d) = (\mathbb{R}^2, d_{\mathrm{euc}})$, $(\mathbb{H}^2, d_{\mathrm{hyp}})$ or $(\mathbb{S}^2, d_{\mathrm{sph}})$. Let $E_1, E_2, \ldots, E_n$ be its edges. In the tiling of X by the images of Δ under the elements of Γ, each edge E_j separates Δ from some other tile $\gamma_j(\Delta)$ and coincides with the image $\gamma_j(E_{i_j})$ of an edge E_{i_j}, possibly with $E_{i_j} = E_j$. Note that E_{i_j} separates Δ from $\gamma_j^{-1}(\Delta)$, and is equal to $\gamma_j^{-1}(E_j)$, so that $\gamma_{j_i} = \gamma_j^{-1}$. As a consequence, the rule $j \mapsto i$ defines a bijection of $\{1, 2, \ldots, n\}$.

In this situation, it is more convenient to write $j = j_i$ and $\varphi_i = \gamma_{j_i} = \gamma_i^{-1}$. In particular, φ_i is an element of Γ, and sends the edge E_i to the edge E_{j_i}.

Theorem 7.11. *Let Δ be a fundamental domain for the isometric action of the group Γ over $(X, d) = (\mathbb{R}^2, d_{\text{euc}})$, $(\mathbb{H}^2, d_{\text{hyp}})$ or $(\mathbb{S}^2, d_{\text{sph}})$. As above, consider for each edge E_i of Δ the element $\varphi_i \in \Gamma$ sending E_i to some other edge E_{j_i} (possibly equal to E_i). Then the group Γ is generated by the φ_i.*

Proof. Let $\gamma \in \Gamma$. Let g be an oriented geodesic arc in (X, d) going from a point P in the interior of Δ to a point Q in the interior of $\gamma(\Delta)$.

We claim that g can meet only finitely may vertices and finitely many edges of the tessellation of X by the images of Δ under the elements of Γ. This follows from the fact that as a bounded closed subset of $X = \mathbb{R}^2$, $\mathbb{H}^2$ or $\mathbb{S}^2$, the geodesic g is compact (see Theorems 6.13, 6.15 or 6.16). If g met infinitely many edges, looking at the intersection points of g with these edges and extracting a converging subsequence, the limit of that subsequence would contradict the Local Finiteness Condition in the definition of tessellations.

In particular, g meets only finitely many vertices of the tessellation. By slightly moving g (and its endpoints P and Q), we can arrange that it actually meets no vertex of the tessellation.

Let Δ, $\gamma_1(\Delta)$, $\gamma_2(\Delta)$, $\ldots$, $\gamma_{m-1}(\Delta)$, $\gamma_m(\Delta) = \gamma(\Delta)$ be the tiles of the tessellation traversed by g, in this order. It is quite possible that two γ_k are equal when the fundamental domain Δ is not convex. Let E_{i_k} and E_{j_k} be the edges such that the geodesic g enters the tile $\gamma_k(\Delta)$ by the edge $\gamma_k(E_{i_k})$ and exits it by the edge $\gamma_k(E_{j_k})$. In particular, $\gamma_k(E_{i_k}) = \gamma_{k-1}(E_{j_{k-1}})$, with the convention that $\gamma_0 = \text{Id}_X$.

By construction, φ_{i_k} is the unique element of Γ that sends Δ to a tile $\varphi_{i_k}(\Delta)$ adjacent to Δ in such a way that $\varphi_{i_k}(E_{i_k})$ coincides with another edge of Δ. It follows that $\varphi_{i_k} = \gamma_{k-1}^{-1} \circ \gamma_k$, namely, $\gamma_k = \gamma_{k-1} \circ \varphi_{i_k}$.

By induction,

$$\gamma = \gamma_m = \varphi_{i_1} \circ \varphi_{i_2} \circ \cdots \circ \varphi_{i_m}.$$

Since this holds for every $\gamma \in \Gamma$, this proves that Γ is generated by the φ_i. $\qquad\square$

The isometries φ_i, sending an edge E_i of the polygon Δ to another edge E_{j_i}, define a gluing data of the type considered in Sections 4.3 and 4.5. However, note that it is quite possible that $j_i = i$, so that the edge E_i may actually be glued to itself as in Section 6.3.4.

Let $(\widehat{\Delta}, \widehat{d}_\Delta)$ be the quotient space obtained from the polygon Δ by performing the edge gluings specified by the isometries φ_i. Here, we are using hats $\widehat{\ }$ instead of bars $\overline{\ }$ to distinguish objects in $(\widehat{\Delta}, \widehat{d}_\Delta)$ from elements of the other quotient metric space $(\bar{X}, \bar{d}) = (X/\Gamma, \bar{d})$, the quotient of (X, d) by the action of the group Γ. In particular, a point $P \in \Delta$ defines points $\widehat{P} \in \widehat{\Delta}$ and $\bar{P} \in \bar{X}$ in each of these two quotient spaces.

Theorem 7.12. *Let the group Γ act by isometries and discontinuously on $(X, d) = (\mathbb{R}^2, d_{\mathrm{euc}})$, $(\mathbb{H}^2, d_{\mathrm{hyp}})$ or $(\mathbb{S}^2, d_{\mathrm{sph}})$, and let Δ be a fundamental domain for the action of Γ. Then, for the above definitions, the space $(\widehat{\Delta}, \widehat{d}_\Delta)$ obtained by gluing edges of Δ is isometric to the quotient space $(X/\Gamma, \bar{d})$ of (X, d) by the action of Γ.*

Proof. By definition of the gluing process, if two points P and $Q \in \Delta$ give the same point $\widehat{P} = \widehat{Q}$ in the quotient space $\widehat{\Delta}$, then Q is the image of P under a composition of gluing maps φ_i (none if P is an interior point of Δ, exactly one φ_i if P and Q belong to edges but are not vertices, and possibly several φ_i when P and Q are vertices). As a consequence, when P and Q define the same point in $\widehat{\Delta}$, there exists an element $\gamma \in \Gamma$ such that $Q = \gamma(P)$, so that P and Q also define the same element $\bar{P} = \bar{Q}$ in X/Γ.

We can therefore define a map $\rho \colon \widehat{\Delta} \to X/\Gamma$ by associating $\bar{P} \in X/\Gamma$ to $\widehat{P} \in \widehat{\Delta}$. Indeed, the above observation shows that $\bar{P} \in X/\Gamma$ does not depend of the point P that we used to represent $\widehat{P} \in \widehat{\Delta}$.

To show that ρ is surjective, consider an element $\bar{P} \in X/\Gamma$, represented by $P \in X$. Because the images $\gamma(\Delta)$ of the fundamental domain Δ under the elements γ of Γ cover all of X, there exists a point $P' \in \Delta$ and a group element $\gamma \in \Gamma$ such that $P = \gamma(P')$. Then,

$\bar{P}$ is equal to $\bar{P}'$ in the quotient space X/Γ, which itself is the image under ρ of the element $\widehat{P}' \in \widehat{\Delta}$. This proves that ρ is surjective.

To prove that ρ is injective, suppose that P, $Q \in \Delta$ are such that $\rho(\widehat{P}) = \rho(\widehat{Q})$, namely, such that $\bar{P} = \bar{Q}$ in X/Γ. This property means that there exists $\gamma \in \Gamma$ such that $P = \gamma(Q)$. In particular, the two tiles Δ and $\gamma(\Delta)$ meet at P. If P is in interior point of Δ, this is possible only if $\gamma = \mathrm{Id}_X$, so that P is equal to Q. If P is in an edge and is not a vertex, then by definition γ is one of the gluing maps φ_i, so that P and $Q = \varphi_i(P)$ define the same point $\widehat{P} = \widehat{Q}$ in $\widehat{\Delta}$. Finally, if P is a vertex, let $\Delta = \gamma_0(\Delta)$, $\gamma_1(\Delta)$, $\gamma_2(\Delta)$, ..., $\gamma_k(\Delta) = \gamma(\Delta)$ be a sequence of tiles going from Δ to $\gamma(\Delta)$ around P. There are two possible such sequences, according to the direction in which one turns around P. By construction, each tile $\gamma_j(\Delta)$ meets $\gamma_{j+1}(\Delta)$ along an edge $\gamma_{j+1}(E_{i_j})$ near P, so that $\gamma_j^{-1} \circ \gamma_{j+1}$ is equal to the gluing map φ_{i_j}. In particular, the two vertices $\gamma_j^{-1}(P)$ and $\gamma_{j+1}^{-1}(P)$ are glued together by the gluing map φ_{i_j}. As a consequence, the element $\widehat{P} \in \widehat{\Delta}$ contains the vertices $P = \gamma_0^{-1}(P)$, $\gamma_1^{-1}(P)$, ..., $\gamma_{k-1}^{-1}(P)$, $\gamma_k^{-1}(P) = \gamma^{-1}(P) = Q$ of Δ, so that $\widehat{P} = \widehat{Q}$.

In all cases we therefore showed that if two points P and $Q \in \Delta$ define the same point $\bar{P} = \bar{Q} \in X/\Gamma$, then they also correspond to the same point $\widehat{P} = \widehat{Q}$ in the quotient space $\widehat{\Delta}$. This proves that ρ is injective, and therefore bijective.

Since Δ is a subset of X, every discrete walk w from $\widehat{P}$ to $\widehat{Q} \in \widehat{\Delta}$ in Δ is also a discrete walk from $\bar{P}$ to $\bar{Q} \in X/\Gamma$ in X, and $\ell_d(w) \leqslant \ell_{d_\Delta}(w)$ since $d(P', Q') \leqslant d_\Delta(P', Q')$ for every P', $Q' \in \Delta$. It follows that $\bar{d}(\bar{P}, \bar{Q}) \leqslant \widehat{d}_\Delta(\widehat{P}, \widehat{Q})$.

Conversely, consider P, $Q \in \Delta \subset X$. For $\varepsilon > 0$, Proposition 7.6 provides a $\gamma \in \Gamma$ such that $d(P, \gamma(Q)) < \bar{d}(\bar{P}, \bar{Q}) + \varepsilon$. As in the proof of Theorem 7.11, let g be a geodesic arc going from P to $\gamma(Q)$ in X, and meeting the tiles $\Delta = \gamma_0(\Delta)$, $\gamma_1(\Delta)$, $\gamma_2(\Delta)$, ..., $\gamma_{m-1}(\Delta)$, $\gamma_m(\Delta) = \gamma(\Delta)$ in this order. This time, g is allowed to cross the vertices of the tiling.

For every i, let P_i and $Q_i \in \Delta$ be such that the piece of g crossing $\gamma_i(\Delta)$ goes from $\gamma_i(P_i)$ to $\gamma_i(Q_i)$. Note that $d_\Delta(P_i, Q_i) = d(P_i, Q_i)$ since these two points are joined by a geodesic contained in Δ. Also

$\gamma_{i+1}(P_{i+1}) = \gamma_i(Q_i)$, so that $\bar{P}_{i+1} = \bar{Q}_i$ in X/Γ; it follows that P_{i+1} and Q_i are glued together in $\widehat{\Delta}$ since we just proved that ρ is injective. Similarly, the last point Q_m is glued to Q in $\widehat{\Delta}$. Therefore, the sequence $P = P_0, Q_0 \sim P_1, Q_1 \sim P_2, \ldots, Q_{m-1} \sim P_m, Q_m \sim Q$ forms a discrete walk w from $\widehat{P}$ to $\widehat{Q}$ in $\widehat{\Delta}$, whose d_Δ-length is

$$\ell_{d_\Delta}(w) = \sum_{i=0}^{m} d_\Delta(P_i, Q_i) = \sum_{i=0}^{m} d(P_i, Q_i) = \sum_{i=0}^{m} d\big(\gamma_i(P_i), \gamma_i(Q_i)\big)$$
$$= \ell(g) = d\big(P, \gamma(Q)\big) < \bar{d}(\bar{P}, \bar{Q}) + \varepsilon,$$

where $\ell(g)$ is the euclidean, hyperbolic or spherical arc length according to whether $X = \mathbb{R}^2$, $\mathbb{H}^2$ or $\mathbb{S}^2$. It follows that $\widehat{d}_\Delta(\widehat{P}, \widehat{Q}) < \bar{d}(\bar{P}, \bar{Q}) + \varepsilon$ for every ε, and therefore that $\widehat{d}_\Delta(\widehat{P}, \widehat{Q}) \leqslant \bar{d}(\bar{P}, \bar{Q})$.

Since we had already proved the reverse inequality, this shows that $\widehat{d}_\Delta(\widehat{P}, \widehat{Q}) = \bar{d}(\bar{P}, \bar{Q})$ for every $\widehat{P}, \widehat{Q} \in \widehat{\Delta}$, and completes the proof that $\rho\colon \widehat{\Delta} \to X/\Gamma$, which associates $\bar{P} \in X/\Gamma$ to $\widehat{P} \in \widehat{\Delta}$, is an isometry from $(\widehat{\Delta}, \widehat{d}_\Delta)$ to $(X/\Gamma, \bar{d})$. $\qquad\square$

7.4. Dirichlet domains

We now prove a converse to Proposition 7.10 by showing that for any discontinuous group Γ of isometries of the euclidean plane $(\mathbb{R}^2, d_{\text{euc}})$, the hyperbolic plane $(\mathbb{H}^2, d_{\text{hyp}})$ or the sphere $(\mathbb{S}^2, d_{\text{sph}})$, there always exists a fundamental domain for the action of Γ. We will not use this property until Section 12.4, where a variation of Dirichlet domains, called Ford domains, will play an important role.

This converse will hold provided we slightly extend our definition of polygons in $(X, d) = (\mathbb{R}^2, d_{\text{euc}})$, $(\mathbb{H}^2, d_{\text{hyp}})$ or $(\mathbb{S}^2, d_{\text{sph}})$. Namely, To page 201 we will allow a polygon Δ to have infinitely many edges and vertices, provided that the families of edges and vertices are *locally finite* in the following sense: For every $P \in \Delta$, there is a ball $B_d(P, \varepsilon)$ centered at P which meets only finitely many edges and vertices of Δ. Such a polygon will be called a *locally finite polygon* to distinguish this notion from the finite polygons considered so far. By compactness of $\mathbb{S}^2$, one easily sees that a locally finite polygon in the sphere $\mathbb{S}^2$ can have only finitely many edges and vertices, and consequently is finite.

If the group Γ of isometries of the metric space (X, d) acts discontinuously on X, the **Dirichlet domain** of Γ centered at the point $P_0 \in X$ is the subset

$$\Delta_\Gamma(P_0) = \{P \in X; d(P, P_0) \leqslant d(P, \gamma(P_0)) \text{ for every } \gamma \in \Gamma\}.$$

In other words, the Dirichlet domain centered at P_0 consists of those points that are at least as close to P_0 as to any other points of its orbit. It is named after Gustav Lejeune Dirichlet (1805–1859), who introduced a similar construction to study the number-theoretic properties of quadratic forms with integer coefficients. Dirichlet domains and their variations are ubiquitous in mathematics, where they arise under many different names. For instance, if we replace the orbit of P_0 by an arbitrary locally finite subset $A \subset X$, the similarly defined subset of X is called the Voronoi domain of A centered at $P_0 \in A$. See also the Ford domains that we will consider in Section 12.4.

Theorem 7.13. *Let the group Γ act by isometries and discontinuously on $(X, d) = (\mathbb{R}^2, d_{\mathrm{euc}})$, $(\mathbb{H}^2, d_{\mathrm{hyp}})$ or $(\mathbb{S}^2, d_{\mathrm{sph}})$. Then, for every $P_0 \in X$, the Dirichlet domain $\Delta_\Gamma(P_0)$ is a locally finite polygon and, as γ ranges over all the elements of the group Γ, the $\gamma(\Delta_\Gamma(P_0))$ form a tessellation of X.*

If, in addition, the point P_0 is fixed by no element of Γ except for the identity, then the Dirichlet domain $\Delta_\Gamma(P_0)$ is a fundamental domain for the action of Γ.

We will split the proof of Theorem 7.13 into several lemmas. The following elementary construction is the key geometric ingredient.

Lemma 7.14. *Let P and Q be two distinct points in $(X, d) = (\mathbb{R}^2, d_{\mathrm{euc}})$, $(\mathbb{H}^2, d_{\mathrm{hyp}})$ or $(\mathbb{S}^2, d_{\mathrm{sph}})$. Then the set of $R \in X$ such that $d(P, R) = d(Q, R)$ is a complete geodesic β_{PQ} of X. In addition, the set of $R \in X$ such that $d(P, R) \leqslant d(Q, R)$ is the closed half-space delimited by β_{PQ} in X and containing P.*

Proof. This is immediate by elementary geometry in the euclidean and spherical case. See Exercise 2.4 for the hyperbolic case.   □

By analogy with the euclidean case, the geodesic β_{PQ} is the **perpendicular bisector** of the points P and Q.

Lemma 7.15. *If Γ acts by isometry and discontinuously on $(X,d) = (\mathbb{R}^2, d_{\mathrm{euc}})$, $(\mathbb{H}^2, d_{\mathrm{hyp}})$ or $(\mathbb{S}^2, d_{\mathrm{sph}})$ then, for every $P \in X$ and $r > 0$, there are only finitely many $\gamma \in \Gamma$ such that $d\big(P, \gamma(P_0)\big) \leqslant r$.*

Proof. Suppose, in search of a contradiction, that there are infinitely many elements $\gamma_1, \gamma_2, \ldots, \gamma_n, \ldots$ in Γ such that all $\gamma_n(P_0)$ belong to the closed ball B of radius r centered at P, consisting of all $Q \in X$ such that $d(P, Q) \leqslant r$.

One easily checks that B is a closed subset of X. Since it is clearly bounded, it follows from Theorems 6.13, 6.15 and 6.16 that B is compact. As a consequence, there exists a subsequence $(\gamma_{n_k})_{k \in \mathbb{N}}$ such that $\gamma_{n_k}(P_0)$ converges to some $P_\infty \in B$ as k tends to ∞.

This implies that for every ε, there exists a k_0 such that $d\big(\gamma_{n_k}(P_0), P_\infty\big) < \frac{\varepsilon}{2}$ for every $k \geqslant k_0$. In particular, for every $k \geqslant k_0$,

$$
\begin{aligned}
d\big(P_0, \gamma_{n_{k_0}}^{-1} \circ \gamma_{n_k}(P_0)\big) &= d\big(\gamma_{n_{k_0}}(P_0), \gamma_{n_k}(P_0)\big) \\
&\leqslant d\big(\gamma_{n_{k_0}}(P_0), P_\infty\big) + d\big(P_\infty, \gamma_{n_k}(P_0)\big) \\
&< \tfrac{\varepsilon}{2} + \tfrac{\varepsilon}{2} = \varepsilon.
\end{aligned}
$$

So, for every $\varepsilon > 0$, we found infinitely many $\gamma \in \Gamma$ for which $\gamma(P_0)$ is in the ball $B_d(P_0, \varepsilon)$, contradicting the fact that the action of Γ is discontinuous.

This proves that our original assumption was false and therefore that B contains only finitely many $\gamma(P)$. $\square$

Lemma 7.16. *The Dirichlet domain $\Delta_\Gamma(P_0)$ is a locally finite polygon.*

Proof. For every $\gamma \in \Gamma$, let $H_{P_0 \gamma(P_0)}$ be the set of points $P \in X$ such that $d(P, P_0) \leqslant d(P, \gamma(P_0))$. Lemma 7.14 shows that if $\gamma(P_0) \neq P_0$, $H_{P_0 \gamma(P_0)}$ is a half-space delimited by the perpendicular bisector $\beta_{P_0 \gamma(P_0)}$ of P_0 and $\gamma(P_0)$. If $\gamma(P_0) = P_0$, $H_{P_0 \gamma(P_0)}$ is of course the whole space X and we set $\beta_{P_0 P_0}$ to be the empty set in this case.

By definition, the Dirichlet domain $\Delta_\Gamma(P_0)$ is the intersection of the half-spaces $H_{P_0 \gamma(P_0)}$ as γ ranges over all elements of Γ. To prove that $\Delta_\Gamma(P_0)$ is a locally finite polygon, it consequently suffices to show that the family of geodesics $\beta_{P_0 \gamma(P_0)}$ bounding the $H_{P_0 \gamma(P_0)}$ is locally finite. Namely, for every $P \in X$, we need to find an ε such that the ball $B_d(P, \varepsilon)$ meets only finitely many $\beta_{P_0\gamma(P_0)}$.

Actually, any $\varepsilon > 0$ will do. If $\beta_{P_0\gamma(P_0)} \cap B_d(P, \varepsilon)$ is nonempty, pick a point Q in this intersection. By the Triangle Inequality, the point Q is at distance $\leqslant d(P, P_0) + \varepsilon$ from P_0. Since $Q \in \beta_\gamma$, it is also at the same distance from $\gamma(P_0)$ as from P_0. Another application of the Triangle Inequality then shows that $\gamma(P_0)$ is at distance $\leqslant 2d(P, P_0) + 2\varepsilon$ from P_0. By Lemma 7.15, this can happen for only finitely many $\gamma \in \Gamma$.

This proves that the family of the perpendicular bisectors $\beta_{P_0\gamma(P_0)}$ is finite, and therefore that the Dirichlet domain $\Delta_\Gamma(P)$ is a locally finite polygon. □

Before proving that the images of the Dirichlet domain $\Delta_\Gamma(P_0)$ under the elements of Γ tessellate X, let us first observe that each of these images is also a Dirichlet domain.

Lemma 7.17. *For every* $\gamma \in \Gamma$, $\gamma\big(\Delta_\Gamma(P_0)\big) = \Delta_\Gamma\big(\gamma(P_0)\big)$.

Proof. The Dirichlet domain $\Delta_\Gamma\big(\gamma(P_0)\big)$ consists of those $P \in X$ which are at least as close to $\gamma(P_0)$ as to any other point of the orbit $\Gamma(\gamma(P_0)) = \Gamma(P_0)$. Since γ is an isometry of X, this property is equivalent to the fact that $\gamma^{-1}(P)$ is at least as close to P_0 as to any other point of its orbit. Therefore, P is in $\Delta_\Gamma\big(\gamma(P_0)\big)$ if and only if $\gamma^{-1}(P)$ is in $\Delta_\Gamma(P_0)$, namely, if and only if P is in $\gamma\big(\Delta_\Gamma(P_0)\big)$. □

Lemma 7.18. *As γ ranges over all the elements of Γ, the Dirichlet domains $\Delta_\Gamma\big(\gamma(P_0)\big)$ form a tessellation of X.*

Proof. A point $P \in X$ belongs to $\Delta_\Gamma\big(\gamma(P_0)\big)$ if and only if $d\big(P, \gamma(P_0)\big)$ is equal to

$$\inf\{d(P, \gamma'(P_0); \gamma' \in \Gamma\} = \bar{d}(\bar{P}, \bar{P}_0),$$

where the equality comes from Proposition 7.6. Lemma 7.15 shows that there are only finitely many $\gamma' \in \Gamma$ such that $d\big(P, \gamma'(P_0)\big) < \bar{d}(\bar{P}, \bar{P}_0) + 1$. This has two consequences.

The first one is that there exists at least one γ such that $d\big(P, \gamma(P_0)\big) = \bar{d}(\bar{P}, \bar{P}_0)$. Indeed, the above observation shows that it suffices to consider the infimum over finitely many elements, so that the infimum is actually a minimum. As a consequence, there exists $\gamma \in \Gamma$ such that $P \in \Delta\big(\gamma(P_0)\big)$.

The second consequence is that the set of $\gamma \in \Gamma$ such that $P \in \Delta\big(\gamma(P_0)\big)$ is a finite set $\{\gamma_1, \gamma_2, \ldots, \gamma_n\}$. In particular, P belongs to only finitely many Dirichlet domains $\Delta_\Gamma\big(\gamma(P_0)\big)$ and, near P, these $\Delta_\Gamma\big(\gamma_i(P_0)\big)$ are delimited by the perpendicular bisectors $\beta_{\gamma_i(P_0)\gamma_j(P_0)}$.

These two properties show that the union of the domains $\Delta_\Gamma\big(\gamma(P_0)\big)$ is equal to X and that when two distinct $\Delta_\Gamma\big(\gamma(P_0)\big)$ and $\Delta_\Gamma\big(\gamma'(P_0)\big)$ meet, they meet along edges and/or vertices. Since any two of these Dirichlet domains are isometric by Lemma 7.17, this completes the proof. $\qquad\square$

The combination of Lemmas 7.17 and 7.18 shows that the images of the Dirichlet domain $\Delta_\Gamma(P_0)$ under the action of Γ form a tessellation of X. This was the main statement of Theorem 7.13.

To prove the second part of this theorem, assume that the stabilizer Γ_{P_0} consists only of the identity map. Then, whenever $\gamma \neq \gamma'$, the two points $\gamma(P_0)$ and $\gamma'(P_0)$ are distinct so that the Dirichlet domains $\Delta_\Gamma\big(\gamma(P_0)\big) = \gamma\big(\Delta_\Gamma(P_0)\big)$ and $\Delta_\Gamma\big(\gamma'(P_0)\big) = \gamma'\big(\Delta_\Gamma(P_0)\big)$ are distinct. This is exactly the additional condition needed to prove that $\Delta_\Gamma(P_0)$ is a fundamental domain.

This concludes the proof of Theorem 7.13. $\qquad\square$

For Theorem 7.13 to provide a (locally finite) fundamental domain for the action of Γ, we need to find a point P_0 whose stabilizer Γ_{P_0} consists only of the identity. This of course is automatic when the action is free. In the general case, almost every $P_0 \in X$ will have the required property. Indeed, the analysis of all possible stabilizers in Exercise 7.12 (suitably extended to the euclidean and spherical contexts) shows that every point of X can be approximated by a point with a trivial stabilizer.

From page 197

We will let you check that the results of Sections 4.3, 4.4, 4.5, 6.3 and 7.3 on edge gluings of polygons immediately extend from finite polygons to locally finite polygons Δ in $X = \mathbb{R}^2$ or $\mathbb{H}^2$, provided that we impose the following additional condition.

Finite Gluing Condition. *Each point of the polygon Δ is glued to only finitely many other points of Δ.*

Exercises for Chapter 7

Exercise 7.1. For every rational number $\frac{p}{q}$, let $\gamma_{\frac{p}{q}}\colon \mathbb{R} \to \mathbb{R}$ be the translation defined by $\gamma_{\frac{p}{q}}(x) = x + \frac{p}{q}$.

  a. Show that the set $\Gamma = \{\gamma_{\frac{p}{q}}; \frac{p}{q} \in \mathbb{Q}\}$ of all such rational translations is a goup of isometries of the metric space $(\mathbb{R}, d_{\mathrm{euc}})$.

  b. Show that the quotient semi-metric $\bar{d}$ induced by $d = d_{\mathrm{euc}}$ on the quotient space $\bar{X} = X/\Gamma$ is not a metric. Actually, compute $\bar{d}(\bar{P}, \bar{Q})$ for every $\bar{P}, \bar{Q} \in \bar{X}$.

Exercise 7.2. This exercise is devoted to a few classical transformation groups that arise in linear algebra.

  a. Let the **general linear group** $\mathrm{GL}_n(\mathbb{R})$ consist of all the linear maps $\mathbb{R}^n \to \mathbb{R}^n$ whose associated matrix M has nonzero determinant $\det(M) \neq 0$. Show that $\mathrm{GL}_n(\mathbb{R})$ is a transformation group of $\mathbb{R}^n$. Hint: It may be useful to remember that any $n \times n$-matrix M with $\det(M) \neq 0$ admits an inverse, and that $\det(MN) = \det(M)\det(N)$.

  b. Let the **special linear group** $\mathrm{SL}_n(\mathbb{R})$ consist of all the linear maps $\mathbb{R}^n \to \mathbb{R}^n$ whose associated matrix M has determinant $\det(M)$ equal to 1. Show that $\mathrm{SL}_n(\mathbb{R})$ is a transformation group of $\mathbb{R}^n$.

  c. Let the **projective space** $\mathbb{RP}^{n-1}$ consist of all the lines passing through the origin in $\mathbb{R}^n$ (compare Exercise 2.12). Every $\varphi \in \mathrm{GL}_n(\mathbb{R})$ induces a map $\bar{\varphi}\colon \mathbb{RP}^{n-1} \to \mathbb{RP}^{n-1}$, which associates to each line $L \in \mathbb{RP}^{n-1}$ the line $\varphi(L) \in \mathbb{RP}^{n-1}$. Show that as φ ranges over all elements of $\mathrm{GL}_n(\mathbb{R})$, the corresponding maps $\bar{\varphi}$ form a transformation group of $\mathbb{RP}^{n-1}$. This group is called the **projective general linear group** and is traditionally denoted by $\mathrm{PGL}_n(\mathbb{R})$.

  d. Let the **projective special linear group** $\mathrm{PSL}_n(\mathbb{R})$ consist of all the maps $\bar{\varphi}\colon \mathbb{RP}^{n-1} \to \mathbb{RP}^{n-1}$ induced as in part c by linear maps $\varphi \in \mathrm{SL}_n(\mathbb{R})$. Show that $\mathrm{PSL}_n(\mathbb{R})$ is a transformation group of $\mathbb{RP}^{n-1}$.

Exercise 7.3 (Abstract groups). A **group law** on a set Γ is a map $\Gamma \times \Gamma \to \Gamma$, denoted by $(\gamma, \gamma') \mapsto \gamma \cdot \gamma' \in \Gamma$ for every $\gamma, \gamma' \in \Gamma$, such that:

  a. $\gamma \cdot (\gamma' \cdot \gamma'') = (\gamma \cdot \gamma') \cdot \gamma''$ for every $\gamma, \gamma', \gamma'' \in \Gamma$;

  b. there exists an element $\iota \in \Gamma$ such that $\gamma \cdot \iota = \iota \cdot \gamma = \gamma$ for every $\gamma \in \Gamma$;

  c. for every $\Gamma \in \Gamma$, there exists an element $\gamma' \in \Gamma$ such that $\gamma \cdot \gamma' = \gamma' \cdot \gamma = \iota$.

Show that if Γ is a transformation group on a set X, the map $\Gamma \times \Gamma \to \Gamma$ defined by $(\gamma, \gamma') \mapsto \gamma \circ \gamma'$ is a group law on Γ.

An **abstract group** is the data of a set Γ and of a group law on Γ.

Exercise 7.4. Let the group Γ act discontinuously and by isometries on the metric space (X, d). Show that if X is compact, Γ is necessarily finite.

Exercise 7.5. Let the group Γ act by isometries on $(X, d) = (\mathbb{R}^2, d_{\mathrm{euc}})$, $(\mathbb{H}^2, d_{\mathrm{hyp}})$ or $(\mathbb{S}^2, d_{\mathrm{sph}})$. Show that if Γ acts discontinuously at *some* point $P_0 \in X$, in the sense that there exists an $\varepsilon > 0$ such that $\gamma(P) \in B_d(P_0, \varepsilon)$ for only finitely many $\gamma \in \Gamma$, then it acts discontinuously at *all* points $P \in X$. Possible hint: Borrow ideas from the proof of Lemma 7.15.

Exercise 7.6. Let the group Γ act discontinuously on the complete metric space (X, d). We want to show that the quotient space $(X/\Gamma, \bar{d})$ is complete. For this, let $(\bar{P}_n)_{n \in \mathbb{N}}$ be a sequence in this quotient space $\bar{X} = X/\Gamma$ with finite length $\sum_{n=1}^{\infty} \bar{d}(\bar{P}_n, \bar{P}_{n+1}) < \infty$.

    a. By induction on n, construct a sequence $(P'_n)_{n \in \mathbb{N}}$ in X such that for every n, $\bar{P}'_n = \bar{P}_n$ and $d(P'_n, P'_{n+1}) \leqslant \bar{d}(\bar{P}_n, \bar{P}_{n+1}) + 2^{-n}$. Hint: Proposition 7.6.

    b. Show that the sequence $(P'_n)_{n \in \mathbb{N}}$ converges to some point $P_\infty \in X$.

    c. Conclude that $(\bar{P}_n)_{n \in \mathbb{N}}$ converges to $\bar{P}_\infty$ in $(X/\Gamma, \bar{d})$.

Exercise 7.7. Let Γ be the group of integral translations of the euclidean plane (R^2, d_{euc}), of the form $(x, y) \mapsto (x + m, y + n)$ with $(m, n) \in \mathbb{Z}^2$.

    a. Let Y be the parallelogram with vertices $(0, 0)$, $(1, 1)$, $(2, 1)$ and $(1, 0)$, as in Figure 6.5. Prove that Y is a fundamental domain for the action of Γ. Possible hint: Glue opposite sides of Y, apply Theorem 6.1, and check that the corresponding tiling group is equal to Γ.

    b. Do the same for the parallelogram whose vertices are $(0, 0)$, $(1, 1)$, $(3, 2)$ and $(2, 1)$.

Exercise 7.8. Let Γ be the group of isometries of $(\mathbb{S}^2, d_{\mathrm{sph}})$ consisting of the identity map $\mathrm{Id}_{\mathbb{S}^2}$ and of the antipode map γ defined by $\gamma(P) = -P$.

    a. Check that Γ is really a transformation group acting discontinuously on $\mathbb{S}^2$.

    b. Show that the quotient space $(\mathbb{S}^2/\Gamma, \bar{d}_{\mathrm{sph}})$ is isometric to the projective plane of Section 5.3. Hint: Find a fundamental domain for the action of Γ and apply Theorem 7.12.

Exercise 7.9. In the euclidean plane, consider the infinite strip $X_w = [0, w] \times (-\infty, +\infty)$ of width $w > 0$ and, for $t \in \mathbb{R}$, let $\bar{X}_{w,t}$ be the cylinder obtained from X_w by gluing the left-hand side $\{0\} \times (-\infty, +\infty)$ to the right-hand side $\{w\} \times (-\infty, +\infty)$ by the translation $\varphi \colon \{0\} \times (-\infty, +\infty) \to \{w\} \times (-\infty, +\infty)$ defined by $\varphi(0, y) = (w, y + t)$. Endow $\bar{X}_{w,t}$ with the quotient metric $\bar{d}_{\mathrm{euc}}$ induced by the euclidean metric d_{euc} of X_w.

    a. Show that $(\bar{X}_{w,t}, \bar{d}_{\mathrm{euc}})$ is isometric to a quotient space $(\mathbb{R}^2/\Gamma_{w,t}, \bar{d}_{\mathrm{euc}})$, where $\Gamma_{w,t}$ is the group of isometries of the euclidean plane $(\mathbb{R}^2, d_{\mathrm{euc}})$

generated by a translation along a certain vector of length $\sqrt{w^2 + t^2}$.
Hint: Theorems 6.1 and 7.12.

b. Suppose that $w_1^2 + t_1^2 = w_2^2 + t_2^2$. Show that the cylinders $(\bar{X}_{w_1,t_1}, \bar{d}_{\text{euc}})$ and $(\bar{X}_{w_2,t_2}, \bar{d}_{\text{euc}})$ are isometric. Possible hint: First find a euclidean isometry $\psi \colon (\mathbb{R}^2, d_{\text{euc}}) \to (\mathbb{R}^2, d_{\text{euc}})$ such that $\Gamma_{w_2,t_2} = \{\psi \circ \gamma \circ \psi^{-1}; \gamma \in \Gamma_{w_1,t_1}\}$, and then use ψ to construct and isometry $\bar{\psi} \colon (\mathbb{R}^2/\Gamma_{w_1,t_1}, \bar{d}_{\text{euc}}) \to (\mathbb{R}^2/\Gamma_{w_2,t_2}, \bar{d}_{\text{euc}})$.

c. Show that the cylinders $(\bar{X}_{w_1,0}, \bar{d}_{\text{euc}})$ and $(\bar{X}_{w_2,0}, \bar{d}_{\text{euc}})$ are not isometric if $w_1 \neq w_2$. Possible hint: Look at closed geodesics in these cylinders.

d. Show that the cylinders $(\bar{X}_{w_1,t_1}, \bar{d}_{\text{euc}})$ and $(\bar{X}_{w_2,t_2}, \bar{d}_{\text{euc}})$ are isometric if and only if $w_1^2 + t_1^2 = w_2^2 + t_2^2$.

Exercise 7.10. Let the group Γ act isometrically and discontinuously on the metric space (X, d). Let Δ be the Dirichlet domain of Γ at P. Show that Δ is invariant under the stabilizer Γ_P of P, namely that $\gamma(\Delta) = \Delta$ for every $\gamma \in \Gamma$ with $\gamma(P) = P$.

Exercise 7.11. For $a \in \mathbb{R}$, let Γ be the group of isometries of the euclidean plane $(\mathbb{R}^2, d_{\text{euc}})$ generated by the translations $\varphi_1 \colon (x, y) \mapsto (x + 1, y)$ and $\varphi_2 \colon (x, y) \mapsto (x + a, y + 1)$. Determine the Dirichlet domain $\Delta_{\Gamma_a}(O)$ of Γ_a centered at the origin $O = (0, 0)$. The answer will depend on a.

Exercise 7.12 (2-dimensional stabilizers). Let Γ act by isometries and discontinuously on the hyperbolic plane. The goal of the exercise is to determine the possible types for a stabilizer Γ_P. It will be convenient to use the ball model $(\mathbb{B}^2, d_{\mathbb{B}^2})$. Using a hyperbolic isometry φ sending P to O and replacing Γ by the group $\{\varphi \circ \gamma \circ \varphi^{-1}; \gamma \in \Gamma\}$, we can assume without loss of generality that P is the euclidean center O of the disk $\mathbb{B}^2$.

a. Show that every element of the stabilizer Γ_O is the restriction to $\mathbb{B}^2$ of either a euclidean rotation around O or a euclidean reflection across a line passing through O.

b. Suppose that every element of Γ_O is a rotation. Show that there exists an integer $n \geqslant 1$ so that Γ_O consists of the n rotations around O of angles $\frac{2k\pi}{n}$, with $k = 0, 1, \ldots, n - 1$. Hint: Consider an element of $\Gamma_O - \{\text{Id}_{\mathbb{H}^2}\}$ whose rotation angle is smallest.

c. Suppose that Γ_O contains at least one reflection. Show that there exists an integer $n \geqslant 1$ and a line L passing through the origin such that Γ_O consists of the n rotations around O of angles $\frac{2k\pi}{n}$, with $k = 0, 1, \ldots, n - 1$, and the n reflections across the lines obtained by rotating L around O by angles $\frac{k\pi}{n}$, with $k = 0, 1, \ldots, n - 1$. Hint: First consider the elements of Γ_O that are rotations.

Exercise 7.13. Let the group Γ act by isometries and discontinuously on the hyperbolic plane $(\mathbb{H}^2, d_{\mathrm{hyp}})$. Let P be a point whose stabilizer Γ_P contains more elements than just the identity map.

   a. Show that there is an angle sector A, delimited by two infinite geodesics issued from P, which is a fundamental domain for the action of Γ_P. It may be convenient to use Exercise 7.12.

   b. Let $\Delta_\Gamma(P)$ be the Dirichlet domain of Γ at P. Show that the intersection $A \cap \Delta_\Gamma(P)$ is a fundamental domain for the action of Γ.

Exercise 7.14. Let Γ act by isometries and discontinuously on the hyperbolic plane $(\mathbb{H}^2, d_{\mathrm{hyp}})$.

   a. Let Γ_P be the stabilizer of a point $P \in \mathbb{H}^2$, and choose an arbitrary $\varepsilon > 0$. Show that the quotient space $(B_{d_{\mathrm{hyp}}}(P, \varepsilon)/\Gamma_P, \bar{d}_{\mathrm{hyp}})$ is isometric, either to a hyperbolic cone of radius ε and cone angle $\frac{2\pi}{n}$ for some integer $n \geqslant 1$ (as defined in Exercise 4.7 for the euclidean context) or to a hyperbolic disk sector of radius ε with angle $\frac{\pi}{n}$ for some integer $n \geqslant 1$. Hint: Use Exercises 7.12 and 7.13.

   b. Assume, in addition, that every element of Γ is a linear fractional map. Show that the quotient space $(\mathbb{H}^2/\Gamma, \bar{d}_{\mathrm{hyp}})$ is a hyperbolic surface with cone singularities, as defined in Exercise 4.8. Hint: Use Theorem 7.8.

Exercises 7.12, 7.13 and 7.14 have immediate generalizations to the euclidean and spherical context.

Chapter 8

The Farey tessellation and circle packing

The Farey tessellation and circle packing rank among some of the most beautiful objects in mathematics. It turns out that they are closely related to the punctured torus that we encountered in Section 5.5, and to the tiling group of the corresponding tessellation constructed in Section 6.6. Some of the exercises at the end of this chapter explore additional connections between the Farey tessellation and various number-theoretic and combinatorial problems.

8.1. The Farey circle packing and tessellation

$$\infty = \tfrac{1}{0}$$

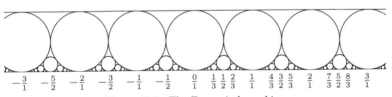

$$-\tfrac{3}{1} \quad -\tfrac{5}{2} \quad -\tfrac{2}{1} \quad -\tfrac{3}{2} \quad -\tfrac{1}{1} \quad -\tfrac{1}{2} \quad \tfrac{0}{1} \quad \tfrac{1}{3}\,\tfrac{1}{2}\,\tfrac{2}{3} \quad \tfrac{1}{1} \quad \tfrac{4}{3}\,\tfrac{3}{2}\,\tfrac{5}{3} \quad \tfrac{2}{1} \quad \tfrac{7}{3}\,\tfrac{5}{2}\,\tfrac{8}{3} \quad \tfrac{3}{1}$$

Figure 8.1. The Farey circle packing

For every rational number $\frac{p}{q} \in \mathbb{Q}$ with p, q coprime and $q > 0$, draw in the plane $\mathbb{R}^2$ the circle $C_{\frac{p}{q}}$ of diameter $\frac{1}{q^2}$ that is tangent to the x-axis at $(\frac{p}{q}, 0)$ and lies above this axis. These circles $C_{\frac{p}{q}}$ fit together to form the pattern illustrated in Figure 8.1.

We can get a better view of this circle pattern by zooming in, as in Figures 8.2 and 8.3.

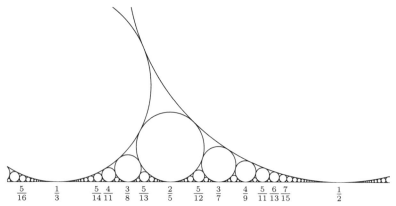

$\frac{5}{16}$ $\frac{1}{3}$ $\frac{5}{14}\frac{4}{11}$ $\frac{3}{8}\frac{5}{13}$ $\frac{2}{5}$ $\frac{5}{12}\frac{3}{7}$ $\frac{4}{9}\frac{5}{11}\frac{6}{13}\frac{7}{15}$ $\frac{1}{2}$

Figure 8.2. Zooming in on the Farey circle packing

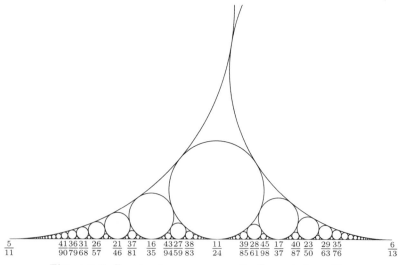

$\frac{5}{11}$ $\frac{41}{90}\frac{36}{79}\frac{31}{68}\frac{26}{57}$ $\frac{21}{46}\frac{37}{81}$ $\frac{16}{35}$ $\frac{43}{94}\frac{27}{59}\frac{38}{83}$ $\frac{11}{24}$ $\frac{39}{85}\frac{28}{61}\frac{45}{98}\frac{17}{37}$ $\frac{40}{87}\frac{23}{50}$ $\frac{29}{63}\frac{35}{76}$ $\frac{6}{13}$

Figure 8.3. Zooming in once more on the Farey circle packing

By inspection, we can make the following experimental observations:

- the circles $C_{\frac{p}{q}}$ have disjoint interiors;
- two circles $C_{\frac{p}{q}}$ and $C_{\frac{p'}{q'}}$ are tangent exactly when

$$pq' - p'q = \pm 1;$$

- three circles $C_{\frac{p}{q}}$, $C_{\frac{p'}{q'}}$ and $C_{\frac{p''}{q''}}$ with $\frac{p}{q} < \frac{p''}{q''} < \frac{p'}{q'}$ are tangent to each other exactly when $\frac{p''}{q''}$ is the **Farey sum** $\frac{p}{q} \oplus \frac{p'}{q'}$ of $\frac{p}{q}$ and $\frac{p'}{q'}$, namely, when

$$\frac{p''}{q''} = \frac{p}{q} \oplus \frac{p'}{q'} = \frac{p+p'}{q+q'}.$$

The same properties hold if we consider, in addition, the infinite rational number $\infty = \frac{1}{0} = \frac{-1}{0}$ and introduce C_∞ as the horizontal

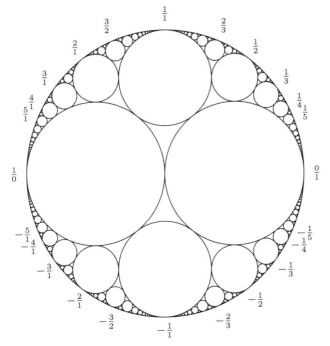

Figure 8.4. The Farey circle packing in the disk model $\mathbb{B}^2$

line of equation $y = 1$, while the interior of C_∞ is defined as the half-space consisting of those points (x, y) with $y > 1$.

The family of the circles $C_{\frac{p}{q}}$ is the **Farey circle packing**.

Some of the beauty of this collection of circles is also visible in Figure 8.4, which represents the image of the Farey circle packing in the disk model $\mathbb{B}^2$ for the hyperbolic plane, under the linear fractional map $\Phi(z) = -\frac{z-\mathrm{i}}{z+\mathrm{i}}$ sending the upper half-plane to the disk $\mathbb{B}^2$. The numbers in that picture label some of the images $\Phi(\frac{p}{q})$ of the corresponding points $\frac{p}{q} \in \mathbb{Q}$.

Quite remarkably, this Farey circle packing is related to the hyperbolic punctured torus that we considered in Section 5.5.

We can get a hint at this relationship if we erase the circles $C_{\frac{p}{q}}$, and if we connect the two points $(\frac{p}{q}, 0)$ and $(\frac{p'}{q'}, 0)$ by a semi-circle centered on the x-axis exactly when the circles $C_{\frac{p}{q}}$ and $C_{\frac{p'}{q'}}$ are tangent. The resulting collection of hyperbolic geodesics is illustrated in Figure 8.5, which bears a strong analogy with the tessellation of Figure 6.12 that we associated to our hyperbolic punctured torus. This connection is made precise in Section 8.2.

Figures 8.6 and 8.7 are obtained by zooming in on this collection of semi-circles, called the **Farey tessellation** of the hyperbolic plane.

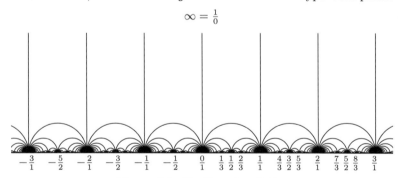

Figure 8.5. The Farey tessellation

The Farey circle packing and the Farey tessellation are named after John Farey (1766–1826), a geologist who experimentally observed (without proof) the following elementary property. For a fixed number $N > 0$, consider all rational numbers $\frac{p}{q}$ with $0 \leqslant p \leqslant q \leqslant N$

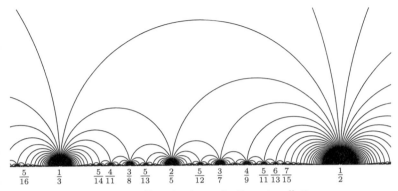

$$\frac{5}{16} \qquad \frac{1}{3} \qquad \frac{5}{14}\frac{4}{11} \quad \frac{3}{8} \quad \frac{5}{13} \quad \frac{2}{5} \quad \frac{5}{12} \quad \frac{3}{7} \quad \frac{4}{9} \quad \frac{5}{11}\frac{6}{13}\frac{7}{15} \qquad \frac{1}{2}$$

Figure 8.6. Zooming in on the Farey tessellation

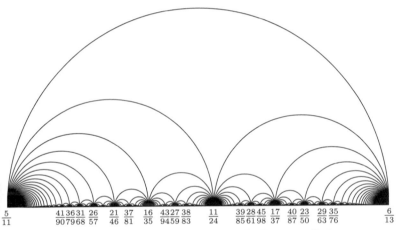

$$\frac{5}{11} \qquad \frac{41}{90}\frac{36}{79}\frac{31}{68}\frac{26}{57} \quad \frac{21}{46} \quad \frac{37}{81} \quad \frac{16}{35} \quad \frac{43}{94}\frac{27}{59}\frac{38}{83} \quad \frac{11}{24} \qquad \frac{39}{85}\frac{28}{61}\frac{45}{98}\frac{17}{37} \quad \frac{40}{87}\frac{23}{50} \quad \frac{29}{63}\frac{35}{76} \qquad \frac{6}{13}$$

Figure 8.7. Zooming in once more on the Farey tessellation

coprime, and list these numbers by order of increasing size

$$0 < \frac{1}{N} < \cdots < \frac{p_{i-1}}{q_{i-1}} < \frac{p_i}{q_i} < \frac{p_{i+1}}{q_{i+1}} < \cdots < \frac{N-1}{N} < 1.$$

Then, $p_{i+1}q_i - p_iq_{i+1} = 1$ for any two consecutive $\frac{p_i}{q_i} < \frac{p_{i+1}}{q_{i+1}}$, and $\frac{p_i}{q_i} = \frac{p_{i-1}}{q_{i-1}} \oplus \frac{p_{i+1}}{q_{i+1}}$ for any three consecutive $\frac{p_{i-1}}{q_{i-1}} < \frac{p_i}{q_i} < \frac{p_{i+1}}{q_{i+1}}$. The same property had actually been discovered, with partial proofs, earlier in 1802 by C. Haros. However, the name of Farey series of order N became attached to the above sequence when Augustin Cauchy, who was only aware of Farey's note [**Farey**], provided a complete proof of these two statements in 1816. A good reference for the Farey series is [**Hardy & Wright**].

Of course, Farey did not know anything about hyperbolic geometry. The first published account of what we call here the Farey circle packing is due to Lester R. Ford (1886–1967) [**Ford₂**]. In particular, these circles are also often called *Ford circles*.

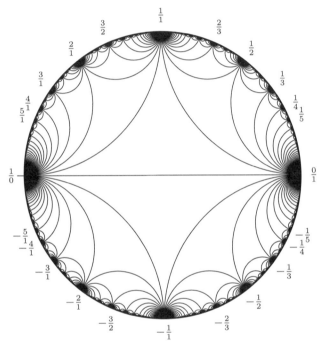

Figure 8.8. The Farey tessellation in the disk model for $\mathbb{H}^2$

8.2. The Farey tessellation and the once-punctured torus

Let us go back to the tessellation of the hyperbolic plane $\mathbb{H}^2$ discussed in Section 6.6. We started with the hyperbolic square X with vertices 0, $+1$, -1 and $\infty \in \mathbb{R} \cup \{\infty\}$ at infinity of $\mathbb{H}^2$. We also considered the transformation group Γ generated by the transformations $\varphi_1(z) = \frac{z+1}{z+2}$ and $\varphi_3(z) = \frac{z-1}{-z+2}$. We then showed that the polygons

$\varphi(X)$, as φ ranges over all the elements of Γ, form a tessellation of $\mathbb{H}^2$.

Let us split the square X along the diagonal 0∞ into the two triangles T^+ and T^-, where T^+ has vertices 0, $+1$ and ∞ and T^- has vertices 0, -1 and ∞. Note that the hyperbolic reflection $z \mapsto -\bar{z}$ across the vertical half-line 0∞ exchanges T^+ and T^-, so that these two triangles are isometric. (Actually, Lemma 8.4 below will show that any two triangles with vertices at infinity are isometric.)

It follows that the collection of the triangles $\varphi(T^+)$ and $\varphi(T^-)$, with $\varphi \in \Gamma$, forms a tessellation of $\mathbb{H}^2$ by ideal triangles. Let $\mathcal{T}$ denote this tessellation.

We will prove that this tessellation is exactly the Farey tessellation, in the sense that its edges are exactly the hyperbolic geodesics joining $\frac{p}{q}$ to $\frac{p'}{q'}$ when $pq' - p'q = \pm 1$.

For this it is convenient to consider, as in Example 7.3, the set $\mathrm{PSL}_2(\mathbb{Z})$ consisting of all linear fractional maps

$$\varphi(z) = \frac{az + b}{cz + d},$$

with a, b, c, $d \in \mathbb{Z}$ and $ad - bc = 1$. We already observed in Example 7.3 that $\mathrm{PSL}_2(\mathbb{Z})$ is a transformation group of $\mathbb{H}^2$. See Exercises 2.12 and 7.2 for an explanation of the notation.

Since the gluing maps φ_1 and φ_3 generating Γ are elements of $\mathrm{PSL}_2(\mathbb{Z})$, we conclude that Γ is contained in $\mathrm{PSL}_2(\mathbb{Z})$.

When we consider a rational number $\frac{p}{q} \in \mathbb{Q} \cup \{\infty\}$, we will use the convention that p and q are coprime and that $q > 0$, with the exception of $\infty = \frac{1}{0} = \frac{-1}{0}$. We say that $\frac{p}{q}$, $\frac{p'}{q'} \in \mathbb{Q} \cup \{\infty\}$ form a **Farey pair** if $pq' - p'q = \pm 1$.

Lemma 8.1. *If $\varphi \in \mathrm{PSL}_2(\mathbb{Z})$ and if $\frac{p}{q}$, $\frac{p'}{q'} \in \mathbb{Q} \cup \{\infty\}$ form a Farey pair, then $\varphi(\frac{p}{q})$ and $\varphi(\frac{p'}{q'})$ form a Farey pair.*

Proof. Immediate computation. □

Note that the pairs $\{0, \infty\}$, $\{1, \infty\}$ and $\{-1, \infty\}$ are Farey pairs. From Lemma 8.1 and the fact that $\Gamma \subset \mathrm{PSL}_2(\mathbb{Z})$, we conclude that the endpoints of each edge of the tessellation $\mathcal{T}$ form a Farey pair.

Conversely, let g be a geodesic of $\mathbb{H}^2$ whose endpoints form a Farey pair.

Lemma 8.2. *Let g_1 and g_2 be two distinct geodesics of $\mathbb{H}^2$ whose endpoints each form a Farey pair. Then g_1 and g_2 are disjoint.*

Proof. If the endpoints of g_1 are $\frac{p_1}{q_1}$ and $\frac{p_1'}{q_1'}$ and if we choose the indexing so that $p_1' q_1 - p_1 q_1' = +1$, the map $\varphi \in \mathrm{PSL}_2(\mathbb{Z})$ defined by $\varphi(z) = \frac{q_1 z - p_1}{-q_1' z + p_1'}$ sends g to the geodesic with endpoints 0 and ∞. Let $\frac{p_2}{q_2}$ and $\frac{p_2'}{q_2'}$ be the endpoints of $\varphi(g_2)$. If g and g' meet each other, then $\varphi(g')$ must meet $\varphi(g)$ so that $\frac{p_2}{q_2}$ and $\frac{p_2'}{q_2'}$ have opposite signs. But this is incompatible with the fact that by Lemma 8.1, they must satisfy the Farey relation $p_2 q_2' - p_2' q_2 = \pm 1$. □

Lemma 8.2 shows that a geodesic g whose endpoints form a Farey pair must be an edge of the tessellation $\mathcal{T}$. Indeed, the tiles of the tessellation $\mathcal{T}$ are ideal triangles, and their interiors consequently cannot contain any complete geodesic. Therefore, g must meet an edge g' of $\mathcal{T}$, and must be equal to g' by Lemma 8.2.

This shows that the edges of the tessellation $\mathcal{T}$ are exactly the complete geodesics of $\mathbb{H}^2$ whose endpoints in $\mathbb{R} \cup \{\infty\}$ form a Farey pair. As a consequence, the tessellation $\mathcal{T}$ coincides with the Farey tessellation. In particular, this proves that the Farey tessellation is indeed a tessellation, something that we had taken for granted so far.

8.3. Horocircles and the Farey circle packing

If $\xi \in \mathbb{R} \cup \{\infty\}$ is a point at infinity of the hyperbolic plane $\mathbb{H}^2$, recall from Section 6.8 that a *horocircle* centered at ξ is a curve $H = C - \{\xi\} \subset \mathbb{H}^2$ where C is a euclidean circle passing through ξ and tangent to $\mathbb{R}$. In particular, when $\xi = \infty$, a horocircle is just a horizontal line contained in $\mathbb{H}^2$.

Note that since linear and antilinear fractional maps send circles to circles, every isometry of $\mathbb{H}^2$ sends horocircle to horocircle.

We had already encountered horocircles when analyzing the punctured torus of Section 5.5. For a given a, we indeed used horocircles C_∞, C_{-1}, C_0, C_1 centered at ∞, -1, 0, 1, respectively, to cut out

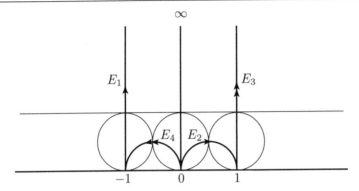

Figure 8.9. A hyperbolic square with horocircles

pieces U_∞, U_{-1}, U_0, U_1 from the hyperbolic square X. If we take the value a to be equal to 1 (which was not allowed in Section 5.5), these horocircles are tangent to each other. See Figure 8.9.

This is the secret behind the Farey circle packing. Indeed, we observed in Section 5.5 that the gluing maps φ_1, $\varphi_2 = \varphi_1^{-1}$, φ_3, $\varphi_4 = \varphi_3^{-1}$ respect this set of four horocircles, in the sense that $\varphi_i(C_\xi) = C_{\xi'}$ whenever $\varphi_i(\xi) = \xi'$. Therefore, as we tessellate the hyperbolic plane $\mathbb{H}^2$ by the tiles $\varphi(X)$ with $\varphi \in \Gamma$, the images of these four horocircles under the transformations $\varphi \in \Gamma$ form a family of horocircles, all centered at points $\frac{p}{q} \in \mathbb{Q} \cup \{\infty\}$. Looking at their intersection with a given tile, it is immediate that two of these horocircles meet only when they are tangent to each other, and that this happens exactly when their centers are the ends of an edge of the Farey tessellation $\mathcal{T}$, namely, when their centers form a Farey pair.

The family of these horocircles, combined with the Farey tessellation $\mathcal{T}$, is illustrated in Figure 8.10 for the upper half-space $\mathbb{H}^2$, and in Figure 8.11 for the disk model $\mathbb{B}^2$ for the hyperbolic plane.

To identify this family of horocircles to the Farey circle packing, it suffices to combine the following lemma with the fact that Γ is contained in $\mathrm{PSL}_2(\mathbb{Z})$.

Lemma 8.3. *Consider the horocircle* $C_\infty = \{z \in \mathbb{H}^2, \mathrm{Im}(z) = 1\}$ *centered at* ∞. *For every* $\varphi \in \mathrm{PSL}_2(\mathbb{Z})$ *of the form*

$$\varphi(z) = \frac{az + b}{cz + d},$$

with a, b, c, $d \in \mathbb{Z}$ and $ad - bc = 1$, the image $\varphi(C_\infty)$ is equal to $C_{\frac{a}{c}} - \{\frac{a}{c}\}$, where $C_{\frac{a}{c}}$ is the euclidean circle of diameter $\frac{1}{c^2}$ tangent to $\mathbb{R}$ at $\frac{a}{c}$.

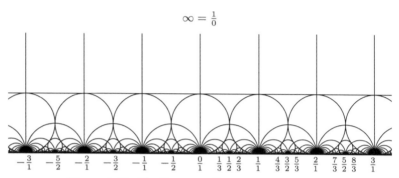

Figure 8.10. The Farey tessellation and circle packing

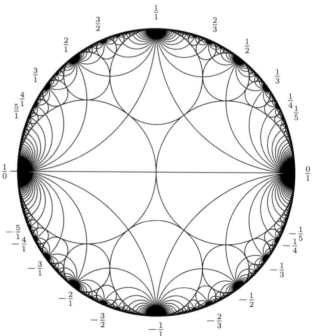

Figure 8.11. The Farey circle packing and tessellation in the disk model $\mathbb{B}^2$

Proof. Since φ sends horocircle to horocircle and since it sends ∞ to $\frac{a}{c}$, the only issue is to compute the (euclidean) diameter of the circle $C_{\frac{a}{c}}$. This can be easily checked by decomposing φ as a composition of horizontal translations, homotheties and inversions as in Lemma 2.12. $\qquad\square$

8.4. Shearing the Farey tessellation

We now revisit the complete and incomplete punctured tori of Sections 5.5 and 6.7.2, and their associated tessellations or partial tessellations of the hyperbolic plane.

8.4.1. Revisiting the partial tessellations associated to incomplete punctured tori. The key property is the following.

Lemma 8.4. *Given a triple of distinct points ξ_1, ξ_2, $\xi_3 \in \mathbb{R} \cup \{\infty\}$ and another such triple of distinct points ξ_1', ξ_2', $\xi_3' \in \mathbb{R} \cup \{\infty\}$, there is a unique isometry φ of the hyperbolic plane $(\mathbb{H}^2, d_{\mathrm{hyp}})$ sending each ξ_i to the corresponding ξ_i'.*

Also, at each ξ_i, there is a unique horocircle C_i centered at ξ_i such that any two C_i and C_j are tangent to each other and meet at a point of the complete hyperbolic geodesic going from ξ_i to ξ_j.

Proof. The first part is a simple algebraic computation, using the fact that every isometry of the hyperbolic plane is a linear or antilinear fractional map; compare Exercise 2.9. Observe that φ is orientation-preserving, namely, it is linear fractional, precisely when $(\xi_1 - \xi_2)(\xi_2 - \xi_3)(\xi_3 - \xi_1)$ and $(\xi_1' - \xi_2')(\xi_2' - \xi_3')(\xi_3' - \xi_1')$ have the same sign (suitably interpreted when one of these points is equal to ∞).

To prove the second statement, the first part shows that we can restrict attention to one specific example, such as the case where the points are 0, 1 and ∞. The result then follows from elementary geometric considerations. Compare Figure 8.9. $\qquad\square$

A consequence of Lemma 8.4 is that every edge of an ideal triangle has a preferred base point, namely, the point where the two horocircles centered at its endpoints and singled out by Lemma 8.4 touch each other.

The hyperbolic punctured tori constructed in Sections 5.5 and 6.7.2 were obtained from the ideal square X with vertices -1, 0, 1 and ∞ by gluing the edge E_1 going from -1 to ∞ to the edge E_2 going from 0 to 1 by the map

$$\varphi_1(z) = \frac{z+1}{z+a},$$

and by gluing the edge E_3 going from 1 to ∞ to the edge E_4 going from 0 to -1 by the map

$$\varphi_3(z) = \frac{z-1}{-z+b}.$$

In the first case that we considered, in Section 5.5, the constants were chosen so that $a = b = 2$ and the construction provided a complete punctured torus.

Split the square X along its diagonal 0∞, namely, along the geodesic going from 0 to ∞. This gives two ideal triangles, one with vertices -1, 0, ∞, and another one with vertices 0, 1 and ∞. Applying Lemma 8.4 to these two triangles now provides a base point on each of the edges of X, namely, $P_1 = -1 + i \in E_1$, $P_2 = \frac{1}{2} + \frac{i}{2} \in E_2$, $P_3 = 1 + i \in E_3$ and $P_4 = -\frac{1}{2} + \frac{i}{2} \in E_4$. Compare Figure 8.9.

When $a = b = 2$, the gluing maps $\varphi_1 \colon E_1 \to E_2$ and $\varphi_3 \colon E_3 \to E_4$ exactly send base point to base point. This is why when we tessellate the hyperbolic plane by the images of X under the elements of the tiling group generated by φ_1 and φ_3, the horocircles fit nicely together, as in Figure 8.10.

However, this is not so in the general case. We now give a geometric interpretation to the constants a and b occurring in the definition of φ_1 and φ_3. First of all, recall the definition of the other gluing maps defined by $\varphi_2 = \varphi_1^{-1}$ and $\varphi_4 = \varphi_3^{-1}$.

In the edge $E_1 = \varphi_2(E_2)$, the base point determined by the horocircles of $\varphi_2(X)$ is the image $\varphi_2(P_2) = -1 + i(a-1)$ of the base point $P_2 = \frac{1}{2} + \frac{i}{2}$. In particular, the hyperbolic distance from this base point to the base point $P_1 = -1 + i$ determined by the horocircles of X is equal to $|\log(a-1)|$. More precisely, as seen from the interior of X, $\varphi_2(P_2)$ is at signed distance $s_1 = -\log(a-1)$ to the left of P_1, where we count the distance as negative when the point is to the right.

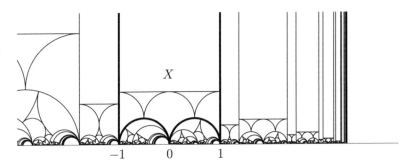

Figure 8.12. Horocircle arcs in the tessellation of Section 6.7.2

Note that this signed distance convention is symmetric. If we look at the edge $E_1 = \varphi_2(E_2)$ from the interior of $\varphi_2(X)$, the base point P_1 marked on $E_1 = \varphi_2(E_2)$ by the horocircles of the "outside" polygon X is also at signed distance s_1 to the left of the base point determined by the horocircles of $\varphi_2(X)$.

Transporting the geometric situation to the edge $E_2 = \varphi_1(E_1)$ by $\varphi_1 = \varphi_2^{-1}$, we see that, as seen from the interior of X, the base point determined by $\varphi_1(X)$ is at signed distance s_1 to the left of the base point P_2 determined by X.

Similarly, on the edge $E_3 = \varphi_4(E_4)$, the base point $\varphi_4(P_4)$ determined by $\varphi_4(X)$ is at signed distance $s_3 = \log(b-1)$ from the base point P_3 determined by the horocircles of X as seen from the interior of X. Also, on the edge $E_4 = \varphi_3(E_3)$, the base point $\varphi_3(P_3)$ is to the left of the base point P_4 as seen from inside X.

Actually, we have the same situation on each edge of the partial tessellation of the hyperbolic plane associated to the square X and to the gluing maps φ_1 and φ_3. Each such edge E is of the form $E = \varphi(E_1)$ or $\varphi(E_3)$ for some element φ of the tiling group Γ generated by φ_1 and φ_3, and separates the polygons $\varphi(X)$ from the polygon $\psi(X)$ associated to another $\psi \in \Gamma$. As seen from inside $\varphi(X)$, the base point determined by $\psi(X)$ is at signed distance s_1 if $E = \varphi(E_1)$, and at signed distance s_3 if $E = \varphi(E_3)$, to the left of the base point determined by $\varphi(X)$. This immediately follows by transporting to $\varphi(X)$ by φ our analysis of the partial tessellation near X.

This is illustrated in Figure 8.12 in the case where $s_1 = 0.25$ and $s_3 = -1$. Compare this figure to Figures 8.9 and 8.10, and to Figures 6.12 and 6.16.

In particular, every tile of this partial tessellation $\mathcal{T}_{s_1 s_3}$ corresponds to a tile of the tessellation $\mathcal{T}$ associated to the case where $s_1 = s_3 = 0$ (and corresponding to a complete punctured torus). Actually, one goes from $\mathcal{T}$ to $\mathcal{T}_{s_1 s_3}$ by progressively sliding all the tiles to the left along the edges, and by a signed distance of s_1 or s_3 according to whether the edge considered is associated to E_1 or E_3 by the tiling group. We say that $\mathcal{T}_{s_1 s_3}$ is obtained by **shearing** $\mathcal{T}$ according to the **shear parameters** s_1 and s_3.

8.4.2. Shearing the Farey tessellation. We can generalize the above construction by introducing an additional edge, namely, the diagonal of X formed by the geodesic joining 0 to ∞. This geodesic splits X into two ideal triangles: the triangle T^+ with vertices 0, 1 and ∞; and the triangle T^- with vertices -1, 0 and ∞. Given an additional shear parameter s_5, we can then deform X by sliding T^- along E_5 by a distance of s_5 to the left of T^+. Namely, we can replace T^- by its image under the hyperbolic isometry $z \mapsto e^{-s_5} z$.

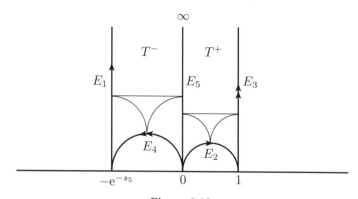

Figure 8.13

The corresponding new hyperbolic quadrilateral (not a square any more) is illustrated in Figure 8.13 in the case where $s_5 = -.25$.

Starting with this sheared quadrilateral, we can then construct a partial tessellation of the hyperbolic plane as before, using shear parameters s_1 and s_3. In practice, one begins by gluing the sides of the new quadrilateral X using the gluing maps $\varphi_1 \colon E_1 \to E_2$ and $\varphi_3 \colon E_3 \to E_4$ defined by

$$\varphi_1(z) = \frac{e^{s_5} z + 1}{e^{s_5} z + e^{-s_1} + 1}$$

and

$$\varphi_3(z) = e^{-s_5} \frac{z - 1}{-z + e^{s_3} + 1}.$$

These formulas are easily obtained from the formulas for the original gluing maps, using the fact that $a = e^{-s_1} + 1$ and $b = e^{s_3} + 1$. One can then consider the partial tessellation of the hyperbolic plane associated to these edge gluings.

Figure 8.14 illustrates the case where $s_1 = 0.25$, $s_3 = -0.75$ and $s_5 = -0.25$.

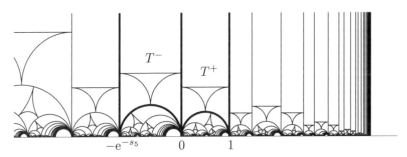

Figure 8.14. Shearing the Farey tessellation

Lemma 8.5. *The images of X under the tiling group Γ generated by the gluing maps φ_1 and φ_3 is the whole hyperbolic plane $\mathbb{H}^2$ exactly when $s_1 + s_3 + s_5 = 0$.*

Proof. We will use Poincaré's Polygon Theorem 6.25. The gluing maps $\varphi_1 \colon E_1 \to E_3$, $\varphi_3 \colon E_3 \to E_4$, $\varphi_2 \colon E_2 \to E_1$ and $\varphi_4 \colon E_4 \to E_2$ form an edge cycle around the ideal vertex $\overline{\infty} = \{\infty, 1, 0, -e^{-s_5}\}$. This is the only edge cycle in the gluing data.

Theorem 6.25 asserts that the quotient space $(\bar{X}, \bar{d}_X)$ is complete if and only if the composition map $\varphi_4 \circ \varphi_2 \circ \varphi_3 \circ \varphi_1$ is horocyclic at

∞. Remembering that $\varphi_2 = \varphi_1^{-1}$ and $\varphi_4 = \varphi_3^{-1}$, an immediate computation yields

$$\varphi_4 \circ \varphi_2 \circ \varphi_3 \circ \varphi_1(z) = e^{2s_1+2s_3+2s_5} z + 1 + e^{s_3} + e^{s_3+s_1}$$
$$+ e^{s_3+s_1+s_5} + e^{2s_3+s_1+s_5} + e^{2s_3+2s_1+s_5}.$$

In particular, this map is horocyclic if and only if $s_1 + s_3 + s_5 = 0$.

If $s_1 + s_3 + s_5 = 0$, Theorem 6.25 guarantees that the quotient space $(\bar{X}, \bar{d}_X)$ is complete. Then Theorem 6.1 shows that the images of X under Γ tessellate $\mathbb{H}^2$. Note that X has no finite vertex, so that the completeness of the quotient space is the only hypothesis that we have to check when applying Theorem 6.1.

Conversely, suppose that the images of X under Γ tessellate $\mathbb{H}^2$. Then, Theorem 7.12 shows that $(\bar{X}, \bar{d}_X)$ is isometric to the quotient $(\mathbb{H}^2/\Gamma, \bar{d}_\Gamma)$ of $\mathbb{H}^2$ under the action of Γ, and Exercise 7.6 implies that $(\mathbb{H}^2/\Gamma, \bar{d}_\Gamma)$ is complete. The converse part of Theorem 6.25 now proves that $s_1 + s_3 + s_5$ must be equal to 0. (We could also have used the analysis of Section 6.7.2.) $\square$

In Chapter 10, we will generalize this construction by allowing, in addition to shearing, bending into a third dimension. For this, we need to consider the 3-dimensional hyperbolic space, which we introduce in the next chapter.

Exercises for Chapter 8

Exercise 8.1. Rigorously show that if the endpoints of an edge of the Farey tessellation are $\frac{p}{q}$, $\frac{p'}{q'} \in \mathbb{Q} \cup \{\infty\}$, this edge separates two tiles of the tessellation which are ideal triangles with respective vertices (at infinity) $\frac{p}{q}$, $\frac{p'}{q'}$ and $\frac{p''}{q''}$ with $\frac{p''}{q''} = \frac{p}{q} \oplus \frac{p'}{q'}$ and $\frac{p''}{q''} = \frac{p}{q} \oplus \frac{-p'}{-q'}$ (abandonning the convention that all fractions have positive denominator, by keeping numerators and denominators coprime), where $\oplus$ denotes the Farey sum.

Exercise 8.2. Consider the hyperbolic square X of Figure 8.9, with vertices -1, 0, 1, ∞. In Section 8.4, we split it along the geodesic 0∞ to obtain two ideal triangles, and applying Lemma 8.4 to these two triangles provided us with base points on each of the edges of X. Compute the base points that one would have obtained if we had instead split X along the other diagonal of the square, namely, along the hyperbolic geodesic going from 1 to -1.

Exercise 8.3 (The Farey Property). For every $N > 0$, consider all rational numbers $\frac{p}{q} \in \mathbb{Q}$ whose denominator is such that $0 < q \leqslant N$. Since there are only finitely many such rational numbers between any two consecutive integers, we can order all these rationals and list them as

$$\cdots < \frac{p_{i-1}}{q_{i-1}} < \frac{p_i}{q_i} < \frac{p_{i+1}}{q_{i+1}} < \cdots$$

with $i \in \mathbb{Z}$, where $q_i > 0$ and p_i and q_i are coprime. This bi-infinite sequence is called the **Farey series** of order N.

a. Show that if $\frac{p}{q}$ and $\frac{p'}{q'} \in \mathbb{Q}$ are such that $p'q - pq' = 1$ and q, $q' > 0$, then $|q''| \geqslant q + q'$ for every $\frac{p''}{q''} \in \mathbb{Q}$ with $\frac{p}{q} < \frac{p''}{q''} < \frac{p'}{q'}$, and equality holds only for $\frac{p''}{q''} = \frac{p+p'}{q+q'}$. Hint: If $\frac{p}{q} < \frac{p''}{q''} < \frac{p+p'}{q+q'}$, then

$$\frac{1}{qq''} \leqslant \frac{p''}{q''} - \frac{p}{q} < \frac{p+p'}{q+q'} - \frac{p}{q} = \frac{1}{q(q+q')}.$$

b. Prove by induction on N that that any two consecutive terms in the Farey series form a Farey pair, namely that $p_{i+1}q_i - p_iq_{i+1} = 1$ for every $i \in \mathbb{Z}$. Hint: Use part a.

c. Show that any three consecutive terms of the Farey series are such that $\frac{p_i}{q_i} = \frac{p_{i-1}}{q_{i-1}} \oplus \frac{p_{i+1}}{q_{i+1}}$, namely that $\frac{p_i}{q_i} = \frac{p_{i-1}+p_{i+1}}{q_{i-1}+q_{i+1}}$ for every $i \in \mathbb{Z}$ (although $p_{i-1} + p_{i+1}$ and $q_{i-1} + q_{i+1}$ are not necessarily coprime).

Exercise 8.4 (Pythagorean triples). A **Pythagorean triple** is a triple (a, b, c) of three coprime integers a, b, $c \geqslant 0$ such that $a^2 + b^2 = c^2$. For instance, $(1, 0, 1)$ and $(3, 4, 5)$ are well-known Pythagorean triples, while $(387, 884, 965)$ is probably a less familiar one.

a. Let $\mathbb{S}^1$ be the circle of radius 1 and center $(0, 0)$ in the plane $\mathbb{R}^2$. Show that the map $\Psi \colon (a, b, c) \mapsto (\frac{a}{c}, \frac{b}{c})$ defines a one-to-one correspondence between Pythagorean triples (a, b, c) and rational points of $\mathbb{S}^1$ located in the first quadrant, namely, points $(x, y) \in \mathbb{S}^1$ whose coordinates x, y are rational and nonnegative.

b. Consider our usual isometry $\Phi(z) = -\frac{z-i}{z+i}$ from the upper half-space $(\mathbb{H}^2, d_{\mathrm{hyp}})$ to the disk model $(\mathbb{B}^2, d_{\mathbb{B}^2})$. Show that Φ sends each rational point $\frac{p}{q} \in \mathbb{Q} \cap [0, 1]$ to a rational point of $\mathbb{S}^1$ in the first quadrant.

c. Conversely, for $(x, y) \in \mathbb{S}^1$ with x, y rational and nonnegative, show that $\Phi^{-1}(x, y)$ is a rational number in the interval $[0, 1]$.

As a consequence, the composition $\Psi^{-1} \circ \Phi$ provides a one-to-one correspondence between rational numbers in the interval $[0, 1]$ and all Pythagorean triples. For instance, $\Psi^{-1} \circ \Phi(\frac{1}{2}) = (3, 4, 5)$, and $\Psi^{-1} \circ \Phi(\frac{17}{26}) = (387, 884, 965)$.

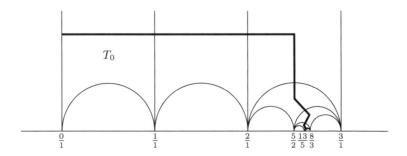

Figure 8.15. Traveling in the Farey tessellation

Exercise 8.5 (Travelling in the Farey tessellation). Consider a Farey triangle T with nonnegative vertices $\frac{p}{q}$, $\frac{p+p'}{q+q'}$, $\frac{p'}{q'} \geqslant 0$. This triangle can be connected to the base triangle T_0 with vertices 0, 1, ∞ by a sequence of Farey triangles T_0, T_1, T_2, ..., T_{n-1}, $T_n = T$ in such a way that each T_{i+1} is adjacent to T_i and that the T_i are all distinct. As one enters T_i from T_{i-1}, since we are not allowed to backtrack to T_{i-1}, there are exactly two possibilities for T_{i+1}, which are respectively to the left or to the right of T_i as seen from T_{i-1}. The above sequence of Farey triangles T_0, T_1, T_2, ..., T_{n-1}, $T_n = T$ can therefore be described by a sequence of symbols $S_1 S_2 \ldots S_n$, where each S_i is the symbol L if T_{i+1} is to the left of T_i, and S_i is the symbol R if it is to the right of T_i. By convention, T_0 is considered to have been entered through the edge 0∞, so that $S_1 = L$ if T_1 is the triangle 12∞, and $S_1 = R$ if T_1 is the triangle $0\frac{1}{2}1$.

For instance, Figure 8.15 illustrates the case of the triangle T with vertices $\frac{13}{5}$, $\frac{21}{8}$ and $\frac{8}{3}$, associated to the symbol sequence $S_1 S_2 S_3 S_4 S_5 S_6 = LLRLRL$. (The vertex $\frac{21}{8}$ is not labeled due to lack of space.)

Finally, consider the matrices $\lambda = \left(\begin{smallmatrix} 1 & 1 \\ 0 & 1 \end{smallmatrix}\right)$ and $\rho = \left(\begin{smallmatrix} 1 & 0 \\ 1 & 1 \end{smallmatrix}\right)$, and let the matrix σ be associated to the symbol sequence $S_1 S_2 \ldots S_n$ by the property that $\sigma = \sigma_1 \sigma_2 \ldots \sigma_n$ where $\sigma_i = \lambda$ when $S_i = L$ and to $\sigma_i = \rho$ when $S_i = R$.

Show that the vertices $\frac{p}{q}$, $\frac{p+p'}{q+q'}$, $\frac{p'}{q'}$ of T can be directly read from the matrix σ, by the property that $\sigma = \left(\begin{smallmatrix} p' & p \\ q' & q \end{smallmatrix}\right)$. Hints: Proof by induction on n. It may also be convenient to consider, as in Exercise 2.12, the linear fractional map $\varphi_\alpha(z) = \frac{az+b}{cz+d}$ associated to the matrix $\alpha = \left(\begin{smallmatrix} a & b \\ c & d \end{smallmatrix}\right)$, and note that φ_λ sends T_0 to the triangle 12∞ and that φ_ρ sends T_0 to the triangle $0\frac{1}{2}1$.

Exercise 8.6 (The Farey tessellation and continued fractions). A *continued fraction* is an expression of the form

$$[a_1, a_2, \ldots, a_n] = \cfrac{1}{a_1 + \cfrac{1}{a_2 + \cfrac{1}{\cdots + \cfrac{1}{a_n}}}}.$$

For instance, $[2, 1, 1, 1] = \frac{8}{3}$ and $[2, 1, 1, 1, 1] = \frac{13}{5}$. With the notation of Exercise 8.5, write the symbol sequence $S_1 S_2 \ldots S_n$ as $L^{m_1} R^{n_1} L^{m_2} R^{n_2} \ldots L^{m_k} R^{n_k}$, where L^{m_i} denotes $m_i \geqslant 0$ copies of the symbol L, and similarly R^{n_i} denotes $n_i \geqslant 0$ copies of R.

Show that

$$\frac{p}{q} = [m_1, n_1, m_2, n_2, \ldots, m_k, n_k]$$

and

$$\frac{p'}{q'} = [m_1, n_1, m_2, n_2, \ldots, n_{k-1}, m_k],$$

using the conventions that $\frac{1}{0} = \infty$, $\frac{1}{\infty} = 0$ and $a + \infty = \infty$. Hint: Induction on k.

Exercise 8.7 (Domino diagrams). In the diagram

(8.1)

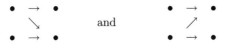

consider, for i, $j \in \{1, 2\}$, the paths that jump from bullet to bullet by following the arrows, and that go from the bullet marked i on the left to the bullet marked j on the right; let n_{ij} be the number of such paths. For instance one relatively easily sees that among the paths starting from the bullet marked 2 on the left, three of them end at the bullet marked 1 while five of them end at the bullet marked 2, so that $n_{21} = 3$ and $n_{22} = 5$. A little more perseverance shows that $n_{11} = 8$ and $n_{12} = 13$.

The above diagram can also be described by chaining together two types of "dominos", namely

• → •   • → •
  ↘   and   ↗
• → •   • → •

If we denote the domino on the left by L and the domino on the right by R, the above diagram corresponds to the chain $LLRLRL$. Comparing with the example discussed in Exercise 8.5, we may find it surprising that $\frac{n_{11}}{n_{21}} = \frac{8}{3}$, and $\frac{n_{12}}{n_{22}} = \frac{13}{5}$. Namely, in the Farey tessellation, the path described by the same symbol $LLRLRL$ leads to the ideal triangle T with vertices $\frac{n_{11}}{n_{21}}$, $\frac{n_{12}}{n_{22}}$, and $\frac{n_{11}+n_{12}}{n_{21}+n_{22}}$. We want to show that this is no coincidence.

Given a symbol sequence $S_1 S_2 \ldots S_n$, where each S_i is L or R, consider the domino diagram obtained by chaining together the dominos L and R associated to the S_i. Let σ be also the matrix associated to $S_1 S_2 \ldots S_n$, as in Exercise 8.5. We will use the notation of that exercise.

a. Write $\sigma = \left(\begin{smallmatrix} s_{11} & s_{12} \\ s_{21} & s_{22} \end{smallmatrix}\right)$ and $\sigma_i = \left(\begin{smallmatrix} s_{11}^{(i)} & s_{12}^{(i)} \\ s_{21}^{(i)} & s_{22}^{(i)} \end{smallmatrix}\right)$. Remember that $\sigma = \sigma_1 \sigma_2 \ldots \sigma_n$. Show that

$$s_{ij} = \sum_{i_1, i_2, \ldots, i_{n-1} \in \{1,2\}} s_{i i_1}^{(1)} s_{i_1 i_2}^{(2)} s_{i_2 i_3}^{(3)} \ldots s_{i_{n-1} j}^{(n)}.$$

It may be useful to first do a few examples with small n.

b. In our case, $\sigma_i = \left(\begin{smallmatrix} 1 & 0 \\ 1 & 1 \end{smallmatrix}\right)$ or $\left(\begin{smallmatrix} 1 & 1 \\ 0 & 1 \end{smallmatrix}\right)$ according to whether the symbol S_i is L or R. Show that each term $s_{i i_1}^{(1)} s_{i_1 i_2}^{(2)} s_{i_2 i_3}^{(3)} \ldots s_{i_{n-1} j}^{(n)}$ is equal to 0 or 1.

c. In the domino diagram associated to $S_1 S_2 \ldots S_n$, label each bullet on the top row by 2 and each bullet on the lower row by 1. Show that $s_{i_0 i_1}^{(1)} s_{i_1 i_2}^{(2)} s_{i_2 i_3}^{(3)} \ldots s_{i_{n-1} i_n}^{(n)} = 1$ if and only if there is a path in the domino diagram where, for each $k = 1, 2, \ldots, n+1$, the kth bullet is labeled by $i_{k-1} \in \{1, 2\}$.

d. Conclude that s_{ij} is equal to the number n_{ij} of paths from i to j in the domino diagram.

e. Let T be the Farey triangle associated to the symbol sequence $S_1 S_2 \ldots S_n$ as in Exercise 8.5. Show that the vertices of T are $\frac{n_{11}}{n_{21}}$, $\frac{n_{12}}{n_{22}}$ and $\frac{n_{11}+n_{12}}{n_{21}+n_{22}}$.

Chapter 9

The 3-dimensional hyperbolic space

We now jump to one dimension higher, to the 3-dimensional hyperbolic space $\mathbb{H}^3$. This space is defined in complete analogy with the hyperbolic plane. Most of the proofs are identical to those that we already used in dimension 2.

9.1. The hyperbolic space

The *3-dimensional hyperbolic space* is the metric space consisting of the upper half-space

$$\mathbb{H}^3 = \{(x, y, u) \in \mathbb{R}^3; u > 0\},$$

endowed with the hyperbolic metric d_{hyp} defined below. Here we are using the letter u for the last coordinate in order to reserve z for the complex number $z = x + iy$.

If γ is a piecewise differentiable curve in $\mathbb{H}^3$ parametrized by the vector-valued function

$$t \mapsto \big(x(t), y(t), u(t)\big), \quad a \leqslant t \leqslant b,$$

its *hyperbolic length* is

$$\ell_{\mathrm{hyp}}(\gamma) = \int_a^b \frac{\sqrt{x'(t)^2 + y'(t)^2 + u'(t)^2}}{u(t)} \, dt.$$

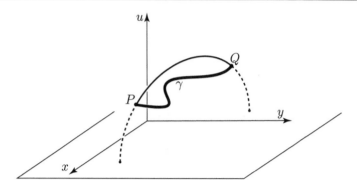

Figure 9.1. The hyperbolic space $\mathbb{H}^3$

The **hyperbolic distance** from P to $Q \in \mathbb{H}^3$ is then defined as

$$d_{\text{hyp}}(P, Q) = \inf \left\{ \ell_{\text{hyp}}(\gamma); \gamma \text{ goes from } P \text{ to } Q \right\}.$$

The proof that $(\mathbb{H}^3, d_{\text{hyp}})$ is a metric space is identical to the proof of the same statement for the hyperbolic plane in Lemma 2.1.

As in dimension 2, the **hyperbolic norm** of a vector $\vec{v}$ based at the point $P = (x, y, u) \in \mathbb{H}^3$ is defined as

$$\|\vec{v}\|_{\text{hyp}} = \frac{1}{u} \|\vec{v}\|_{\text{euc}}$$

where $\|\vec{v}\|_{\text{euc}} = \sqrt{a^2 + b^2 + c^2}$ is the usual euclidean norm of $\vec{v} = (a, b, c)$.

The hyperbolic space admits several "obvious" isometries $\varphi : \mathbb{H}^3 \to \mathbb{H}^3$, which are immediate extensions of their 2-dimensional counterparts.

These include the **horizontal translations** defined by

$$\varphi(x, y, u) = (x + x_0, y + y_0, u)$$

for x_0, $y_0 \in \mathbb{R}$, and the **homotheties**

$$\varphi(x, y, u) = (\lambda x, \lambda y, \lambda u)$$

for $\lambda > 0$.

A new, but not very different, type of symmetry consists of the *rotations* around the u-axis, defined by

$$\varphi(x, y, u) = (x \cos \theta - y \sin \theta, x \sin \theta + y \cos \theta, u)$$

for $\theta \in \mathbb{R}$.

Finally, the 3-dimensional analogue of the inversion across the unit circle is the *inversion across the unit sphere*. This is the map $\varphi \colon \mathbb{R}^3 \cup \{\infty\} \to \mathbb{R}^3 \cup \{\infty\}$ defined by

$$\varphi(x, y, u) = \left(\frac{x}{x^2 + y^2 + u^2}, \frac{y}{x^2 + y^2 + u^2}, \frac{u}{x^2 + y^2 + u^2} \right).$$

This inversion clearly sends the upper half-space $\mathbb{H}^3$ to itself.

Lemma 9.1. *If $\varphi \colon \mathbb{H}^3 \to \mathbb{H}^3$ is the inversion across the unit sphere, and if $\vec{v}$ is a vector based at $P \in \mathbb{H}^3$, then its image $D_P\varphi(\vec{v})$ under the differential of φ, which is a vector based at $\varphi(P)$, is such that*

$$\|D_P\varphi(\vec{v})\|_{\mathrm{hyp}} = \|\vec{v}\|_{\mathrm{hyp}}.$$

As a consequence, φ is an isometry of $(\mathbb{H}^3, d_{\mathrm{hyp}})$.

Proof. This is an immediate computation, identical to the one used in the proof of Lemma 2.3 to show that the inversion across the unit circle is an isometry of the hyperbolic plane. $\square$

Consider the vertical half-plane

$$H = \{(x, 0, u) \in \mathbb{R}^3; u > 0\} \subset \mathbb{H}^3.$$

Replacing the letter u by the letter y, this half-plane has a natural identification with the hyperbolic plane $\mathbb{H}^2$. For this identification, the 3-dimensional hyperbolic length of a curve in $H \subset \mathbb{H}^3$ is the same as the 2-dimensional hyperbolic length of a curve in $\mathbb{H}^2$. The same holds for the hyperbolic norm $\|\vec{v}\|_{\mathrm{hyp}}$ of a vector $\vec{v}$ tangent to H at $P \in H \subset \mathbb{H}^3$.

We can therefore identify the hyperbolic plane $(\mathbb{H}^2, d_{\mathrm{hyp}})$ to the half-plane $H \subset \mathbb{H}^3$, endowed with the metric d_H for which $d_H(P, Q)$ is the infimum of the hyperbolic lengths of all curves that join P to Q in H. Theorem 9.4 below, which identifies the shortest curves joining P to Q in $\mathbb{H}^3$, shows that d_H is just the restriction to H of the hyperbolic metric d_{hyp} of $\mathbb{H}^3$.

Theorem 9.2. *The hyperbolic space* $(\mathbb{H}^3, d_{\mathrm{hyp}})$ *is homogeneous and isotropic. Namely, for every vector $\vec{v}$ at the point P and every $\vec{w}$ at Q with $\|\vec{v}\|_{\mathrm{hyp}} = \|\vec{w}\|_{\mathrm{hyp}}$, there exists an isometry φ of $(\mathbb{H}^3, d_{\mathrm{hyp}})$ such that $\varphi(P) = Q$ and $D_P\varphi(\vec{v}) = \vec{w}$.*

Proof. Modifying $\vec{v}$, P, $\vec{w}$ and Q by suitable horizontal translations and rotations about the u-axis (which are isometries of $\mathbb{H}^3$), we can assume without loss of generality that P and Q are both in the half-plane $\mathbb{H}^2 \subset \mathbb{H}^3$, and that $\vec{v}$ and $\vec{w}$ are both tangent to that half-plane.

Then, because the hyperbolic plane $\mathbb{H}^2$ is isotropic (Proposition 2.20), there exists an isometry $\varphi \colon \mathbb{H}^2 \to \mathbb{H}^2$ which sends P to Q and whose differential map sends $\vec{v}$ to $\vec{w}$. This isometry is a composition of translations along the x-axis, homotheties, and inversions across the unit circle in the plane. Extending the inversions by inversions across the unit sphere, all of these factors extend to isometries of $\mathbb{H}^3$, so that φ extends to an isometry $\varphi \colon \mathbb{H}^3 \to \mathbb{H}^3$ which has the desired properties. $\qquad\square$

Theorem 9.3. *The hyperbolic space* $(\mathbb{H}^3, d_{\mathrm{hyp}})$ *is complete.*

Proof. The proof is identical to that of Theorem 6.10. $\qquad\square$

9.2. Shortest curves in the hyperbolic space

Theorem 9.4. *The shortest curve from P to $Q \in \mathbb{H}^3$ is the circle arc joining P to Q that is contained in the vertical circle passing through P and Q and centered on the xy-plane (possibly a vertical line).*

Proof. The proof is identical to the one we used to prove the same result in dimension 2, namely, Theorem 2.7. Namely, first consider the case where P and Q lie on the same vertical line as in Lemma 2.4. Then, for the general case, use a suitable composition of horizontal translations, homotheties and inversions to send P and Q to the same vertical line, and then use the previous case. We leave the details as an exercise. $\qquad\square$

9.3. Isometries of the hyperbolic space

To list all isometries of the hyperbolic space $\mathbb{H}^3$, it is convenient to identify the xy-plane to the complex plane $\mathbb{C}$, in the usual manner.

Lemma 9.5. *Every linear or antilinear fractional map of* $\widehat{\mathbb{C}} = \mathbb{C} \cup \{\infty\}$ *continuously extends to a map* $\widehat{\varphi} \colon \mathbb{H}^3 \cup \widehat{\mathbb{C}} \to \mathbb{H}^3 \cup \widehat{\mathbb{C}}$ *whose restriction to* $\mathbb{H}^3$ *is an isometry of* $(\mathbb{H}^3, d_{\mathrm{hyp}})$.

Proof. By Lemma 2.12, the linear or antilinear map $\varphi \colon \widehat{\mathbb{C}} \to \widehat{\mathbb{C}}$ is a composition of translations, rotations, homotheties, and inversions across the unit circle. We observed that all of these factors extend to continuous transformations of $\mathbb{H}^3 \cup \widehat{\mathbb{C}}$ inducing isometries of $(\mathbb{H}^3, d_{\mathrm{hyp}})$. $\qquad \square$

A priori, the extension $\widehat{\varphi}$ of φ provided by the proof of Lemma 9.5 might depend on the choice of the decomposition of φ as a composition of translations, rotations, homotheties, and inversions across the unit circle. To show that this is not the case, we use a simple geometric observation.

Lemma 9.6. *The inversion across the unit sphere sends any sphere S centered on the xy-plane to a sphere centered on the xy-plane, possibly a vertical plane.*

Proof. The inversion and the sphere S are both symmetric with respect to rotations around the line joining the origin O to the center of the sphere S. The property consequently follows from the 2-dimensional fact that the inversion across the unit circle is an isometry of $\mathbb{H}^2$ and therefore sends a circle centered on the x-axis to another circle centered on the x-axis (possibly a vertical line). $\qquad \square$

Lemma 9.7. *The extension $\widehat{\varphi} \colon \mathbb{H}^3 \to \mathbb{H}^3$ of the linear or antilinear map $\varphi \colon \widehat{\mathbb{C}} \to \widehat{\mathbb{C}}$ provided by the proof of Lemma 9.5 is independent of choices.*

Proof. If $P \in \mathbb{H}^3$, arbitrarily pick three spheres S_1, S_2 and S_3 centered on the xy-plane, passing through P, and sufficiently generic that P is the only point of the intersection $\mathbb{H}^3 \cap S_1 \cap S_2 \cap S_3$. There are of course many spheres like this.

The spheres S_1, S_2, S_3 intersect the xy-plane $\mathbb{C}$ in three circles C_1, C_2, C_3. By Proposition 2.18, φ sends the circles C_1, C_2, C_3 to circles C_1', C_2', C_3' in $\widehat{\mathbb{C}}$. Then there exist unique spheres S_1', S_2', S_3' centered on the xy-plane which intersect this xy-plane along C_1', C_2', C_3'. By Lemma 9.6 and similar (and obvious) statements for horizontal translations, rotations about the z-axis and homotheties, the extension of $\widehat{\varphi}$ must send the sphere S_1 to S_1', S_2 to S_2' and S_3 to S_3'. As a consequence, $\varphi(P)$ is the unique point of the intersection $\mathbb{H}^3 \cap S_1' \cap S_2' \cap S_3'$.

In particular, $\widehat{\varphi}(P)$ is independent of the decomposition of φ as a composition of translations, rotations, homotheties, and inversions that was used in the proof of Lemma 9.5.   □

Theorem 9.8. *Every linear or antilinear fractional map* $\varphi \colon \widehat{\mathbb{C}} \to \widehat{\mathbb{C}}$ *has a unique continuous extension* $\widehat{\varphi} \colon \mathbb{H}^3 \cup \widehat{\mathbb{C}} \to \mathbb{H}^3 \cup \widehat{\mathbb{C}}$ *whose restriction to* $\mathbb{H}^3$ *is an isometry of* $(\mathbb{H}^3, d_{\mathrm{hyp}})$.

Conversely, every isometry of $(\mathbb{H}^3, d_{\mathrm{hyp}})$ *is obtained in this way.*

Proof. We just proved the existence of the extension in Lemma 9.5.

Conversely, let ψ be an isometry of $(\mathbb{H}^3, d_{\mathrm{hyp}})$. We want to find a linear or antilinear fractional map φ whose isometric extension to $\mathbb{H}^3$ coincides with ψ.

Since ψ is an isometry, it sends the oriented geodesic 0∞ to another complete geodesic of $\mathbb{H}^3$, going from z_1 to $z_2 \in \widehat{\mathbb{C}}$. By elementary algebra, there is a linear fractional map φ sending 0 to z_1 and ∞ to z_2. If $\widehat{\varphi}$ is its isometric extension to $\mathbb{H}^3$, the isometry $\psi \circ \widehat{\varphi}^{-1}$ now sends the oriented geodesic 0∞ to itself.

Replacing ψ by $\psi \circ \widehat{\varphi}^{-1}$ if necessary, we can therefore assume that ψ sends the geodesic 0∞ to itself, without loss of generality. Composing ψ with a homothety if necessary, we can even assume that it fixes some point of this geodesic. Since ψ is an isometry, it now fixes every point of the geodesic 0∞.

Let g be a complete geodesic contained in the vertical euclidean half-plane $\mathbb{H}^2 \subset \mathbb{H}^3$ and crossing 0∞ at some point P_0. Then, $\psi(g)$ is a complete geodesic passing through P_0. Note that $\mathbb{H}^2$ is the union of all complete geodesics that meet both 0∞ and $g-\{P_0\}$. Therefore, $\psi(\mathbb{H}^2)$

is the union of all complete geodesics meeting 0∞ and $\psi(g) - \{\psi(P_0)\}$, and consequently is the vertical euclidean half-plane containing 0∞ and $\psi(g)$. Composing ψ with a rotation around 0∞ if necessary, we can therefore assume that $\psi(\mathbb{H}^2) = \mathbb{H}^2$.

In particular, ψ now restricts to an isometry of $\mathbb{H}^2$. By the classification of isometries of $(\mathbb{H}^2, d_{\mathrm{hyp}})$ in Theorem 2.11, there consequently exists a linear or antilinear fractional map φ, with real coefficients, whose isometric extension $\widehat{\varphi}$ coincides with ψ on $\mathbb{H}^2$. Then, $\psi \circ \widehat{\varphi}^{-1}$ is a hyperbolic isometry of $\mathbb{H}^3$ which fixes every point of $\mathbb{H}^2$. The same argument as in the proof of Lemma 2.10 then shows that $\psi \circ \widehat{\varphi}^{-1}$ is either the identity map or the euclidean reflection across $\mathbb{H}^2$.

This concludes the proof that every isometry of $(\mathbb{H}^3, d_{\mathrm{hyp}})$ coincides with the isometric extension $\widehat{\varphi}$ of some linear or antilinear fractional map φ.

Lemma 9.7 shows that the extension is unique. $\qquad\square$

It is possible to provide explicit formulas for the extension of a given linear or antilinear fractional map $z \mapsto \frac{az+b}{cz+d}$ or $z \mapsto \frac{c\bar{z}+d}{a\bar{z}+b}$ to an isometry of $\mathbb{H}^3$. See Excercise 9.2. However, these expressions are not very illuminating.

In general, we will use the same letter to denote a linear or antilinear fractional map $\varphi \colon \widehat{\mathbb{C}} \to \widehat{\mathbb{C}}$, the associated hyperbolic isometry $\varphi \colon \mathbb{H}^3 \to \mathbb{H}^3$, and the continuous map $\varphi \colon \mathbb{H}^3 \cup \widehat{\mathbb{C}} \to \mathbb{H}^3 \cup \widehat{\mathbb{C}}$ made up of these two maps. We will say that $\varphi \colon \mathbb{H}^3 \to \mathbb{H}^3$ is the **isometric extension** of $\varphi \colon \widehat{\mathbb{C}} \to \widehat{\mathbb{C}}$ to an isometry of $(\mathbb{H}^3, d_{\mathrm{hyp}})$ which, although *stricto sensu* mathematically incorrect (since the actual extension is $\varphi \colon \mathbb{H}^3 \cup \widehat{\mathbb{C}} \to \mathbb{H}^3 \cup \widehat{\mathbb{C}}$), is a convenient short-hand terminology.

For future reference, we indicate here the following property.

Lemma 9.9. *Let $B_{d_{\mathrm{hyp}}}(P_0, r)$ and $B_{d_{\mathrm{hyp}}}(P_0', r)$ be two balls with the same radius $r > 0$ in $(\mathbb{H}^3, d_{\mathrm{hyp}})$. Endow each of these two balls with the restriction of the hyperbolic metric d_{hyp}. Then every isometry $\varphi \colon B_{d_{\mathrm{hyp}}}(P_0, r) \to B_{d_{\mathrm{hyp}}}(P_0', r)$ has a unique extension to an isometry $\varphi \colon \mathbb{H}^3 \to \mathbb{H}^3$.*

Proof. We just follow the lines of the proof of Theorem 9.8.

Transporting everything by an isometry of $\mathbb{H}^3$ if necessary, we can assume without loss of generality that P_0 is on the geodesic 0∞. Then, by the arguments used in the proof of Theorem 9.8, we can progressively compose φ with various isometries of $(\mathbb{H}^3, d_{\mathrm{hyp}})$ to reduce the problem to the case where φ sends P_0 to P_0, and then sends the intersection of $B_{d_{\mathrm{hyp}}}(P_0, r)$ with 0∞ to itself, and then sends $B_{d_{\mathrm{hyp}}}(P_0, r) \cap \mathbb{H}^2$ to itself, and then fixes every point of $B_{d_{\mathrm{hyp}}}(P_0, r) \cap \mathbb{H}^2$, and then fixes every point of $B_{d_{\mathrm{hyp}}}(P_0, r)$.

At this point, the existence of the extension to a global isometry (namely, the identity map) of $\mathbb{H}^3$ is immediate. The uniqueness of the extension is immediate once we notice that it must send each geodesic emanating from P_0 to itself. □

9.4. Hyperbolic planes and horospheres

9.4.1. Hyperbolic planes.
A *hyperbolic plane* in the hyperbolic space $\mathbb{H}^3$ is the intersection H of $\mathbb{H}^3$ with a euclidean sphere centered on the xy-plane or with a vertical euclidean plane. In particular, a special case of *a* hyperbolic plane is that of *the* hyperbolic plane $\mathbb{H}^2 \subset \mathbb{H}^3$.

Proposition 9.10. *Every isometry of* $(\mathbb{H}^3, d_{\mathrm{hyp}})$ *sends hyperbolic plane to hyperbolic plane.*

Proof. The property holds for horizontal translations, homotheties, rotations around vertical axes, and inversions across spheres centered on the xy-plane (including vertical planes) by Lemma 9.6 and elementary observations. The result then follows from Theorem 9.8 and Lemma 2.12, which prove that every isometry is a composition of transformations of this type. □

Conversely, using an appropriate composition of horizontal translations and homotheties, one easily checks that every hyperbolic plane $H \subset \mathbb{H}^3$ is the image of $\mathbb{H}^2$ under an isometry of $(\mathbb{H}^3, d_{\mathrm{hyp}})$.

9.4.2. Horospheres.
A *horosphere centered at* $z \in \mathbb{C}$ is the intersection with $\mathbb{H}^3$ of a euclidean sphere S which is tangent to the xy-plane $\mathbb{C} \subset \mathbb{R}^3$ at z, and lies above this xy-plane. A horosphere

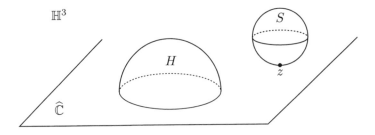

Figure 9.2. A hyperbolic plane H and a horosphere S

centered at the point ∞ is just a horizontal euclidean plane contained in $\mathbb{H}^3$.

Proposition 9.11. *Every isometry of* $(\mathbb{H}^3, d_{\mathrm{hyp}})$ *sends horosphere to horosphere.*

Proof. As in Lemma 9.6, the inversion across the unit sphere must send a horosphere to a sphere, which must be tangent to $\widehat{\mathbb{C}}$ since the inversion respects $\widehat{\mathbb{C}}$. Therefore, the standard inversion sends horospheres to horospheres. Since the same property clearly holds for horizontal translations, homotheties and rotations around vertical axes, the result immediately follows from Theorem 9.8 and Lemma 2.12. □

Exercises for Chapter 9

Exercise 9.1 (Dihedral angles). A *(hyperbolic) dihedron* is the region D of $\mathbb{H}^3$ which is delimited by two hyperbolic half planes Π_1 and Π_2 meeting along a complete geodesic $g = \Pi_1 \cap \Pi_2$, where a hyperbolic half-plane is one of the two regions of a hyperbolic plane delimited by a complete geodesic. If P is a point of this geodesic g, let Π_P be the unique hyperbolic plane which is orthogonal to g at P. Then, $D \cap \Pi_P$ is a angular sector in Π_P delimited by the two semi-infinite geodesics $\Pi_P \cap \Pi_1$ and $\Pi_P \cap \Pi_2$ issued from P. Show that the angle of this angular sector $D \cap \Pi_P$ at P is independent of the choice of $P \in g$. Possible hint: Use an isometry of $(\mathbb{H}^3, d_{\mathrm{hyp}})$ sending g to the geodesic going from 0 to ∞.

This angle is the *dihedral angle* of the dihedron D along its edge g. Dihedral angles will play an important role when we consider hyperbolic polyhedra in Chapter 10.

Exercise 9.2 (Explicit formulas for hyperbolic isometries).

    a. Show that the extension of the linear fractional map $\varphi(z) = \frac{az+b}{cz+d}$, with $ad - bc = 1$, to an isometry of $\mathbb{H}^3 \subset \mathbb{R}^3 = \mathbb{C} \times \mathbb{R}$ is given by

$$\varphi(z, u) = \left(\frac{az + b}{cz + d} + \frac{|cu|^2}{c(cz + d)(|cz + d|^2 + |cu|^2)}, \frac{u}{|cz + d|^2 + |cu|^2} \right)$$

for every $z \in \mathbb{C}$ and $u > 0$. Hint: Decompose φ as a composition of horizontal translations, rotations, homotheties and inversions, and then take the composition of the isometric extensions of these factors.

    b. Give a similar formula for the isometric extension of the antilinear fractional map $\varphi(z) = \frac{c\bar{z}+d}{a\bar{z}+b}$.

Exercise 9.3 (The ball model for the hyperbolic space). Let

$$\mathbb{B}^3 = \{(x, y, u) \in \mathbb{R}^3; x^2 + y^2 + u^2 < 1\}$$

be the open unit ball in $\mathbb{R}^3$. For every piecewise differentiable curve γ in $\mathbb{B}^3$, parametrized by $t \mapsto (x(t), y(t), u(t))$, $a \leqslant t \leqslant b$, define its $\mathbb{B}^3$-length as

$$\ell_{\mathbb{B}^3}(\gamma) = 2 \int_a^b \frac{\sqrt{x'(t)^2 + y'(t)^2 + u'(t)^2}}{1 - x(t)^2 - y(t)^2 - u(t)^2} \, dt.$$

Finally, let the $\mathbb{B}^3$-distance from P to $Q \in \mathbb{B}^3$ be

$$d_{\mathbb{B}^3}(P, Q) = \inf\{\ell_{\mathbb{B}^3}(\gamma); \gamma \text{ goes from } P \text{ to } Q\},$$

where the infimum is taken over all piecewise differentiable curves going from P to Q in $\mathbb{B}^3$.

    a. Let $\Phi \colon \widehat{\mathbb{R}}^3 \to \widehat{\mathbb{R}}^3$ be the inversion across the euclidean sphere of radius $\sqrt{2}$ centered at the point $(0, 0, -1)$. Show that it sends the upper half-space $\mathbb{H}^3$ to $\mathbb{B}^3$.

    b. Show that the restriction of Φ to $\mathbb{H}^3$ defines an isometry from $(\mathbb{H}^3, d_{\mathrm{hyp}})$ to $(\mathbb{B}^3, d_{\mathbb{B}^3})$. Possible hint: If $\vec{v}$ is a vector based at $P \in \mathbb{B}^3$, compute $\|D_P\Phi^{-1}(\vec{v})\|_{\mathrm{hyp}}$.

    c. Let S be a euclidean sphere which orthogonally meets the unit sphere $\mathbb{S}^2$ bounding $\mathbb{B}^3$. Show that the inversion ρ across S respects $\mathbb{B}^3$, and restricts to an isometry of $(\mathbb{B}^3, d_{\mathbb{B}^3})$. Possible hint: For Φ as in part a, it may be easier to prove that $\Phi^{-1} \circ \rho \circ \Phi$ is an isometry of $(\mathbb{H}^3, d_{\mathrm{hyp}})$.

    d. Show that every composition of inversions across euclidean spheres orthogonal to $\mathbb{S}^2$ restricts to an isometry of $(\mathbb{B}^3, d_{\mathbb{B}^3})$. Hint: part c.

    e. Conversely, show that every isometry of $(\mathbb{B}^3, d_{\mathbb{B}^3})$ is a composition of inversions across euclidean spheres which orthogonally meet $\mathbb{S}^2$. Hint: First consider isometries which fix the origin O.

f. Show that every geodesic of $(\mathbb{B}^3, d_{\mathbb{B}^3})$ is a circle arc contained in a euclidean circle which orthogonally meets the sphere $\mathbb{S}^2$ in two points. Hint: Use the isometry Φ of part b.

Exercise 9.4. Let $S = S_{d_{\mathrm{hyp}}}(P_0, r)$ be a hyperbolic sphere of radius r in $\mathbb{H}^3$, consisting of those $P \in \mathbb{H}^3$ such that $d_{\mathrm{hyp}}(P, P_0) = r$. Endow S with the metric d_S for which $d_S(P, Q)$ is the infimum of the hyperbolic lengths of all piecewise differentiable curves joining P to Q in S. Show that the metric space (S, d_S) is isometric to $(\mathbb{S}^2, (\sinh r)d_{\mathrm{sph}})$ where $(\sinh r)d_{\mathrm{sph}}$ is the metric on the standard euclidean sphere $\mathbb{S}^2$ obtained by multiplying the spherical metric d_{sph} by $\sinh r$. Possible hint: It may be useful to consider the ball model $(\mathbb{B}^3, d_{\mathbb{B}^3})$ of Exercise 9.3, and hyperbolic spheres centered at the origin O in this ball model.

Exercise 9.5 (The Möbius group). Let the *Möbius group* $\mathcal{M}_3$ be the group of transformations of $\widehat{\mathbb{R}}^3 = \mathbb{R}^3 \cup \{\infty\}$ generated by all inversions across euclidean spheres (including reflections across euclidean planes). The elements of $\mathcal{M}_3$ are *Möbius transformations*. Show that every Möbius transformation $\varphi \in \mathcal{M}_3$ sends every euclidean sphere to a euclidean sphere (including planes among spheres), and every euclidean circle to a euclidean circle (including lines among circles). Hint: Compare the proof of Lemma 9.6, and note that every circle is the intersection of two circles.

Exercise 9.6. Show that if $\varphi \in \mathcal{M}_3$ is a Möbius transformation and if $\vec{v}$ and $\vec{w}$ are vectors based at the same point $P \in \mathbb{R}^3$, the angle between the vectors $D_P\varphi(\vec{v})$ and $D_P\varphi(\vec{v})$ is equal to the angle between $\vec{v}$ and $\vec{w}$. In particular, isometries of $(\mathbb{H}^3, d_{\mathrm{hyp}})$ are angle-preserving, in this sense. Hint: First consider the case of inversions.

Exercise 9.7 (A characterization of Möbius transformations). The goal is to prove the converse of the statement of Exercise 9.5. Let φ be a homeomorphism of $\widehat{\mathbb{R}}^3$ which sends sphere to sphere.

a. Show that there is a Möbius transformation $\psi \in \mathcal{M}_3$ such that for the usual identification of $\mathbb{R}^3$ with $\mathbb{C} \times \mathbb{R}$ and the standard inclusions $\mathbb{R} \subset \mathbb{C} = \mathbb{C} \times \{0\} \subset \mathbb{C} \times \mathbb{R}$, $\psi(0) = \varphi(0)$, $\psi(1) = \varphi(1)$, $\psi(\infty) = \varphi(\infty)$, $\psi(\mathbb{R}) = \varphi(\mathbb{R})$, $\psi(\mathbb{C}) = \varphi(\mathbb{C})$, $\psi(\mathbb{H}^3) = \varphi(\mathbb{H}^3)$, ψ sends each half-plane of $\mathbb{C}$ delimited by $\mathbb{R}$ to the same half-plane as φ.

b. Set $\varphi' = \varphi \circ \psi^{-1}$. Show that φ' sends sphere to sphere, fixes the points $0, 1, \infty$, and respects $\mathbb{C}, \mathbb{R}, \mathbb{H}^3$ and each of the two half-planes delimited by $\mathbb{R}$ in $\mathbb{C}$.

c. Show that, for any $x, y \in \mathbb{R}$, the midpoint $m = \frac{1}{2}(x + y)$ is uniquely determined by the fact that there exists two circles C_1, C_2 and two lines L_1, L_2 in $\mathbb{C}$ with the following properties: The points m and x belong to C_1; the points m, y belong to C_2; the lines L_1, L_2 are each tangent to both of C_1 and C_2; the three lines L_1, L_2 and $\mathbb{R}$ are disjoint.

(Draw a picture). Conclude that $\varphi'(m)$ is the midpoint of $\varphi'(x)$ and $\varphi'(y)$.

d. Use part c to show, by induction on $|p|$, that $\varphi'(p) = p$ for every integer $p \in \mathbb{Z}$. Then show, by induction on n, that $\varphi'(\frac{p}{2^n}) = \frac{p}{2^n}$ for every $p \in \mathbb{Z}$ and $n \in \mathbb{N}$. Conclude that $\varphi'(x) = x$ for every $x \in \mathbb{R}$.

e. Show that φ' sends every circle in $\mathbb{C}$ with center in $\mathbb{R}$ to another circle centered on $\mathbb{R}$. Hint: Such a circle C is characterized by the property that the lines tangent to C at each of the two points of $C \cap \mathbb{R}$ are disjoint.

f. Use part e to show that φ' fixes every point of $\mathbb{C}$. Hint: Compare the proof of Lemma 9.7.

g. Show that φ' sends each sphere of $\mathbb{R}^3$ centered on $\mathbb{C}$ to a sphere centered on $\mathbb{C}$. Hint: Look at the planes tangent to S at the points of $S \cap \mathbb{C}$.

h. Use part g to show that φ' is the identity map of $\widehat{\mathbb{R}}^3$. Hint: Compare the proof of Lemma 9.7.

i. Conclude that every homeomorphism of $\widehat{\mathbb{R}}^3$ which sends sphere to sphere is a Möbius transformation.

Exercise 9.8. Let $\varphi \in \mathcal{M}_3$ be a Möbius transformation of $\widehat{\mathbb{R}}^3$ which respects the upper half-space $\mathbb{H}^3$. Show that the restriction of φ to $\mathbb{H}^3$ is an isometry of $(\mathbb{H}^3, d_{\mathrm{hyp}})$. Hint: Adapt the steps of Exercise 9.7; more precisely show that, in part a of that exercise, we can take ψ so that it induces an isometry of $\mathbb{H}^3$, and then follow the steps in parts b–i to conclude that $\varphi = \psi$.

Exercise 9.9. Remember that if φ is a function of three variables valued in $\mathbb{R}^3$, its *jacobian* at the point P is the determinant $\det D_P\varphi$. Show that if $\varphi \colon \mathbb{H}^3 \to \mathbb{H}^3$ is the isometric extension of a linear or antilinear fractional map $\widehat{\mathbb{C}} \to \widehat{\mathbb{C}}$, the jacobian $\det D_P\varphi$ is positive at every $P \in \mathbb{H}^3$ for a linear fractional map $\widehat{\mathbb{C}} \to \widehat{\mathbb{C}}$, and it is negative at every $P \in \mathbb{H}^3$ for an antilinear fractional map. Hint: First consider an inversion, and remember that $D_P(\varphi \circ \psi) = D_{\psi(P)}\psi \circ D_P\varphi$.

In other words, the extension of a linear fractional map to an isometry of $(\mathbb{H}^3, d_{\mathrm{hyp}})$ is *orientation-preserving*, and the isometric extension of an antilinear fractional map to $\mathbb{H}^3$ is *orientation-reversing*. See our discussion of 3-dimensional orientation in Section 12.1.1.

Exercise 9.10 (Classification of orientation-preserving isometries of $\mathbb{H}^3$). Let $\varphi \colon \mathbb{H}^3 \to \mathbb{H}^3$ be the isometric extension of a linear fractional map $\varphi \colon \widehat{\mathbb{C}} \to \widehat{\mathbb{C}}$. Assume in addition that φ is not the identity map.

a. Show that $\varphi \colon \widehat{\mathbb{C}} \to \widehat{\mathbb{C}}$ fixes exactly one or two points of $\widehat{\mathbb{C}}$.

b. If φ fixes only one point $z \in \widehat{\mathbb{C}}$, let ψ be a hyperbolic isometry sending z to ∞. Show that $\psi \circ \varphi \circ \psi^{-1}$ is a horizontal translation; in other words, transporting all points of $\mathbb{H}^3$ by ψ replaces φ by a horizontal translation. In this case, φ is said to be **parabolic**.

c. Now consider the case where φ fixes two distinct points z_1 and z_2 of $\widehat{\mathbb{C}}$. Let ψ be a hyperbolic isometry sending z_1 to 0 and z_2 to ∞. Show that $\psi \circ \varphi \circ \psi^{-1}$ is a homothety-rotation $z \mapsto az$ with $a \in \mathbb{C} - \{0\}$. Conclude that φ respects the complete hyperbolic geodesic g going from z_1 to z_2, and acts on g by a translation of constant hyperbolic distance $|\log|a||$. In this case, φ is said to be **elliptic** if it fixes every point of g (namely, if $|a| = 1$), and **loxodromic** otherwise.

d. Show that a parabolic or loxodromic isometry of $(\mathbb{H}^3, d_{\text{hyp}})$ fixes no point of $\mathbb{H}^3$. Show that the fixed points of an elliptic isometry form a complete geodesic in $\mathbb{H}^3$.

Exercise 9.11 (Classification of orientation-reversing isometries of $\mathbb{H}^3$). Let $\varphi \colon \mathbb{H}^3 \to \mathbb{H}^3$ be the isometric extension of an antilinear fractional map $\varphi \colon \widehat{\mathbb{C}} \to \widehat{\mathbb{C}}$. Consider its square $\varphi^2 = \varphi \circ \varphi$.

a. Show that $\varphi^2 \colon \widehat{\mathbb{C}} \to \widehat{\mathbb{C}}$ is a linear fractional map, and that φ sends each fixed point of φ^2 to a fixed point of φ^2.

b. With the terminology of Exercise 9.10, suppose that φ^2 is parabolic and is not the identity. Show that there exists a hyperbolic isometry ψ such that $\psi \circ \varphi \circ \psi^{-1}$ is the composition of the euclidean reflection across a vertical euclidean plane with the horizontal translation along a nonzero vector parallel to that plane. In this case, φ is **orientation-reversing parabolic**.

c. Suppose that φ^2 fixes exactly two points z_1, $z_2 \in \widehat{\mathbb{C}}$, and suppose that φ fixes both z_1 and z_2 (compare part a). Show that there exists a hyperbolic isometry ψ such that $\psi \circ \varphi \circ \psi^{-1}$ is the composition of a homothety with the reflection across a vertical euclidean plane passing through the point 0. Show that φ respects the complete geodesic g going from z_1 to z_2, and acts on g by a nontrivial hyperbolic translation. In this case, φ is a **orientation-reversing loxodromic**.

d. Suppose that φ^2 fixes exactly two points z_1, $z_2 \in \widehat{\mathbb{C}}$, and suppose that φ exchanges z_1 and z_2 (compare part a). Show that there exists a hyperbolic isometry ψ such that $\psi \circ \varphi \circ \psi^{-1}$ is the composition of the inversion across a sphere centered at the origin 0 with the rotation around the vertical line passing through 0 (namely, our u-axis) by an angle θ which is not an integer multiple of π. In this case, φ is **orientation-reversing elliptic**.

e. Finally, suppose that φ^2 is the identity map. Choose an arbitrary point $P_0 \in \mathbb{H}^3$ which is not fixed by φ (which must exist since φ is not the identity), and let g be a unique complete hyperbolic geodesic passing through P_0 and $\varphi(P_0)$. Let ψ be an isometry of $\mathbb{H}^3$ sending g to the vertical half-line going from 0 to ∞. Show that $\psi \circ \varphi \circ \psi^{-1}$ is either the inversion across a euclidean sphere centered at 0 or the composition of the inversion across a sphere centered at the origin 0 with the rotation of angle π around the vertical line passing through 0. Then φ is a **hyperbolic reflection** in the first case, and it is again **orientation-reversing elliptic** in the second case.

f. Show that an orientation-reversing parabolic or loxodromic isometry of $(\mathbb{H}^3, d_{\mathrm{hyp}})$ fixes no point of $\mathbb{H}^3$, that an orientation-reversing elliptic isometry fixes exactly one point of $\mathbb{H}^3$, and that the fixed points of a hyperbolic reflection form a hyperbolic plane $\Pi \subset \mathbb{H}^3$.

Exercise 9.12 (Hyperbolic volume). If D is a region in the hyperbolic space $\mathbb{H}^3 = \{(x, y, u) \in \mathbb{R}^3; u > 0\}$, let its **hyperbolic volume** be defined as

$$\mathrm{vol}_{\mathrm{hyp}}(D) = \iiint_D \frac{1}{u^3}\, dx\, dy\, du \in [0, \infty].$$

Note that this volume may be infinite if D is unbounded in $\mathbb{H}^3$.

a. Let $\varphi \colon \mathbb{H}^3 \to \mathbb{H}^3$ be a horizontal translation, a rotation around the u-axis, a homothety or an inversion across the unit sphere. Show that $\mathrm{vol}_{\mathrm{hyp}}\big(\varphi(D)\big) = \mathrm{vol}_{\mathrm{hyp}}(D)$. As for the hyperbolic area in Exercise 2.14, you may need to remember the formula for changes of variables in triple integrals.

b. Conclude that $\mathrm{vol}_{\mathrm{hyp}}\big(\varphi(D)\big) = \mathrm{vol}_{\mathrm{hyp}}(D)$ for every isometry of $(\mathbb{H}^3, d_{\mathrm{hyp}})$.

Exercise 9.13 (Hyperbolic volume in the ball model). Let $\mathbb{B}^3$ be the ball model of Exercise 9.3. If D is a region of $\mathbb{B}^3$, define its hyperbolic volume as

$$\mathrm{vol}_{\mathbb{B}^3}(D) = \mathrm{vol}_{\mathrm{hyp}}\big(\varphi(D)\big),$$

where $\varphi \colon \mathbb{B}^3 \to \mathbb{H}^3$ is an arbitrary isometry from $(\mathbb{B}^3, d_{\mathbb{B}^3})$ to $(\mathbb{H}^3, d_{\mathrm{hyp}})$.

a. Show that this volume is independent of the choice of the isometry φ. Hint: Exercise 9.12.

b. Show that

$$\mathrm{vol}_{\mathbb{B}^3}(D) = \iiint_D \frac{8}{(1 - x^2 - y^2 - z^2)^3}\, dx\, dy\, dz.$$

Hint: Remember the change of variable formula for triple integral, and use a convenient isometry φ.

Exercise 9.14. Show that the hyperbolic volume of a hyperbolic ball $B_{d_{\mathrm{hyp}}}(P, r) \subset \mathbb{H}^3$ of radius r is equal to $\pi \sinh 2r - 2\pi r$. Hint: It might be convenient to use the ball model $\mathbb{B}^3$, and to consider the case where the ball is centered at the origin O, as in Exercise 2.13a.

Chapter 10

Kleinian groups

This chapter is devoted to kleinian groups, which are certain groups of isometries of the hyperbolic space $\mathbb{H}^3$. We begin with an experimental investigation, in Section 10.1, of the tiling groups that one obtains by suitably bending the Farey tessellation into the 3-dimensional hyperbolic space $\mathbb{H}^3$. The following sections develop the basic definitions and properties of kleinian groups. We then return to the examples of Section 10.1 by rigorously proving, at least in some cases, that the phenomena we had experimentally observed correspond to real mathematical facts. This includes a few surprising properties, such as the natural occurrence of continuous curves in the Riemann sphere $\widehat{\mathbb{C}}$ which are differentiable at very few points.

10.1. Bending the Farey tessellation

10.1.1. Crooked Farey tessellations. In Section 8.4.2, we associated to real numbers s_1, s_3, $s_5 \in \mathbb{R}$ with $s_1 + s_3 + s_5 = 0$ a tessellation of the hyperbolic plane $\mathbb{H}^2$, defined as follows. We started with the ideal triangle T^+ with vertices 0, 1 and ∞, and the hyperbolic triangle T^- with vertices $-\mathrm{e}^{-s_5}$, 0 and ∞. We also considered the linear fractional maps

$$\varphi_1(z) = \frac{\mathrm{e}^{s_5} z + 1}{\mathrm{e}^{s_5} z + \mathrm{e}^{-s_1} + 1}, \quad \varphi_2(z) = \varphi_1^{-1}(z) = \mathrm{e}^{-s_5} \frac{-(\mathrm{e}^{-s_1} + 1)z + 1}{z - 1},$$

$$\varphi_3(z) = e^{-s_5}\frac{z-1}{-z+e^{s_3}+1}, \qquad \varphi_4(z) = \varphi_3^{-1}(z) = \frac{e^{s_5}(e^{s_3}+1)z+1}{e^{s_5}z+1}.$$

The tessellation then consisted of all triangles of the form $\varphi(T^+)$ or $\varphi(T^-)$ as φ ranges over all elements of the transformation group Γ generated by φ_1 and φ_3. Recall that Γ consists of all the linear fractional maps of the form

$$\varphi = \varphi_{i_1} \circ \varphi_{i_2} \circ \cdots \circ \varphi_{i_k},$$

where each φ_{i_j} is equal to φ_1, φ_2, φ_3 or φ_4. The standard Farey tessellation corresponded to the case where $s_1 = s_2 = s_3 = 0$.

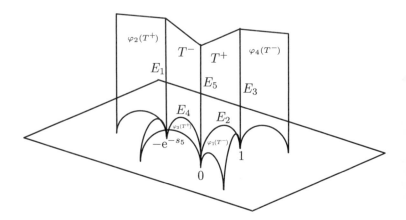

Figure 10.1

We now make the key observation that this setup makes sense even when s_1, s_3, $s_5 \in \mathbb{C}$ are complex numbers, provided that we consider ideal triangles in the hyperbolic space $\mathbb{H}^3$. See Section T.4 in the TOOL KIT for the definition of the complex exponential e^s when $s \in \mathbb{C}$ is a complex number.

Indeed, given complex numbers s_1, s_3, $s_5 \in \mathbb{C}$, let T^+ and T^- be the ideal triangles in the hyperbolic space $\mathbb{H}^3$ with vertices 0, 1, $\infty \in \widehat{\mathbb{C}} = \mathbb{C} \cup \{\infty\}$, and $-e^{-s_5}$, 0 and $\infty \in \widehat{\mathbb{C}}$, respectively. Let E_1, E_2, E_3, E_4, E_5 be their edges, where E_1 goes from $-e^{-s_5}$ to ∞, E_2 from 0 to 1, E_3 from 1 to ∞, E_4 from 0 to $-e^{-s_5}$, and E_5 from 0 to ∞. See Figure 10.1.

The linear fractional maps $\varphi_1(z) = \frac{e^{s_5}z+1}{e^{s_5}z+e^{-s_1}+1}$ and $\varphi_3(z) = e^{-s_5}\frac{z-1}{-z+e^{s_3}+1}$ define isometries of the hyperbolic space $(\mathbb{H}^3, d_{\text{hyp}})$, which we denote by the same letter. We can then consider the group Γ of isometries on $\mathbb{H}^3$ that is generated by φ_1, φ_2, φ_3 and φ_4, and the family $\mathcal{T}$ of all ideal triangles $\varphi(T^\pm)$ as φ ranges over all elements of Γ.

Of course, because the triangles $\varphi(T^\pm)$ are 2-dimensional, there is no way they can cover the whole 3-dimensional space $\mathbb{H}^3$. However, we will see that these triangles still lead to some very interesting geometric objects. We will call the family $\mathcal{T}$ of the triangles $\varphi(T^\pm)$ a (2-dimensional) **crooked Farey tessellation** in $\mathbb{H}^3$.

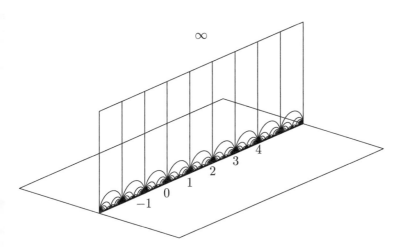

Figure 10.2. Side view of the Farey tessellation

10.1.2. Experimental examples. Let us look at a few examples before going any further. In the real case, the situation where $s_1 + s_3 + s_5 = 0$ was better behaved, as it provided a tessellation of the whole hyperbolic plane $\mathbb{H}^2$. For the same reasons (both mathematical and aesthetic), we will restrict our attention to complex s_1, s_3, $s_5 \in \mathbb{C}$ with $s_1 + s_3 + s_5 = 0$. In particular, this guarantees that the isometry of $\mathbb{H}^3$ defined by

$$\varphi_4 \circ \varphi_2 \circ \varphi_3 \circ \varphi_1(z) = \mathrm{e}^{2s_1+2s_3+2s_5} z + 1 + \mathrm{e}^{s_3} + \mathrm{e}^{s_3+s_1}$$
$$+ \mathrm{e}^{s_3+s_1+s_5} + \mathrm{e}^{2s_3+s_1+s_5} + \mathrm{e}^{2s_3+2s_1+s_5}$$
$$= z + 2 + 2\mathrm{e}^{s_3} + 2\mathrm{e}^{s_3+s_1}$$

of Γ is a horizontal translation. Note that we already encountered this element $\varphi_4 \circ \varphi_2 \circ \varphi_3 \circ \varphi_1$ of Γ in Sections 5.5 and 6.7.2, where it was used to glue together the two sides of the vertical half-strip V.

Since $\varphi_4 \circ \varphi_2 \circ \varphi_3 \circ \varphi_1$ is an element of Γ, the crooked tessellation $\mathcal{T}$ is invariant under this horizontal translation. Indeed, it sends the triangle $\varphi(T^\pm)$ of $\mathcal{T}$ to $\varphi_4 \circ \varphi_2 \circ \varphi_3 \circ \varphi_1 \circ \varphi(T^\pm)$, which is another triangle of $\mathcal{T}$.

First consider the case where $s_1 = s_3 = s_5 = 0$. In this situation, there is no shearing or bending, and we just get the standard Farey tessellation in the hyperbolic plane $\mathbb{H}^2 \subset \mathbb{H}^3$. See Figure 10.2 for a 3-dimensional view.

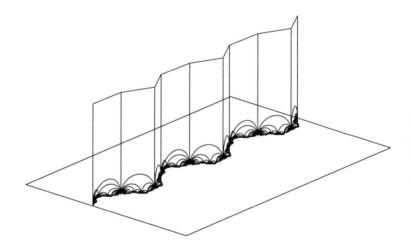

Figure 10.3. Shearing and bending the Farey tessellation a little

The situation becomes more interesting if we slightly move the s_i away from 0. Figure 10.3 represents the case where $s_1 \approx -0.19+0.55\mathrm{i}$, $s_3 \approx 0.15 + 0.42\mathrm{i}$ and $s_5 = 0.04 - 0.97\mathrm{i}$. Here, we have rotated the picture, so that the translation $\varphi_4 \circ \varphi_2 \circ \varphi_3 \circ \varphi_1 \in \Gamma$, which leaves the crooked Farey tessellation $\mathcal{T}$ invariant, appears to be parallel to the x-axis in the standard coordinate frame for $\mathbb{R}^3$.

Note the bending along the edges going to ∞. The bending along the other edges is a little harder to see, but results in the wiggly shape of the crooked tessellation as it approaches the Riemann sphere $\widehat{\mathbb{C}} = \mathbb{C} \cup \{\infty\}$ bounding the hyperbolic space $\mathbb{H}^3$.

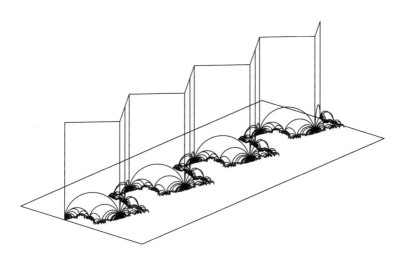

Figure 10.4. Shearing and bending the Farey tessellation much more

Figure 10.4 offers another example where $s_1 = -0.5 + 1.4i$, $s_3 = 0.3 - 1.4i$ and $s_5 = 0.2$. Note that there is no bending between T^+ and T^- because s_5 is real. On the other hand, the picture clearly appears more intricate than Figure 10.3, which is consistent with the fact that s_1 and s_3 are "less real" in this case, in the sense that their imaginary parts are larger.

10.1.3. Footprints. In both Figures 10.3 and 10.4, the ideal triangles of the crooked tessellation appear to draw a relatively complex curve in the sphere $\widehat{\mathbb{C}} = \mathbb{C} \cup \{\infty\}$ bounding the hyperbolic space $\mathbb{H}^3$. More precisely, let us plot the vertices of all ideal triangles. For the cases of Figures 10.2, 10.3 and 10.4, these "footprints" of the crooked tessellation in $\widehat{\mathbb{C}}$ are represented in Figures 10.5, 10.6 and 10.7, respectively.

Figure 10.5. The vertices of the ideal triangles of Figure 10.2 (a boring straight line)

Figure 10.6. The vertices of the ideal triangles of Figure 10.3

Figure 10.7. The vertices of the ideal triangles of Figure 10.4

One can construct many examples in this way. We cannot help including one more example in Figure 10.8, corresponding to $s_1 = 1.831 - 2.16355i$, $s_3 = -s_1$ and $s_5 = 0$, which is particularly pretty.

The pictures of Figures 10.5–10.8 (and Figure 11.2 in Chapter 11) were drawn using the wonderful piece of software *OPTi*, developed by Masaaki Wada [**Wada**]. Figures 10.2–10.4 (and many other complex pictures in this book) were drawn using *Mathematica*® programs written by the author.

What is very clearly apparent in Figures 10.5, 10.6 and 10.7 is that these footprints appear to form a continuous curve, going from the left-hand side of the picture to the right-hand side, and invariant

Figure 10.8. Another pretty example

under a horizontal translation. Upon closer inspection, the same actually holds for Figure 10.8, where the intricacy of the curve is much more amazing. Except for the straight line of Figure 10.5, these continuous curves do not appear to have a very well-defined tangent line at most points.

Experimentally, as one moves the parameters s_1, s_3 and s_5 away from 0, one observes the following phenomenon. For a while, the crooked tessellation and the limit set look very similar to the examples that we have seen so far, except that the pictures are getting more complex. Then, suddenly, complete chaos occurs: The triangles of the crooked tessellation are everywhere in the space $\mathbb{H}^3$ and in all directions, and its footprints cover the whole Riemann sphere $\widehat{\mathbb{C}}$.

10.2. Kleinian groups and their limit sets

The tiling groups Γ of the crooked tessellations of the previous section are examples of kleinian groups, and the curves drawn by their footprints in $\widehat{\mathbb{C}}$ are the limit sets of these kleinian groups. Let us give the precise definition of these concepts.

A ***kleinian group*** is a group Γ of isometries of the hyperbolic space $(\mathbb{H}^3, d_{\text{hyp}})$ whose action on $\mathbb{H}^3$ is discontinuous.

Arbitrarily choose a point $P_0 \in \mathbb{H}^3$, and consider all limit points of its orbit in $\mathbb{R}^3 \cup \{\infty\}$. More precisely, a ***limit point*** of the orbit $\Gamma(P_0)$ is a point $P \in \mathbb{R}^3 \cup \{\infty\}$ for which there exists a sequence $(\gamma_n)_{n \in \mathbb{N}}$ of elements of Γ such that $\gamma_n(P_0) \neq P$ for every n and, more importantly, such that

$$P = \lim_{n \to \infty} \gamma_n(P_0)$$

in $\mathbb{R}^3 \cup \{\infty\}$ and for the euclidean metric d_{euc}. This means that

$$\lim_{n \to \infty} d_{\text{euc}}(P, \gamma_n(P_0)) = 0$$

when $P \neq \infty$, and that

$$\lim_{n \to \infty} d_{\text{euc}}(P_1, \gamma_n(P_0)) = +\infty$$

for an arbitrary base point P_1 when $P = \infty$.

The ***limit set*** of the kleinian group Γ is the set Λ_Γ of all limit points of the orbit $\Gamma(P_0)$ in $\mathbb{R}^3 \cup \{\infty\}$.

Lemma 10.1. *The limit set Λ_Γ of a kleinian group Γ is contained in the Riemann sphere $\widehat{\mathbb{C}} = \mathbb{C} \cup \{\infty\}$ bounding the hyperbolic space $\mathbb{H}^3$ and is independent of the point $P_0 \in \mathbb{H}^3$ chosen.*

Proof. Clearly, a point $P = (x, y, u) \in \mathbb{R}^3$ with $u < 0$ cannot be a limit point, since all points of the orbit $\Gamma(P_0) \subset \mathbb{H}^3$ have positive u-coordinate.

Suppose that a point $P = (x, y, u) \in \mathbb{H}^3$, with $u > 0$, is in the limit set. Then $\lim_{n \to \infty} d_{\text{euc}}(P, \gamma_n(P_0)) = 0$ for some sequence $(\gamma_n)_{n \in \mathbb{N}}$ in Γ with $\gamma_n(P_0) \neq P$ for every n. A simple comparison of the euclidean and hyperbolic metrics, similar to the one used at the end of our proof of Theorem 6.10 or to the estimates that we use below in this proof, shows that $\lim_{n \to \infty} d_{\text{hyp}}(P, \gamma_n(P_0)) = 0$ as well. In particular, for every ε, there are infinitely many γ_n such that $d_{\text{hyp}}(P, \gamma_n(P_0)) < \varepsilon$, contradicting our hypothesis that Γ acts discontinuously on $\mathbb{H}^3$ (compare Lemma 7.15). Therefore, no $P \in \mathbb{H}^3$ can be a limit point of the orbit $\Gamma(P_0)$ either.

The combination of these two arguments shows that the limit set is contained in the set of $(x, y, u) \in \mathbb{R}^3$ with $u = 0$ plus the point ∞, namely, that the limit set is contained in $\widehat{\mathbb{C}}$.

To show that Λ_Γ is independent of the choice of base point, consider another point $P_0' \in \mathbb{H}^3$. Let $\xi = \lim_{n \to \infty} \gamma_n(P_0) \in \widehat{\mathbb{C}}$ be a point on the limit set of $\Gamma(P_0)$.

First consider the case where $\xi \neq \infty$, so that $\xi = (x, y, 0) \in \mathbb{C}$. Write $\gamma_n(P_0) = (x_n, y_n, u_n)$, $d_n = d_{\text{euc}}(\xi, \gamma_n(P_0))$ and $D = d_{\text{hyp}}(P_0, P_0')$ to simplify the notation. Then, for every point $P' = (x', y', u')$ of the hyperbolic geodesic g joining $\gamma_n(P_0)$ to $\gamma_n(P_0')$,

$$\log \frac{u'}{u_n} \leqslant d_{\text{hyp}}\big(\gamma_n(P_0), P'\big) \leqslant d_{\text{hyp}}\big(\gamma_n(P_0), \gamma_n(P_0')\big) = d_{\text{hyp}}(P_0, P_0') = D$$

by Lemma 2.5, and since γ_n is a hyperbolic isometry. As a consequence, for every point (x', y', u') of the geodesic g going from $\gamma_n(P_0)$ to $\gamma_n(P_0')$, its third coordinate is such that

$$u' \leqslant \mathrm{e}^D u_n \leqslant \mathrm{e}^D d_{\text{euc}}\big(\xi, \gamma_n(P_0)\big) = \mathrm{e}^D d_n.$$

From the formula for the hyperbolic length, we conclude that

$$\begin{aligned} d_{\text{euc}}\big(\gamma_n(P_0), \gamma_n(P_0')\big) &\leqslant \ell_{\text{euc}}(g) \leqslant \mathrm{e}^D d_n\, \ell_{\text{hyp}}(g) \\ &\leqslant \mathrm{e}^D d_n\, d_{\text{hyp}}\big(\gamma_n(P_0), \gamma_n(P_0')\big) \\ &\leqslant \mathrm{e}^D d_n\, d_{\text{hyp}}(P_0, P_0') = \mathrm{e}^D d_n\, D. \end{aligned}$$

Since $\lim_{n \to \infty} d_n = 0$, this proves that $d_{\text{euc}}\big(\gamma_n(P_0), \gamma_n(P_0')\big)$ converges to 0 as n tends to ∞, so that

$$\lim_{n \to \infty} \gamma_n(P_0') = \lim_{n \to \infty} \gamma_n(P_0) = \xi$$

for the metric d_{euc}. As a consequence, ξ is also a limit point of the orbit $\Gamma(P_0')$.

The argument is similar when $\xi = \infty$. Then

$$\lim_{n \to \infty} d_{\text{euc}}\big(O, \gamma_n(P_0)\big) = \infty$$

for the origin $O = (0, 0, 0)$. If the coordinates u_n admit an upper bound U, namely, if $u_n \leqslant U$ for every n, then the same argument as before shows that

$$d_{\text{euc}}\big(\gamma_n(P_0), \gamma_n(P_0')\big) \leqslant \mathrm{e}^D U D,$$

so that by the Triangle Inequality,

$$d_{\text{euc}}\big(O, \gamma_n(P_0')\big) \geqslant d_{\text{euc}}\big(O, \gamma_n(P_0)\big) - d_{\text{euc}}\big(\gamma_n(P_0), \gamma_n(P_0')\big)$$
$$\geqslant d_{\text{euc}}\big(O, \gamma_n(P_0')\big) - \mathrm{e}^D U D$$

tends to ∞ as n tends to infinity. Otherwise, there is a subsequence $\big(\gamma_{n_k}\big)_{k \in \mathbb{N}}$ such that $\lim_{k \to \infty} u_{n_k} = \infty$. Then, another application of Lemma 2.5 shows that the third coordinate of $\gamma_{n_k}(P_0') = (x_{n_k}', y_{n_k}', u_{n_k}')$ is such that $u_{n_k}' \geqslant u_{n_k} \mathrm{e}^{-D}$, so that $d_{\text{euc}}\big(O, \gamma_{n_k}(P_0')\big) \geqslant u_{n_k}'$ tends to ∞ as k tends to ∞. Therefore, ∞ is a limit point of the orbit $\Gamma(P_0')$ in both cases.

This proves that every limit point of $\Gamma(P_0)$ is also a limit point of $\Gamma(P_0')$. By symmetry, it follows that $\Gamma(P_0)$ and $\Gamma(P_0')$ have the same limit points, so that Λ_Γ does not depend on the choice of the base point $P_0 \in \mathbb{H}^3$. $\square$

Lemma 10.2. *The limit set Λ_Γ of a kleinian group Γ is closed in the Riemann sphere $\widehat{\mathbb{C}}$, and is invariant under the action of Γ on $\widehat{\mathbb{C}}$ in the sense that it is respected by every $\gamma \in \Gamma$.*

Proof. If ξ_∞ is the limit of a sequence of points $\xi_n \in \Lambda_\Gamma$, then, picking for each n a point $\gamma_n(P_0)$ of the orbit of the base point $P_0 \in \mathbb{H}^3$ that is sufficiently close to ξ_n, we see that ξ_∞ is also the limit of a sequence of points $\gamma_n(P_0)$ of the orbit $\Gamma(P_0)$. (Exercise: Write a complete proof of this statement with the appropriate ε's, and distinguish the cases according to whether $\xi = \infty$ or not.) As a consequence, ξ_∞ is in the limit set Λ_Γ. Since this holds for any such converging sequence in Λ_Γ, this proves that the limit set Λ_Γ is closed.

If $\xi \in \Lambda_\Gamma$ and $\gamma \in \Gamma$, consider a sequence of points $\gamma_n(P_0)$ of the orbit $\Gamma(P_0)$ converging to ξ. Then the points $\gamma \circ \gamma_n(P_0)$ are also in $\Gamma(P_0)$, and converge to $\gamma(\xi)$ by continuity of γ. As a consequence, $\gamma(\xi)$ is also in the limit set Λ_Γ. Since this holds for every $\xi \in \Lambda_\Gamma$ and $\gamma \in \Gamma$, this proves that Λ_Γ is invariant under the action of Γ. $\square$

The following result shows that except in a few degenerate cases, the limit set Λ_Γ is the smallest subset of $\widehat{\mathbb{C}}$ satisfying the conclusions of Lemma 10.2.

To
page 252

Proposition 10.3. *Let K be a closed subset of the Riemann sphere $\widehat{\mathbb{C}}$ which is invariant under the action of a kleinian group Γ and which has at least two points. Then K contains the limit set Λ_Γ.*

Proof. Since K has at least two points, there exists a complete geodesic g of $\mathbb{H}^3$ whose endpoints are in $K \subset \widehat{\mathbb{C}}$. Pick a base point P_0 on this geodesic g.

If ξ is a point of the limit set Λ_Γ, first consider the case where ξ is different from ∞. Then, for every ε, the euclidean ball $B_{d_{\mathrm{euc}}}(\xi, \varepsilon)$ contains an element $\gamma(P_0)$ of the orbit $\Gamma(P_0)$, with $\gamma \in \Gamma$. In particular, the geodesic $\gamma(g)$ meets this ball $B_{d_{\mathrm{euc}}}(\xi, \varepsilon)$. Because g is also a euclidean semi-circle orthogonal to $\mathbb{R}^2$ in $\mathbb{R}^3$, elementary euclidean geometry shows that at least one of the endpoints ξ' of $\gamma(g)$ is contained in $B_{d_{\mathrm{euc}}}(\xi, \varepsilon)$. Note that ξ' is in the subset K, since the endpoints of g are in K and since this set is invariant under the action of Γ. Therefore, for every ε, we found a point $\xi' \in K$ such that $d_{\mathrm{euc}}(\xi, \xi') < \varepsilon$. Since K is closed, this shows that ξ belongs to K.

The argument is similar in the case where $\xi = \infty$. The combination of both cases proves that every point of Λ_γ also belongs to K, so that Λ_Γ is contained in K. $\qquad\square$

See Exercise 10.1 for an example showing that in Proposition 10.3 the hypothesis that K has at least two elements is necessary.

Proposition 10.4. *Suppose that the quotient $(\mathbb{H}^3/\Gamma, \bar{d}_{\mathrm{hyp}})$ of the hyperbolic space $\mathbb{H}^3$ by the kleinian group Γ is compact. Then the limit set Λ_Γ is the whole Riemann sphere $\widehat{\mathbb{C}}$.*

Proof. Pick a base point $P_0 \in \mathbb{H}^3$.

We claim that there is a number D such that $\bar{d}_{\mathrm{hyp}}(\bar{P}, \bar{P}_0) < D$ for every $\bar{P} \in \mathbb{H}^3/\Gamma$. Indeed, we would otherwise be able to construct, by induction, a sequence $(\bar{P}_n)_{n \in \mathbb{N}}$ in $\mathbb{H}^3/\Gamma$ such that $d_{\mathrm{hyp}}(\bar{P}_n, \bar{P}_0) \geqslant n$ for every n. By the compactness hypothesis on $\mathbb{H}^3/\Gamma$, there exists a subsequence $\left(\bar{P}_{n_k}\right)_{k\in\mathbb{N}}$ converging to some $\bar{P}_\infty$, so that

$$\lim_{k\to\infty} d_{\text{hyp}}(\bar{P}_{n_k}, \bar{P}_0) = d_{\text{hyp}}(\bar{P}_\infty, \bar{P}_0)$$

(for instance, use the inequality of Exercise 1.4). But this would contradict the fact that $\lim_{n\to\infty} d_{\text{hyp}}(\bar{P}_{n_k}, \bar{P}_0) = \infty$ by construction of the sequence $(\bar{P}_n)_{n\in\mathbb{N}}$. This proves the existence of a constant D as required.[1]

By Proposition 7.6, $\bar{d}_{\text{hyp}}(\bar{P}, \bar{P}_0)$ is the infimum of the distances $d_{\text{hyp}}(P, \gamma(P_0))$ as γ ranges over all elements of Γ. Therefore, for every $P \in \mathbb{H}^3$, there exists a point $\gamma(P_0)$ of the orbit $\Gamma(P_0)$ such that $d(P, \gamma(P_0)) < D$.

Given a point ξ of the sphere at infinity $\widehat{\mathbb{C}}$ which is not ∞, namely, given $\xi \in \mathbb{C}$, apply the above property to any point $P \in \mathbb{H}^3$ which is at euclidean distance $< \varepsilon$ from ξ. This provides a point $\gamma(P_0) \in \Gamma(P_0)$ such that $d_{\text{hyp}}(P, \gamma(P_0)) < D$. The same estimates as in the proof of Lemma 10.1 then show that $d_{\text{euc}}(P, \gamma(P_0)) < e^D \varepsilon D$, so that $d_{\text{euc}}(\xi, \gamma(P_0)) < \varepsilon(e^D D + 1)$. Taking ε sufficiently small, we conclude that there are points of the orbit $\Gamma(P_0)$ which are at arbitrarily small euclidean distance from ξ. Namely, this shows that $\xi \in \mathbb{C}$ is in the limit set Λ_Γ of Γ.

From page 251

For the point ∞, pick a point $z \in \mathbb{R}$ that is very close to ∞, namely, very far from $O = (0,0,0)$ for the euclidean metric. By the previous case, the orbit $\Gamma(P_0)$ contains points which are very close to z, again for the euclidean metric. It follows that there are points of $\Gamma(P_0)$ which are arbitrarily close to ∞ for the euclidean metric. Namely, ∞ is in the limit set Λ_Γ.   □

10.3. First rigorous example: fuchsian groups

To page 257

In Chapter 6, our examples of various tessellations of the hyperbolic plane $\mathbb{H}^2$ provided us with many examples of groups acting by isometry and discontinuously on $\mathbb{H}^2$, arising as the tiling groups of these tessellations (Proposition 7.10 guarantees discontinuity). Let us see how these also lead to examples of kleinian groups.

---

[1]If you have taken a course in analysis or topology, you may recognize here the proof that every continuous function on a compact space is bounded, as applied to the function $f(\bar{P}) = \bar{d}(\bar{P}, \bar{P}_0)$.

An isometry φ of $(\mathbb{H}^2, d_{\mathrm{hyp}})$ is determined by its extension to the circle at infinity $\widehat{\mathbb{R}} = \mathbb{R} \cup \{\infty\}$, which is a linear or antilinear map $\varphi(x) = \frac{ax+b}{cx+d}$ of $\frac{cx+d}{ax+b}$ with a, b, c, $d \in \mathbb{R}$ and $ad - bc = 1$. Replacing $x \in \widehat{\mathbb{R}}$ by $z \in \widehat{\mathbb{C}} = \mathbb{C} \cup \{\infty\}$ in the above formula yields a linear or antilinear transformation $\varphi \colon \widehat{\mathbb{C}} \to \widehat{\mathbb{C}}$ defined by $\varphi(z) = \frac{az+b}{cz+d}$ in the first case, and $\frac{c\bar{z}+d}{a\bar{z}+b}$ in the second case. This linear or antilinear transformation of $\widehat{\mathbb{C}}$ uniquely extends to an isometry $\varphi \colon \mathbb{H}^3 \to \mathbb{H}^3$ of $(\mathbb{H}^3, d_{\mathrm{hyp}})$.

Since $\varphi \colon \widehat{\mathbb{C}} \to \widehat{\mathbb{C}}$ respects the circle $\widehat{\mathbb{R}} \subset \widehat{\mathbb{C}}$, its extension $\varphi \colon \mathbb{H}^3 \to \mathbb{H}^3$ respects the hyperbolic plane that is bounded by this circle, which is

$$\mathbb{H}^2 = \{(x, u) \in \mathbb{R}^2; u > 0\} = \{(x, 0, u) \in \mathbb{R}^2; u > 0\} \subset \mathbb{H}^3.$$

The restriction of $\varphi \colon \mathbb{H}^3 \to \mathbb{H}^3$ to $\mathbb{H}^2$ then is exactly the original isometry $\varphi \colon \mathbb{H}^2 \to \mathbb{H}^2$. Indeed, the property is immediate when φ is a homothety, a horizontal translation of an inversion across a circle centered on $\widehat{\mathbb{C}}$ (in which case the 3-dimensional extension is the inversion across the sphere with the same center and the same radius), and the general case follows from these special examples since Lemma 2.12 (or, more precisely, the proof of Lemma 2.9) provides a decomposition of every isometry of $\mathbb{H}^2$ as a composition of homotheties, translations and inversions.

In this way, each isometry φ of the hyperbolic plane $\mathbb{H}^2$ has a natural extension to the hyperbolic space $\mathbb{H}^3$.

We can give a less algebraic and more geometric description of this extension in the following way. There is a natural **orthogonal projection** $p \colon \mathbb{H}^3 \to \mathbb{H}^2$ defined as follows (compare Exercise 2.5). For $P \in \mathbb{H}^3$, there is a unique complete geodesic g that passes through P and is orthogonal to $\mathbb{H}^2$; then $p(P)$ is the point where g meets $\mathbb{H}^2$. Elementary geometry shows that in cartesian coordinates, $p(P) = (x, 0, \sqrt{y^2 + u^2})$ if $P = (x, y, u)$. This construction also provides a **signed distance function** $q \colon \mathbb{H}^3 \to \mathbb{R}$ defined by $q(P) = \pm d_{\mathrm{hyp}}(P, p(P))$ where $\pm$ is the sign of the y-coordinate of $P = (x, y, u)$. See Figure 10.9.

Lemma 10.5. *If φ is an isometry of the hyperbolic plane $(\mathbb{H}^2, d_{\mathrm{hyp}})$, its extension to an isometry of the hyperbolic space $(\mathbb{H}^3, d_{\mathrm{hyp}})$ sends*

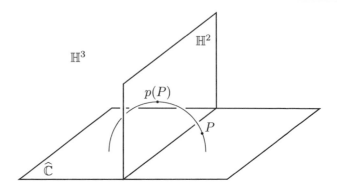

Figure 10.9. The orthogonal projection p

the point $P \in \mathbb{H}^3$ to the point of $\mathbb{H}^3$ that is at the same signed distance $q(\psi(P)) = q(P)$ from $\mathbb{H}^2$ as P, and that projects to the point
$p(\psi(P)) = \varphi(p(P)) \in \mathbb{H}^2$ image of $p(P)$ under φ.

Proof. This immediately follows from the fact that the extension
of φ to $\mathbb{H}^3$ is an isometry and respects each of the two hyperbolic
half-spaces delimited by $\mathbb{H}^2$ in $\mathbb{H}^3$.   □

Lemma 10.6. Let the group Γ act by isometries and discontinuously
on the hyperbolic plane $(\mathbb{H}^2, d_{\mathrm{hyp}})$. Extend this action to an isometric
action on the hyperbolic space $(\mathbb{H}^3, d_{\mathrm{hyp}})$ as above. Then, the action
of Γ on $\mathbb{H}^3$ is discontinuous.

Proof. Consider a point $P \in \mathbb{H}^3$ and its projection $p(P) \in \mathbb{H}^2$. By
hypothesis, the action of Γ on $\mathbb{H}^2$ is discontinuous at $p(P)$. This
means that there exists a small radius $\varepsilon > 0$ such that there are only
finitely many $\gamma \in \Gamma$ for which $\gamma(p(P))$ is in the ball $B_{d_{\mathrm{hyp}}}(p(P), \varepsilon)$.

By continuity of the orthogonal projection $p \colon \mathbb{H}^3 \to \mathbb{H}^2$, there
exists an η such that $d_{\mathrm{hyp}}(p(P), p(Q)) < \varepsilon$ whenever $d_{\mathrm{hyp}}(P, Q) < \delta$.
In other words, the image of the ball $B_{d_{\mathrm{hyp}}}(P, \delta)$ under p is contained
in the ball $B_{d_{\mathrm{hyp}}}(p(P), \varepsilon)$.

If $\gamma(P)$ is in the ball $B_{d_{\text{hyp}}}(P, \delta)$, then $p(\gamma(P))$ is in the ball $B_{d_{\text{hyp}}}(p(P), \varepsilon)$. Lemma 10.5 shows that $p(\gamma(P)) = \gamma(p(P))$. Therefore, $\gamma(P) \in B_{d_{\text{hyp}}}(P, \delta)$ only when $\gamma(p(P)) \in B_{d_{\text{hyp}}}(p(P), \varepsilon)$. By choice of ε, this occurs only for finitely many $\gamma \in \Gamma$.

This proves that the action of Γ is discontinuous at every $P \in \mathbb{H}^3$.   □

By Lemma 10.6, every discontinuous isometric group action on the hyperbolic plane $\mathbb{H}^2$ extends to a group Γ of isometries of $\mathbb{H}^3$ whose action is also discontinuous on $\mathbb{H}^3$. Namely, this group Γ is a kleinian group. A kleinian group obtained in this way is called a *fuchsian group*.

The limit set Λ_Γ of a fuchsian group Γ is clearly contained in the circle $\widehat{\mathbb{R}} = \mathbb{R} \cup \{\infty\}$. Indeed, if we choose the base point P_0 in $\mathbb{H}^2$, its orbit $\Gamma(P_0)$ is completely contained in $\mathbb{H}^2$ so that its limit points must be in $\widehat{\mathbb{R}}$.

If the quotient space $\mathbb{H}^2/\Gamma$ is compact, Proposition 10.4 (adapted to one dimension lower) shows that Λ_Γ is the whole circle $\widehat{\mathbb{R}}$. For instance, this will occur when Γ is the tiling group of a tessellation associated to the gluing of edges of a bounded polygon in $\mathbb{H}^2$, as in the hyperbolic examples of Section 6.5.

The following lemma enables us to generalize this to more tiling groups.

Lemma 10.7. *In the hyperbolic plane $\mathbb{H}^2$, let X be a polygon with edge gluing data satisfying the conditions of Theorem 6.1, so that the images of X under the element of the tiling group Γ generated by the gluing maps form a tessellation of $\mathbb{H}^2$. Suppose in addition that X touches the circle at infinity $\widehat{\mathbb{R}}$ at only finitely many points (possibly none). Then, if we extend Γ to a fuchsian group, the limit set Λ_Γ is equal to the whole circle $\widehat{\mathbb{R}}$.*

Proof. Let $(\bar{X}, \bar{d}_X)$ be the quotient space obtained by performing the prescribed gluings on the edges of X. By Theorem 6.25, we can choose a horocircle C_ξ at each ideal vertex ξ of X so that whenever a gluing map sends ξ to ξ', it also sends C_ξ to $C_{\xi'}$. In addition, by Complement 6.24, these horocircles can be chosen small enough that they are disjoint and that the only edges met by C_ξ are the two edges leading to ξ. For each ideal vertex ξ, let B_ξ be the horodisk bounded by the horocircle C_ξ.

Then the images of $X \cap \bigcup_\xi B_\xi$ under all elements of the tiling group Γ is the union B of a family of (usually infinitely many) disjoint horodisks in $\mathbb{H}^3$. Also, the hypothesis that X touches the circle at infinity only at the ideal vertices ξ implies that if we clip the B_ξ off X, what is left is bounded in $\mathbb{H}^2$. Namely, for an arbitrary base point $P_0 \in X$, the complement $X - \bigcup_\xi B_\xi$ is contained in a large ball $B_{d_{\mathrm{hyp}}}(P_0, D)$.

The argument is then very similar to that of Proposition 10.4. Let $x \in \mathbb{R}$ be a point of the circle at infinity $\widehat{\mathbb{R}}$ which is different from ∞. Pick a point $P \in \mathbb{H}^2$ which is at euclidean distance $< \varepsilon$ from x. If P happens to be in one of the horodisks of B, we can move it out of B while keeping it at euclidean distance $< \varepsilon$ from x (since B consists of disjoint euclidean disks tangent to $\mathbb{R}$). As a consequence, we can always choose the point P outside of B. Since the images of X under the elements of Γ tessellate $\mathbb{H}^2$, there exists $\gamma \in \Gamma$ be such that P is in the tile $\gamma(X)$. Then, P is at hyperbolic distance $\leqslant D$ from the point $\gamma(P_0)$ of the orbit $\Gamma(P_0)$ and, as in the proofs of Lemma 10.1 and Proposition 10.4, at euclidean distance $< \mathrm{e}^D \varepsilon D$ from this point $\gamma(P_0)$. Taking ε sufficiently small, it follows that the orbit $\Gamma(P_0)$ contains points which are at arbitrarily small euclidean distance from x. In other words, $x \in \mathbb{R}$ is contained in the limit set Λ_Γ.

For the point ∞, pick a point $x \in \mathbb{R}$ which is very close to ∞, namely, very far from the origin O for the euclidean metric. By the previous case, there are points of $\Gamma(P_0)$ that are very close to x, again for the euclidean metric. It follows that there are points of $\Gamma(P_0)$ that are arbitrarily close to ∞ for the euclidean metric. Namely, ∞ is in the limit set Λ_Γ. $\square$

According to a traditional terminology, a fuchsian group is said to be **of the first type** if its limit set is the whole circle at infinity $\widehat{\mathbb{R}}$; otherwise, it is **of the second type**. Lemma 10.7 and Chapter 5 provide many examples of fuchsian groups of the first type.

Simple examples of fuchsian groups of the second type include the following: the group Γ_1 generated by the inversion $z \mapsto \frac{1}{\bar{z}}$, in which case the limit set Λ_{Γ_1} is empty (Γ_1 is finite with only two elements); the group Γ_2 generated by the translation $z \mapsto z + 1$, for which Λ_{Γ_2} consists of the single point ∞; the group Γ_3 generated by the homothety $z \mapsto 2z$, for which $\Lambda_{\Gamma_3} = \{0, \infty\}$ has two elements. See Exercise 10.11 for a fuchsian group of the second type whose limit set is not finite.

Fuchsian groups are convenient examples of kleinian groups, but not very exciting because of their intrinsically 2-dimensional nature. The results of the next section will enable us to construct more interesting examples.

From page 252

10.4. Poincaré's Polyhedron Theorem

Poincaré's Polyhedron Theorem is the natural generalization to three dimensions of the Tessellation Theorem 6.1 and of Poincaré's Polygon Theorem 6.25. It will enable us to construct tessellations of the hyperbolic space $\mathbb{H}^3$ by hyperbolic polyhedra, and will provide many examples of kleinian groups.

10.4.1. Gluing the faces of a hyperbolic polyhedron. We define polyhedra in the 3-dimensional hyperbolic space $\mathbb{H}^3$. In order to do this, we first need to fix conventions about polygons.

A *polygon* in $\mathbb{H}^3$ is a subset F of a hyperbolic plane $\Pi \subset \mathbb{H}^3$ which is delimited in Π by finitely many geodesics called its *edges*. Recall that the hyperbolic plane Π is isometric to the standard hyperbolic plane $\mathbb{H}^2$. Under any such isometry, the polygon contained in Π then corresponds to a polygon in $\mathbb{H}^2$, as defined in Section 4.5.1. As in the case of $\mathbb{H}^2$, we require that edges can only meet at their endpoints, called *vertices*, and that each vertex is adjacent to exactly two edges. Some of these edges are allowed to be infinite and lead to a point of the sphere at infinity $\mathbb{C} \cup \{\infty\}$; such a point is a *vertex at infinity* or an *ideal vertex* of the polygon. Also, the polygon F is required to contain all its edges and vertices, so that F is a closed subset of $\mathbb{H}^3$.

A **polyhedron** in the hyperbolic space $\mathbb{H}^3$ is a region X in $\mathbb{H}^3$ delimited by finitely many polygons, called its **faces**. By convention, faces can only meet along some of their edges and vertices, and an edge is adjacent to exactly two faces. As for polygons, we require that a polygon contains all of its edges and vertices, so that it is a closed subset of $\mathbb{H}^3$.

The **edges** and **vertices** of a polyhedron are the edges and vertices of its faces. An **ideal vertex** or **vertex at infinity** is a point of the sphere at infinity $\widehat{\mathbb{C}}$ which touches two distinct faces of the polyhedron. There ideal vertices can be of two types. The most common one is that of an ideal vertex of a face, also corresponding to an endpoint of an infinite edge. However, it is also possible for two faces F_1 and F_2 to touch at a point $\xi \in \widehat{\mathbb{C}}$ without ξ being the endpoint of an edge; this happens when F_1 and F_2 touch $\widehat{\mathbb{C}}$ along two circle arcs that are tangent to each other at ξ. To avoid unnecessary complications, we assume that the polygon X approaches an ideal vertex ξ from only one direction, in the sense that there exists a small euclidean ball $B_{d_{\mathrm{euc}}}(\xi, \varepsilon)$ in $\mathbb{R}^3$ such that the intersection $X \cap B_{d_{\mathrm{euc}}}(\xi, \varepsilon)$ is connected.

We now introduce face gluing data in complete analogy with the case of polygons in Sections 4.3 and 4.5. Namely, we group the faces of the polyhedron X into disjoint pairs $\{F_1, F_2\}$, $\{F_3, F_4\}$, ..., $\{F_{2p-1}, F_{2p}\}$. For each such pair $\{F_{2k-1}, F_{2k}\}$, we are given an isometry $\varphi_{2k-1} \colon F_{2k-1} \to F_{2k}$ with respect to the restriction of the hyperbolic metric d_{hyp} to these faces, and we define $\varphi_{2k} \colon F_{2k} \to F_{2k-1}$ to be the inverse $\varphi_{2k} = \varphi_{2k-1}^{-1}$.

Let $\bar{X}$ be the partition of X defined by the property that for every $P \in X$ the corresponding element $\bar{P} \in \bar{X}$ containing P consists of all points $Q = \varphi_{i_k} \circ \varphi_{i_{k-1}} \circ \cdots \circ \varphi_{i_1}(P)$, where the indices i_1, i_2, ..., i_k are such that $\varphi_{i_{j-1}} \circ \cdots \circ \varphi_{i_1}(P) \in F_{i_j}$ for every j.

Endow X with the metric d_X defined by the property that $d_X(P, Q)$ is the infimum of the hyperbolic lengths of all piecewise differentiable curves joining P to Q in X. We can then consider the quotient semi-metric $\bar{d}_X$ on $\bar{X}$.

The proof of Theorem 4.3 automatically extends to this context to prove:

Lemma 10.8. *The semi-metric $\bar{d}_X$ is a metric, in the sense that $\bar{d}_X(\bar{P}, \bar{Q}) > 0$ whenever $\bar{P} \neq \bar{Q}$.* $\qquad\qquad\Box$

10.4.2. Poincaré's Polyhedron Theorem. We now consider a 3-dimensional version of the Tessellation Theorem 6.1 and of Poincaré's Polygon Theorem 6.25. It is convenient to combine the two results into a single statement.

Using the homogeneity and isotropy of $\mathbb{H}^3$ (Theorem 9.2), Lemma 4.8 immediately extends to the 3-dimensional context, and provides a unique extension of each gluing map $\varphi_i \colon F_i \to F_{i\pm1}$ to a hyperbolic isometry $\varphi_i \colon \mathbb{H}^3 \to \mathbb{H}^3$ that along F_i, sends X to the side of $F_{i\pm1}$ that is opposite X.

As before, let the **tiling group** associated to the above gluing data be the group Γ of isometries of $(\mathbb{H}^3, d_{\mathrm{hyp}})$ generated by these extended gluing maps $\varphi_i \colon \mathbb{H}^3 \to \mathbb{H}^3$.

As in dimension 2, the images of the polyhedron X under the elements of Γ form a **tessellation** of the hyperbolic space $\mathbb{H}^3$ if:

(1) as γ ranges over all elements of the tiling group Γ, the tiles $\gamma(X)$ cover the whole space $\mathbb{H}^3$, in the sense that their union is equal to $\mathbb{H}^3$;

(2) the intersection of any two distinct tiles $\gamma(X)$ and $\gamma'(X)$ consists only of vertices, edges and faces of $\gamma(X)$, which are also vertices, edges or faces of $\gamma'(X)$;

(3) (Local Finiteness) for every point $P \in \mathbb{H}^3$, there exists a ball $B_{d_{\mathrm{hyp}}}(P, \varepsilon)$ which meets only finitely many tiles $\gamma(X)$.

If, in addition, distinct $\gamma \in \Gamma$ give distinct tiles $\gamma(X)$, the polyhedron X is a **fundamental domain** for the action of Γ on $\mathbb{H}^3$.

The bending of the boundary of a hyperbolic polyhedron X along an edge E is measured by its dihedral angle along E, which is defined as in euclidean geometry. More precisely, for a point $P \in E$ which is not a vertex and for $\varepsilon > 0$ sufficiently small, the hyperbolic plane Π orthogonal to E at P cuts the intersection $X \cap B_{d_{\mathrm{hyp}}}(P, \varepsilon)$ along a disk sector of radius ε and of angle θ. The **dihedral angle** of the polyhedron X along the edge E is exactly this angle θ. One easily checks that it does not depend on the point $P \in E$ (see Exercise 9.1).

Theorem 10.9 (Poincaré's Polyhedron Theorem). *For a connected polyhedron $X \subset \mathbb{H}^3$ with face gluing data as in Section 10.4.1, suppose in addition that the following three conditions hold:*

(1) *(Dihedral Angle Condition) for every edge E of the polyhedron X, the dihedral angles of X along the edges that are glued to E add up to $\frac{2\pi}{n_E}$ for some integer $n_E \geqslant 1$ depending on E;*

(2) *(Edge Orientation Condition) the edges of the polyhedron X can be oriented in such a way that whenever a gluing map $\varphi_i \colon F_i \to F_{i\pm 1}$ sends an edge E to an edge E', it sends the orientation of E to the orientation of E';*

(3) *(Horosphere Condition) for every ideal vertex ξ of X, we can select a horosphere S_ξ such that whenever the gluing map $\varphi_i \colon F_i \to F_{i\pm 1}$ sends the ideal vertex to the ideal vertex ξ', it also sends S_ξ to $S_{\xi'}$.*

Then, as γ ranges over all the elements of the tiling group Γ generated by the extended gluing maps $\varphi_i \colon \mathbb{H}^3 \to \mathbb{H}^3$, the tiles $\gamma(X)$ form a tessellation of the hyperbolic space $\mathbb{H}^3$.

In addition, the tiling group Γ acts discontinuously on $\mathbb{H}^3$, the two quotient spaces $(\mathbb{H}^3/\Gamma, \bar{d}_{\mathrm{hyp}})$ and $(\bar{X}, \bar{d}_X)$ are isometric, and these two metric spaces are complete.

In the Edge Orientation Condition (2), an orientation for the edge E is the choice of a traveling direction along E, usually indicated by an arrow. This is easily ensured by adding a few vertices, provided the Horosphere Condition (3) holds. Indeed, for a semi-infinite edge E going from a finite vertex $P \in \mathbb{H}^3$ to an ideal vertex $\xi \in \widehat{\mathbb{C}}$, we can orient E in the direction from P to ξ. For a finite edge E joining two finite vertices, we can introduce a new vertex M at its midpoint, split E accordingly, and orient the resulting two edges E' and E'' away from M. For a bi-infinite edge E joining two ideal vertices ξ' and ξ'', we can again introduce a new vertex M at the midpoint of the two points $E \cap S_{\xi'}$ and $E \cap S_{\xi''}$ determined by the horospheres provided by the Horosphere Condition (3), and orient the resulting two new semi-infinite edges E' and E'' away from M. The setup is then specially designed so that the gluing maps $\varphi_i \colon F_i \to F_{i\pm 1}$ respect these new vertices, and also respect the edge orientations so defined.

As in the 2-dimensional case, the Dihedral Angle Condition (1) and the Horosphere Condition (3) are much more critical. The Edge Orientation Condition (2) was essentially introduced to simplify the statement of the Dihedral Angle Condition (1).

Proof of Theorem 10.9. The proof is essentially identical to the 2-dimensional case, with only a minor twist indicated in Lemma 10.10 To page 263 below. Consequently, we only give a sketch of the arguments.

Our proof of Poincaré's Polygon Theorem 6.25 immediately extends to the 3-dimensional context to show that because of the Horosphere Condition (3) (to be compared with condition (1) of Proposition 6.23), the quotient space $(\bar{X}, \bar{d}_X)$ is complete.

To prove Theorem 10.9, we follow the strategy of the proof of the Tessellation Theorem 6.1, by setting one tile after the other. In particular, a tile $\varphi(X)$ is **adjacent** to X at P if there exists a sequence of gluing maps $\varphi_{i_1}, \varphi_{i_2}, \ldots, \varphi_{i_l}$ such that $\varphi_{i_{j-1}} \circ \cdots \circ \varphi_{i_1}(P)$ belongs to the face F_{i_j} for every $j \leqslant l$ (including the fact that $P \in F_{i_1}$) and

$$\varphi = \varphi_{i_1}^{-1} \circ \varphi_{i_2}^{-1} \circ \cdots \circ \varphi_{i_l}^{-1}.$$

More generally, the tiles $\varphi(X)$ and $\psi(X)$ are **adjacent** at the point $P \in \varphi(X) \cap \psi(X)$ if $\psi^{-1} \circ \varphi(X)$ is adjacent to X at $\psi^{-1}(P)$ in the above sense.

The only point that really requires some thought is the following analogue of Lemma 6.2 and Complement 6.3. We need some terminology. Let a **polyhedral ball sector** with center P and radius ε in $\mathbb{H}^3$ be a region in the ball $B_{d_{\text{hyp}}}(P, \varepsilon)$ delimited on its sides by finitely many hyperbolic planes passing through the point P; we allow the case where the polyhedral ball sector is the whole ball $B_{d_{\text{hyp}}}(P, \varepsilon)$. This is the 3-dimensional analogue of disk sectors, specially designed so that every point of X is the center of some ball $B_{d_{\text{hyp}}}(P, \varepsilon)$ such that $X \cap B_{d_{\text{hyp}}}(P, \varepsilon)$ is a polyhedral ball sector.

Lemma 10.10. *There are only finitely many tiles $\varphi(X)$ that are adjacent to X at the point P. In addition, these adjacent tiles decompose a small ball $B_{d_{\text{hyp}}}(P, \varepsilon)$ into finitely polyhedral ball sectors with disjoint interiors.*

Proof. First, we observe that there are only finitely many points $P = P_1$, P_2, ..., P_n of X that are glued to P. This is automatic when P is an interior point, since it is only glued to itself. Similarly, a ***face point***, namely, a point of a face which is not contained in an edge, is only glued to one other point, and a ***vertex*** can only be glued to some of the finitely many vertices. The case of ***edge points***, namely, points of edges that are not vertices, will require more thought and critically relies on the Horosphere Condition (3) of Theorem 10.9.

We claim that no two distinct edge points P' and P'' of the same edge E can be glued together. Indeed, if two such points P' and $P'' = \varphi(P')$ are glued together by the gluing map $\varphi = \varphi_{i_1} \circ \varphi_{i_2} \circ \cdots \circ \varphi_{i_k}$, then φ sends the edge E to itself, and respects its orientation by the Edge Orientation Condition (2). In addition, it must send at least one point of E to itself: If both endpoints of E are finite vertices, φ must fix these two endpoints since it respects the orientation of E; if exactly one endpoint of E is a finite vertex, φ must clearly fix this endpoint; finally, when E is a bi-infinite edge joining two ideal vertices ξ and ξ', φ must fix the intersection points $E \cap S_\xi$ and $E \cap S_{\xi'}$ determined by the horospheres S_ξ and $S_{\xi'}$ associated to ξ and ξ' by the Horosphere Condition (3). As a consequence, the restriction of φ to E is an orientation-preserving isometry fixing at least one point, and is consequently the identity. But this would contradict our hypothesis that the edge points P' and $P'' = \varphi(P')$ are distinct.

This proves that the set of points that are glued to an edge point P can include at most one point of each edge, and is consequently finite.

Therefore, in all cases, there are only finitely many points $P = P_1$, P_2, ..., P_n of X that are glued to P.

For ε small enough, the intersection of X with each of the balls $B_{d_{\text{hyp}}}(P_i, \varepsilon)$ is a polyhedral ball sector. Let $S_{d_{\text{hyp}}}(P_j, \varepsilon)$ denote the hyperbolic sphere consisting of the points that are at distance ε from P_i, and let S_j be the intersection $X \cap S_{d_{\text{hyp}}}(P_j, \varepsilon)$. As usual, endow S_j with the metric d_{S_j} where $d_{S_j}(Q, R)$ is the infimum of the hyperbolic lengths of all curves joining Q to R in S_j.

If we rescale the metric d_{S_j} by a factor $\sinh \varepsilon$, Exercise 9.4 shows that the metric space $(S_j, \frac{1}{\sinh \varepsilon} d_{S_j})$ is isometric to a polygon in the sphere $(\mathbb{S}^2, d_{\text{sph}})$, whose angle at each vertex P is equal to the dihedral angle of X along the edge containing P. The restriction of the gluing maps φ_i to the edges of the S_j then defines isometric gluing data for these spherical polygons. Because of the Dihedral Angle Condition (1) of Theorem 10.9 and because no two distinct points of the same edge are glued together, the Tessellation Theorem 6.1 (as extended to disconnected polygons in Section 6.3.4, and using Proposition 6.20 to guarantee completeness of the quotient space) shows that this gluing data provides a tessellation of the sphere $\mathbb{S}^2$.

In particular, if we start from the polygon $S_1 \subset S_{d_{\text{hyp}}}(P, \varepsilon)$ and proceed with the tiling procedure using this gluing data, we obtain a tessellation of the sphere $S_{d_{\text{hyp}}}(P, \varepsilon)$. It immediately follows from definitions that the tiles of this tessellation are exactly the polygons $\varphi(S_j)$, where $\varphi \in \Gamma$ is such that the polyhedron $\varphi(X)$ is adjacent to X at P and where $\varphi(P_j) = P$.

By compactness of $\mathbb{S}^2$, this tessellation of $S_{d_{\text{hyp}}}(P, \varepsilon)$ has finitely many tiles (compare Exercise 6.3). Lemma 10.10 now immediately follows from these observations. $\qquad\square$

Substituting Lemma 10.10 for Lemma 6.2 and Complement 6.3, the proof of Theorem 6.1 now immediately extends to show that the polyhedra $\gamma(X)$ with $\gamma \in \Gamma$ tessellate the hyperbolic space $\mathbb{H}^3$. In other words, X is a fundamental domain for the action of the tiling group Γ on $\mathbb{H}^3$.

Once we have shown that X is a fundamental domain for the action of Γ, the proofs of Proposition 7.10 and Theorem 7.12 immediately extend to three dimensions to show that the action of Γ is discontinuous, and that the quotient space $(\mathbb{H}^3/\Gamma, \bar{d}_{\text{hyp}})$ is isometric to $(\bar{X}, \bar{d}_X)$. $\qquad\square$

From page 261

Similarly, we have the following analogue of Theorem 4.10.

Theorem 10.11. *Under the hypotheses of Theorem 10.9, suppose in addition that:*

(1) *the Dihedral Angle Condition* (1) *of Theorem 10.9 is strengthened so that for every edge E of X, the sum of the dihedral angles of X along the edges that are glued to E is equal to 2π;*

(2–3) *Conditions* (2) *and* (3) *are unchanged;*

(4) *the 3-dimensional extensions $\varphi_i \colon \mathbb{H}^3 \to \mathbb{H}^3$ of the gluing maps are orientation-preserving.*

Then, the quotient space $(\bar{X}, \bar{d}_X)$ is locally isometric to the hyperbolic space $(\mathbb{H}^3, d_{\mathrm{hyp}})$.

To page 265

Proof. We will prove that the action of the tiling group Γ is free, namely, that $\varphi(P) \neq P$ for every $P \in \mathbb{H}^3$ and every $\varphi \in \Gamma$ different from the identity.

Suppose, in search of a contradiction, that some $\varphi \in \Gamma - \{\mathrm{Id}_{\mathbb{H}^3}\}$ fixes a point P of $\mathbb{H}^3$. Then Exercise 9.10 shows that because φ is orientation-preserving by condition (4), it fixes each point of a whole complete geodesic g of $\mathbb{H}^3$.

The geodesic g cannot meet the interior of a tile $\gamma(X)$, with $\gamma \in \Gamma$, of the tessellation provided by Theorem 10.9. Indeed, φ would otherwise send this tile to itself, which is excluded by the fact that X is a fundamental domain for the action of Γ. Therefore, g contains an edge E of some tile $\gamma(X)$.

Replacing φ by $\gamma^{-1} \circ \varphi \circ \gamma$ (which fixes each point of the edge $\gamma^{-1}(E)$ of X) if necessary, we can assume without loss of generality that γ is the identity. In particular, φ now fixes every point of an edge E of X.

Let $X_1 = X$, $X_2 = \gamma_2(X)$, $X_3 = \gamma_3(X)$, ..., $X_n = \gamma_n(X)$ be the tiles sitting around the edge E, and let θ_i be the dihedral angle of the polyhedron X_i along this edge E. Then $\sum_{i=1}^{n} \theta_i = 2\pi$.

By the second half of Theorem 10.9, the inclusion map $X \to \mathbb{H}^3$ induces an isometry $\bar{X} \to \mathbb{H}^3/\Gamma$. In particular, the fact that this map is bijective implies that two points $P, Q \in X$ are glued together if and only if there exists an element $\gamma \in \Gamma$ such that $Q = \gamma(P)$. Therefore, the edges of X that are glued to E are the edges E, $\gamma_2^{-1}(E)$, $\gamma_3^{-1}(E)$, ..., $\gamma_n^{-1}(E)$. Note that the dihedral angle of X along the edge $\gamma_i^{-1}(E)$ is also equal to θ_i.

However, there are duplications in this list. Indeed, the tile $\varphi(X)$ is one of these X_i, so that $\varphi = \gamma_i$ and $\gamma_i^{-1}(E) = \varphi^{-1}(E) = E$ since φ fixes every point of E. Therefore, the sum of the dihedral angles of X along the edges that are glued to E is strictly less than $\sum_{i=1}^{n} \theta_i = 2\pi$, contradicting the stronger Dihedral Angle Condition (1) in the hypotheses of Theorem 10.11.

This proves that the action of Γ on $\mathbb{H}^3$ is free. Since this action is discontinuous by Theorem 10.9, Corollary 7.9 shows that the quotient space $(\mathbb{H}^3/\Gamma, \bar{d}_{\mathrm{hyp}})$ is locally isometric to $(\mathbb{H}^3, d_{\mathrm{hyp}})$. $\qquad\square$

It can actually be shown using Lemma 10.10 that conditions (2–3) of Theorem 10.9 are not really necessary for Theorem 10.11 to hold. Similarly, condition (4) of Theorem 10.11 can be significantly relaxed, but at the expense of making the precise result somewhat more cumbersome to state.

From page 264

Exercise 10.8 shows that Theorem 10.11 fails when only its conditions (1–3) are realized, without condition (4) or any similar additional hypothesis.

10.5. More examples of kleinian groups

We return to the examples of crooked Farey tessellations of Section 10.1. We now have the tools to rigorously justify some of our observations.

10.5.1. Kleinian groups associated to crooked Farey tessellations. We will prove that when the shear-bend parameters s_1, s_3, s_5 are sufficiently small, the tiling group Γ of the crooked Farey tessellation associated to these parameters act discontinuously on the hyperbolic space $\mathbb{H}^3$. We will restrict our attention to the case where s_5 is real, as the arguments are a little simpler with this condition.

First consider our original punctured torus example of Sections 5.5 and 6.6, corresponding to $s_1 = s_3 = s_5 = 0$. Then, the square $T^+ \cup T^-$ is a fundamental domain for the action of Γ on $\mathbb{H}^2$. If we extend the action of Γ on $\mathbb{H}^2$ to a fuchsian action on $\mathbb{H}^3$, Lemma 10.5 shows that this action is well behaved with respect to the orthogonal projection $p\colon \mathbb{H}^3 \to \mathbb{H}^2$. As a consequence, the preimage $X = p^{-1}(T^+ \cup T^-)$ is a fundamental domain for the action of Γ on $\mathbb{H}^3$.

Let us analyze this fundamental domain a little better. As usual, let E_1, E_2, E_3 and E_4 be the edges of the square $T^+ \cup T^-$, where E_1 goes from -1 to ∞, E_2 from 0 to 1, E_3 from 1 to ∞, and E_4 from 0 to -1. Then X is bounded by the preimages $H_i = p^{-1}(E_i)$, and each H_i is a hyperbolic plane in $\mathbb{H}^3$. More precisely, H_1 and H_3 are vertical euclidean half-planes, while H_2 and H_4 are euclidean half-spheres of radius $\frac{1}{2}$ centered at $\pm\frac{1}{2}$. In addition, when we extend the gluing maps to hyperbolic isometries $\varphi_i\colon \mathbb{H}^3 \to \mathbb{H}^3$, φ_1 sends H_1 to H_2 and φ_3 sends H_3 to H_4.

We will now create a very similar fundamental domain in the more general case where s_5 is real and where s_1 and s_3 have an imaginary part which is not too large (and where, as usual, the relation $s_1 + s_3 + s_5 = 0$ is satisfied).

Recall that the tiling group Γ of the corresponding crooked Farey tessellation is generated by

$$\varphi_1(z) = \frac{e^{s_5}z + 1}{e^{s_5}z + e^{-s_1} + 1} \quad \text{and} \quad \varphi_3(z) = e^{-s_5}\frac{z - 1}{-z + e^{s_3} + 1}.$$

Consider the four circles C_1, C_2, C_3, C_4 in $\widehat{\mathbb{C}}$ defined as follows: C_3 is the euclidean line passing through the point 1 and making an angle of $\theta = \frac{1}{2}(\pi - \operatorname{Im}(s_1))$ with the x-axis, counting angles counterclockwise; C_1 is the line parallel to C_3 passing through $-e^{-s_5}$; C_2 is the circle passing through the points 0 and 1 and tangent to the line C_3 at 1; C_4 is the circle passing through 0 and $-e^{-s_5}$ and tangent to the line C_1 at $-e^{-s_5}$. See Figure 10.10.

Lemma 10.12. *If s_5 is real and if $\operatorname{Im}(s_1)$ is sufficiently small that $\cos(\operatorname{Im}(s_1)) > \frac{e^{|s_5|}-1}{e^{|s_5|}+1}$, any two distinct C_i and C_j meet only at 0 or 1 of the points $\{0, 1, -e^{-s_5}, \infty\}$ and are tangent to each other at their intersection point when they meet.*

In addition, φ_1 sends C_1 to C_2, and φ_3 sends C_3 to C_4.

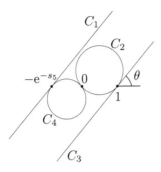

Figure 10.10. A fundamental group for the tiling group of a crooked Farey tessellation

Proof. By elementary euclidean geometry, the two circles C_2 and C_4 have the same tangent line at 0, which makes an angle of $\pi - \theta$ with the x-axis. Therefore, for the first statement, we only need to check that the circle C_2 is disjoint from the line C_1 and that the circle C_4 is disjoint from the line C_3. This is equivalent to the property that the distance between the two lines C_1 and C_3 is greater than the diameters of both C_2 and C_4. Computing these quantities by elementary trigonometry in Figure 10.10 gives the condition stated.

To prove the second statement, note that the differential of φ_3 at the point 1 is the complex multiplication by $e^{-s_5 - s_3}$, by a computation using Proposition 2.15. It follows that the line tangent to $\varphi_3(C_3)$ at $0 = \varphi_3(1)$ makes an angle of $\theta + \text{Im}(-s_5 - s_3)$ with the x-axis. By choice of $\theta = \frac{1}{2}(\pi - \text{Im}(s_1))$ and because of our hypothesis that $-s_5$ is real, this angle is equal to $\pi - \theta$. Since a linear fractional map sends circle to circle (Proposition 2.18), it follows that $\varphi_3(C_3)$ is a circle passing through the points $0 = \varphi_3(1)$ and $-e_{-s_5} = \varphi_3(\infty)$ and whose tangent line at 0 makes an angle of $\pi - \theta$ with the x-axis. There is only one circle with these properties, namely, C_4. Therefore, $\varphi_3(C_3) = C_4$.

The proof that $\varphi_1(C_1) = C_2$ follows from very similar considerations. $\qquad\square$

Proposition 10.13. *Let* Γ *be the tiling group of the crooked Farey tessellation associated to the parameters* s_1, s_3, s_5 *with* s_5 *real and* $\cos(\mathrm{Im}(s_1)) > \frac{e^{|s_5|}-1}{e^{|s_5|}+1}$ *(and* $s_1 + s_3 + s_5 = 0$*). Then, the action of* Γ *on* $\mathbb{H}^3$ *is discontinuous. In particular,* Γ *is a kleinian group.*

Proof. For $i = 1$, 2, 3, 4, let H_i be the hyperbolic plane in $\mathbb{H}^3$ that touches the sphere at infinity $\widehat{\mathbb{C}}$ along the circle C_i. Let X be the polyhedron bounded by these four hyperbolic planes. Namely, X consists of those points in the upper half-space $\mathbb{H}^3$ that are between the euclidean vertical half-planes H_1 and H_3 and above the euclidean half-spheres H_2 and H_4. Note that X is a hyperbolic polyhedron with four faces H_1, H_2, H_3, H_4, with no edge or finite vertex, and with four ideal vertices 0, 1, ∞, $-e^{-s_5}$.

By Lemma 10.12, the isometry $\varphi_1 \colon \mathbb{H}^3 \to \mathbb{H}^3$ sends H_1 to H_2. By considering the image of an additional point, it is also immediate that it sends X to the side of H_3 that is opposite X. Similarly, $\varphi_3 \colon \mathbb{H}^3 \to \mathbb{H}^3$ sends H_3 to H_4 and sends X to the side of H_3 that is opposite X. Consequently, we are in the situation of Poincaré's Polyhedron Theorem 10.9.

The polyhedron X has no edge. Therefore, in order to apply Theorem 10.9, we only need to check the Horosphere Condition (3) of that statement at the ideal vertices of X. As in the 2-dimensional setup of Section 8.4, the hypothesis that $s_1 + s_3 + s_5 = 0$ will be critical here.

We need to find horospheres S_0, S_1, S_∞, $S_{-e^{-s_5}}$ centered at 0, 1, ∞, $-e^{-s_5}$, respectively, such that $\varphi_1(S_\infty) = S_1$, $\varphi_1(S_{-e^{-s_5}}) = S_0$, $\varphi_3(S_\infty) = S_{-e^{-s_5}}$ and $\varphi_3(S_1) = S_0$. We could construct these horospheres "by hand", but it will be easier to use an argument similar to that of Proposition 6.23.

Start with an arbitrary horosphere S_∞ centered at ∞, namely, with an arbitrary horizontal euclidean plane. Then set $S_1 = \varphi_1(S_\infty)$ and $S_{-e^{-s_5}} = \varphi_3(S_\infty)$. We will then be done if we can choose S_0 to be simultaneously equal to $\varphi_3(S_1) = \varphi_3 \circ \varphi_1(S_\infty)$ and to $\varphi_1(S_{-e^{-s_5}}) = \varphi_1 \circ \varphi_3(S_\infty)$, namely, if these two horospheres centered at 0 are equal. The condition that $\varphi_3 \circ \varphi_1(S_\infty) = \varphi_1 \circ \varphi_3(S_\infty)$ is equivalent to the property that $\varphi_3^{-1} \circ \varphi_1^{-1} \circ \varphi_3 \circ \varphi_1 = \varphi_4 \circ \varphi_2 \circ \varphi_3 \circ \varphi_1$ respects the

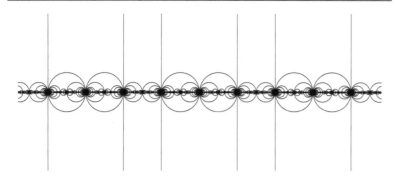

Figure 10.11. The images of the fundamental domain X° under the tiling group Γ° of the standard Farey tessellation

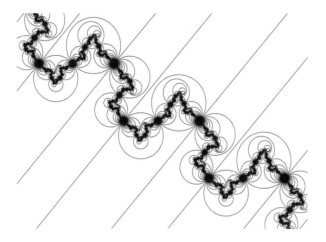

Figure 10.12. The images of the fundamental domain X under the tiling group Γ of a crooked Farey tessellation

horosphere S_∞. We already computed that because $s_1 + s_3 + s_5 = 0$,

$$\varphi_4 \circ \varphi_2 \circ \varphi_3 \circ \varphi_1(z) = e^{2s_1 + 2s_3 + 2s_5} z + 1 + e^{s_3} + e^{s_3 + s_1}$$
$$+ e^{s_3 + s_1 + s_5} + e^{2s_3 + s_1 + s_5} + e^{2s_3 + 2s_1 + s_5}$$
$$= z + 2 + 2e^{s_3} + 2e^{s_3 + s_1}$$

is a translation of $\widehat{\mathbb{C}}$, so that its isometric extension to $\mathbb{H}^3$ is a horizontal translation. In particular, this horizontal translation respects the horizontal plane S_∞, thus we are done.

We can therefore apply Poincaré's Polyhedron Theorem 10.9, and conclude that X is a fundamental domain for the action of Γ on $\mathbb{H}^3$ and that this action is discontinuous. □

The images of the fundamental domain X under the tiling groups are illustrated in Figure 10.11 for the standard case where $s_1 = s_3 = s_5 = 0$, and in Figure 10.12 for a more general example satisfying the hypotheses of Proposition 10.13, in this case $s_1 = -0.5 + 1.4\mathrm{i}$, $s_3 = 0.3 - 1.4\mathrm{i}$ and $s_5 = 0.2$. More precisely, these figures represent in $\widehat{\mathbb{C}}$ the images of the circles C_1, C_2, C_3, C_4 (bounding the hyperbolic planes H_1, H_2, H_3, H_4 delimiting the polyhedron X) under the elements of the tiling group Γ.

We already encountered in Figures 10.4 and 10.7 the crooked Farey tessellation corresponding to the kleinian group of Figure 10.12. When comparing these figures, it is important to remember that Figures 10.4 and 10.7 have been rotated so as to be invariant by a translation parallel to the x-axis.

10.5.2. Limit sets. We now consider the limit sets Λ_Γ of the kleinian groups Γ provided by Proposition 10.13.

The set of ideal vertices of the standard Farey triangulation $\mathcal{T}^\circ$ is the set $\widehat{\mathbb{Q}} = \mathbb{Q} \cup \{\infty\}$ of all rational points in the circle at infinity $\widehat{\mathbb{R}}$. Given a crooked Farey tessellation $\mathcal{T}$, we indicated in Section 10.1 that there is a one-to-one correspondence between the faces, edges and vertices of $\mathcal{T}$ and the faces, edges and vertices of the standard Farey tessellation $\mathcal{T}^\circ$. In particular, each ideal vertex $x \in \widehat{\mathbb{Q}}$ of $\mathcal{T}^\circ$ is associated to an ideal vertex $\lambda(x) \in \widehat{\mathbb{C}}$ of $\mathcal{T}$.

This correspondence defines a map on the set of all rational points $\widehat{\mathbb{Q}}$ in $\widehat{\mathbb{R}}$. We will show that under the hypotheses of Proposition 10.13 (so that the tiling group Γ of $\mathcal{T}$ is a kleinian group), it has a continuous extension to all real numbers.

Proposition 10.14. *Let $\mathcal{T}$ be the crooked Farey tessellation associated to parameters s_1, s_2, s_3 such that s_5 is real and $\cos(\mathrm{Im}(s_1)) > \frac{\mathrm{e}^{|s_5|}-1}{\mathrm{e}^{|s_5|}+1}$ (and $s_1 + s_3 + s_5 = 0$). The above map $x \mapsto \lambda(x)$ extends to a homeomorphism*

$$\lambda \colon \widehat{\mathbb{R}} \to \Lambda_\Gamma.$$

between the circle at infinity $\widehat{\mathbb{R}}$ and the limit set Λ_Γ of the tiling group Γ of $\mathfrak{T}$.

The proof will take a while and is split into several partial steps. Our strategy is reasonably well illustrated by the comparison of Figures 10.11 and 10.12. We will use the family of circles $\gamma(C_i)$, with $\gamma \in \Gamma$ to capture the structure of the limit set, and compare this data to the data similarly associated to the standard Farey tessellation.

To
page 276

With this strategy in mind, it will be convenient to denote by Γ° the tiling group corresponding to $s_1 = s_3 = s_5 = 0$, whereas Γ will denote the tiling group of the crooked Farey tessellation $\mathfrak{T}$ associated to arbitrary s_1, s_3 and s_5 with $s_1 + s_3 + s_5 = 0$, s_5 real, and $\cos(\mathrm{Im}(s_1)) > \frac{e^{|s_5|}-1}{e^{|s_5|}+1}$. Proposition 10.13 guarantees that Γ is a kleinian group, namely, it acts discontinuously on $\mathbb{H}^3$. Remember from Section 10.1 that there is a natural correspondence between elements of Γ and elements of Γ°, and between faces, edges and vertices of the crooked Farey tessellation $\mathfrak{T}$ and faces, edges and vertices of the standard Farey tessellation $\mathfrak{T}^\circ$ of $\mathbb{H}^2 \subset \mathbb{H}^3$. Given any object associated to Γ or $\mathfrak{T}$, we will try as much as possible to denote the corresponding object of Γ° or $\mathfrak{T}^\circ$ by adding a superscript $^\circ$.

Let $\mathcal{C}$ denote the collection of the circles of Figure 10.12, namely, the images of C_1, C_2, C_3, C_4 under the elements of the tiling group Γ. Following the above convention, let $\mathcal{C}^\circ$ denote the corresponding collection of circles in the case where $s_1 = s_3 = s_5 = 0$, as in Figure 10.11.

To fix some terminology, let us say that a point is ***surrounded*** by an actual circle C (namely, not a line) if it is located on the inside of C or on C itself. More generally, a subset of the plane is surrounded by the circle C if every point in this subset is surrounded by C.

Lemma 10.15. *For every $\varepsilon > 0$ and every $C \in \mathcal{C}$ that is an actual circle (namely, not a line), there are only finitely many circles $C' \in \mathcal{C}$ that are surrounded by C and whose euclidean diameter is greater than ε.*

Proof. For each such circle C', consider the hyperbolic plane H' in $\mathbb{H}^3$ bounded by this circle.

Pick an arbitrary base point $P_0 = (x_0, y_0, u_0) \in \mathbb{H}^3$ whose projection $x_0 + iy_0 \in \mathbb{C}$ is surrounded by C. We claim that the hyperbolic distance from P_0 to H' is uniformly bounded in terms of the euclidean diameter D of C, of the coordinate u_0 and of ε. Indeed, if P is the apex of the euclidean half-sphere H', namely, the point of H' with the largest third coordinate, we can join P to P_0 by a curve made up of a vertical line segment of hyperbolic length $\leqslant \log \frac{u_0}{\varepsilon}$ and of a horizontal line segment whose hyperbolic length is $\leqslant \frac{D}{u_0}$. It follows that the hyperbolic distance from P_0 to H' is bounded by $r = \log \frac{u_0}{\varepsilon} + \frac{D}{u_0}$.

By local finiteness of the tessellation of $\mathbb{H}^3$ by the polyhedra $\varphi(X)$ with $\varphi \in \Gamma$, the closed ball $\bar{B}_{d_{\mathrm{hyp}}}(P_0, r)$ can only meet finitely many $\varphi(X)$, and therefore only finitely many such H'. (Compare the proof of Lemma 7.15). This concludes the proof.   □

We can now describe the image of a point $x \in \widehat{\mathbb{R}}$ under the map $\lambda \colon \widehat{\mathbb{R}} \to \Lambda_\Gamma$.

If $x = \infty$, we just set $\lambda(\infty) = \infty$. Otherwise, let $\mathcal{C}_x^\circ$ be the set of those circles in $\mathcal{C}^\circ$ that surround x. Let $\mathcal{C}_x \subset \mathcal{C}$ be associated to $\mathcal{C}_x^\circ$ by the correspondence between circles in $\mathcal{C}$ and circles in $\mathcal{C}^\circ$.

The circles of $\mathcal{C}^\circ$ are also the boundaries of the hyperbolic planes $p^{-1}(E)$, where E ranges over the edges of the tessellation of $\mathbb{H}^2$ by hyperbolic squares constructed in Section 6.6, and associated to our standard punctured torus. In particular, all these edges E are also edges of the Farey tessellation. By inspection of the Farey tessellation, we consequently see two different patterns according to whether x is rational or not.

If x is irrational, it cannot belong to any circle of $\mathcal{C}_x^\circ$. Therefore, the circles of $\mathcal{C}_x^\circ$ are nested together and can be listed as K_1°, K_2°, ..., K_n°, ... in such a way that each K_n° is surrounded by K_{n-1}°.

If x is rational, the elements of $\mathcal{C}_x^\circ$ can be listed into three families. First there is a common stalk (possibly empty) consisting of finitely many circles K_1°, K_2°, ..., K_p° which do not contain the point x, with each K_n° surrounded by K_{n-1}°. Then, $\mathcal{C}_x^\circ$ splits into two infinite families K_{p+1}°, K_{p+2}°, ..., K_n°, ... and L_{p+1}°, L_{p+2}°, ..., $L_n'^\circ$, ... of circles which do contain x, with K_n° to the right of x and L_n° to the left; again, each K_n° is surrounded by K_{n-1}° and L_n° is surrounded by

L_{n-1}° for every $n \geqslant p+1$. In addition, K_p° surrounds all K_n° and L_n° with $n \geqslant p+1$.

Let K_n and L_n be the circles of $\mathcal{C}_x$ associated to K_n° and L_n°, respectively, by the correspondence between $\mathcal{C}_x^\circ$ and $\mathcal{C}_x$.

Lemma 10.16. *If $x \in \mathbb{R} \subset \Lambda_{\Gamma^\circ}$, there is a unique point $\xi \in \Lambda_\Gamma$ that is surrounded by all the circles in $\mathcal{C}_x$.*

Proof. Let $K_1, K_2, \ldots, K_n, \ldots$ be the circles of $\mathcal{C}_x$ defined above. Apply Lemma 10.15 to construct, by induction, an increasing sequence $(n_k)_{k \in \mathbb{N}}$ such that the diameter of each K_{n_k} is less than $\frac{1}{2^k}$. If we pick an arbitrary point ξ_k surrounded by K_{n_k}, the convergence of the series $\sum_{k=1}^\infty \frac{1}{2^k}$ implies that the sequence $(\xi_k)_{k \in N}$ has finite length in $\mathbb{C}$, and therefore converges to some ξ by completeness of $\mathbb{C} = \mathbb{R}^2$ (Theorem 6.8). Every K_n surrounds ξ_k for every k large enough that $n_k > n$, and therefore surrounds ξ by passing to the limit.

By Lemma 10.15, the diameter of K_n tends to 0 as n tends to ∞. It easily follows that ξ is the only point that is surrounded by all the K_n.

When x is irrational, this proves that ξ is surrounded by every circle of $\mathcal{C}_x$. When x is rational we have to worry about the remaining circles L_n. However, in this case, note that the K_n all surround the vertex of the crooked Farey tessellation $\mathcal{T}$ that corresponds to the vertex x of the standard Farey tessellation $\mathcal{T}^\circ$. Therefore, ξ is equal to this vertex of $\mathcal{T}$, which also belongs to all the L_n. As a consequence, ξ is surrounded by all circles of $\mathcal{C}_x$ when x is rational as well.

To conclude the proof, we need to show that ξ is in the limit set Λ_Γ. Let $P_0 \in X$ be an arbitrary base point. For every circle $K_n \in \mathcal{C}_x$ let Π_n be the hyperbolic plane that is bounded by K_n and, among the tiles of the tessellation of $\mathbb{H}^3$ by images of X under elements $\gamma \in \Gamma$, let $\gamma_n(X)$ be the tile that is just below Π_n. The euclidean distance $d_{\mathrm{euc}}(\xi, \gamma_n(P_0))$ is bounded by the euclidean diameter of K_n, and consequently converges to 0 as n tends to $+\infty$ by Lemma 10.15. Therefore, $\xi = \lim_{n \to \infty} \gamma_n(P_0)$ is in the limit set Λ_Γ. □

Define $\lambda \colon \widehat{\mathbb{R}} \to \Lambda_\Gamma$ by the property that $\lambda(\infty) = \infty$ and that the image $\lambda(x)$ of $x \in \mathbb{R}$ is the point ξ provided by Lemma 10.16. Note that this is consistent with our earlier definition when x is rational.

Lemma 10.17. *The map* $\lambda \colon \widehat{\mathbb{R}} \to \Lambda_{\Gamma}$ *is continuous.*

Proof. We will use the notation of Lemma 10.16 and of its proof.

For $x \in \mathbb{R}$ irrational, take an arbitrary $\varepsilon > 0$. By Lemma 10.15, there exists at least one (and in fact infinitely many) n such that the circle $K_n \in \mathcal{C}_x$ has diameter $< \varepsilon$. Since x is irrational, it does not belong to the corresponding circle $K_n^{\circ} \in \mathcal{C}_x^{\circ}$. Therefore, there exists an $\eta > 0$ such that every $y \in \mathbb{R}$ with $d_{\mathrm{euc}}(x, y) < \eta$ is also surrounded by K_n°. The image $\lambda(y)$ of such a point y is surrounded by K_n by construction of λ, so that $d_{\mathrm{euc}}\big(\lambda(x), \lambda(y)\big) < \varepsilon$. Therefore, for every $\varepsilon > 0$, we found an $\eta > 0$ such that $d_{\mathrm{euc}}\big(\lambda(x), \lambda(y)\big) < \varepsilon$ whenever $d_{\mathrm{euc}}(x, y) < \eta$. Namely, λ is continuous at x.

The argument is very similar for x rational, except that we have to also use the circles L_n. For every $\varepsilon > 0$, Lemma 10.15 again provides an n such that K_n and L_n both have diameter $< \varepsilon$. Now, there is an η such that every $y \in \mathbb{R}$ with $d_{\mathrm{euc}}(x, y) < \eta$ is surrounded by K_n° if $y \geqslant x$ and by L_n° if $y \leqslant x$. In both cases this guarantees that $d_{\mathrm{euc}}\big(\lambda(x), \lambda(y)\big) < \varepsilon$. This proves the continuity at every rational x.

For the continuity at $x = \infty$, it will be convenient to consider the translation $\tau = \varphi_4 \circ \varphi_2 \circ \varphi_3 \circ \varphi_1 \in \Gamma$, and the corresponding element $\tau^{\circ} \in \Gamma^{\circ}$. We already computed that $\tau(z) = z + 2 + e^{s_3} + 2e^{s_3 + s_1}$ and that $\tau^{\circ}(z) = z + 6$. Note that λ is well behaved with respect to τ and τ°, in the sense that $\lambda\big(\tau^{\circ}(x)\big) = \tau\big(\lambda(x)\big)$ for every $x \in \mathbb{R}$. If $x \in \mathbb{R}$ is very close to ∞, then, for the euclidean line $C_1^{\circ} \in \mathcal{C}^{\circ}$ passing through 1, it is outside of the vertical strip delimited by the lines $(\tau^{\circ})^n(C_1^{\circ})$ and $(\tau^{\circ})^{-n}(C_1^{\circ})$ for $n > 0$ very large; it follows that $\lambda(x)$ is outside of the strip delimited by the lines $\tau^n(C_1)$ and $\tau^{-n}(C_1)$ and is consequently close to ∞. Formalizing this reasoning with the appropriate quantifiers proves that $\lambda(x)$ tends to $\infty = \lambda(\infty)$ as x tends to ∞. Namely, λ is continuous at ∞. $\qquad\square$

We are now ready to complete the proof of Proposition 10.14, which we restate here as:

Lemma 10.18. *The map* $\lambda \colon \widehat{\mathbb{R}} \to \Lambda_{\Gamma}$ *is a homeomorphism.*

Proof. We have to show that λ is injective, surjective, and that its inverse $\lambda^{-1} \colon \Lambda_{\Gamma} \to \widehat{\mathbb{R}}$ is continuous.

If $x \neq y \in \mathbb{R}$, we can find a circle $C^\circ \in \mathcal{C}^\circ$ which surrounds x but not y. Then, by construction, the corresponding circle $C \in \mathcal{C}$ surrounds $\lambda(x)$ but not $\lambda(y)$, so that $\lambda(x) \neq \lambda(y)$. Also, $\lambda(x) \neq \infty$ when $x \neq \infty$. This proves that λ is injective.

To prove that λ is surjective, consider a point $\xi \neq \infty$ in the limit set Λ_Γ. Pick a base point P_0 in the triangle T^+ of the crooked Farey tessellation $\mathcal{T}$. Then for every $\varepsilon > 0$ there exists a $\gamma \in \Gamma$ such that $d_{\mathrm{euc}}(\gamma(P_0), \xi) < \varepsilon$. We claim that the image of at least one of the vertices of T^+ under γ is at euclidean distance $< \varepsilon$ from ξ. Indeed, the part of $\mathbb{H}^3$ that lies outside of the euclidean ball $B_{d_{\mathrm{euc}}}(\xi, \varepsilon)$ is a hyperbolic half-space bounded by a hyperbolic plane, and is therefore convex in the hyperbolic sense. It follows that if the three vertices of an ideal triangle are outside of $B_{d_{\mathrm{euc}}}(\xi, \varepsilon)$, the whole triangle is outside of that ball. In our case, since the ideal triangle $\gamma(T^+)$ meets $B_{d_{\mathrm{euc}}}(\xi, \varepsilon)$ in P_0, at least one of its vertices must be in $B_{d_{\mathrm{euc}}}(\xi, \varepsilon)$.

As a consequence, for every n, there is a vertex ξ_n of the crooked Farey tessellation $\mathcal{T}$ with $d_{\mathrm{euc}}(\xi_n, \xi) < \frac{1}{n}$. The definition of the map λ shows that $\xi_n = \lambda(x_n)$, where $x_n \in \mathbb{Q} \cup \{\infty\}$ is the vertex of the standard Farey tessellation $\mathcal{T}$ corresponding to ξ_n. If the sequence $(x_n)_{n \in \mathbb{N}}$ is bounded in $\mathbb{R}$, it admits a converging subsequence by Theorem 6.13; otherwise, one can extract from $(x_n)_{n \in \mathbb{N}}$ a subsequence converging to the point ∞ in $\widehat{\mathbb{R}}$. Therefore, we can always find a subsequence $(x_{n_k})_{k \in \mathbb{N}}$ which converges to some $x_\infty \in \widehat{\mathbb{R}}$. Then, ξ is the limit of $\xi_{n_k} = \lambda(x_{n_k})$ as k tends to ∞, which is equal to $\lambda(x_\infty)$ by continuity of λ.

For every $\xi \in \Lambda - \{\infty\}$, we consequently found an $x_\infty \in \Lambda_{\Gamma^\circ}$ such that $\lambda(x_\infty) = \xi$. For $\xi = \infty$, note that $\lambda(\infty) = \infty$. This proves that λ is surjective. In particular, the inverse map $\lambda^{-1} \colon \Lambda_\Gamma \to \widehat{\mathbb{R}}$ is now well defined.

To show that λ^{-1} is continuous, we use a proof by contradiction. Suppose that λ^{-1} is not continuous at $\xi \in \Lambda_\Gamma$ different from ∞. This means that there exists an $\varepsilon > 0$ such that for every $\eta > 0$, there exists $\xi' \in \Lambda_\Gamma$ such that $d_{\mathrm{euc}}(\xi, \xi') < \eta$ but $d_{\mathrm{euc}}\big(\lambda^{-1}(\xi), \lambda^{-1}(\xi')\big) \geqslant \varepsilon$. Applying this to each $\eta = \frac{1}{n}$ provides a sequence $(\xi_n)_{n \in \mathbb{N}}$ which converges to ξ but such that the points $x_n = \lambda^{-1}(\xi_n)$ stay at distance $\geqslant \varepsilon$ from $x = \lambda^{-1}(\xi)$ in $\widehat{\mathbb{R}}$. As above, extract a subsequence $(x_{n_k})_{k \in \mathbb{N}}$

which converges to some $x_\infty \in \widehat{\mathbb{R}}$. Then, by continuity, $\lambda(x_\infty) = \xi = \lambda(x)$ but $d_{\mathrm{euc}}(x, x_\infty) \geqslant \varepsilon$, contradicting the fact that λ is injective.

From page 271
Therefore, λ^{-1} is continuous at every $\xi \in \Lambda_\Gamma$ different from ∞. The continuity of λ^{-1} at ∞ is proved by an almost identical argument.

This concludes the proof that λ is a homeomorphism.   □

10.5.3. Nondifferentiability of the limit set. Proposition 10.14 justifies our observation that many of the limit sets of Section 10.1 form a closed continuous curve with no self-intersection points in the Riemann sphere $\widehat{\mathbb{C}}$. We also noted that the curves in Section 10.1 do not appear very differentiable. We now prove that this is indeed the case, at least under the hypotheses of Proposition 10.14.

A subset Λ of $\mathbb{C}$ has a **tangent line** L at the point $\xi \in \Lambda$ if for every $\eta > 0$ there exists an $\varepsilon > 0$ such that for every $\xi' \in \Lambda \cap B_{d_{\mathrm{euc}}}(\xi, \varepsilon)$, the line $\xi\xi'$ makes an angle $< \eta$ with L.

Proposition 10.19. *Under the hypotheses of Proposition 10.14, so that the homeomorphism $\lambda\colon \widehat{\mathbb{R}} \to \Lambda_\Gamma$ is well defined, suppose in addition that the parameter s_1 is not real. Then, for every irrational point $x \in \mathbb{R} - \mathbb{Q}$, the limit set Λ_Γ admits no tangent line at $\lambda(x)$.*

Remember that "most" real numbers are irrational, in a sense which can be made precise in many mathematical ways; for instance, there are uncountably many irrational numbers but only countably many rationals. See Exercise 10.12 for a proof that Λ_Γ does admit a tangent line at each point $\lambda(x)$ with $x \in \mathbb{Q}$ rational. When s_1 is real, it is immediate from definitions that the limit set Λ_Γ is equal to $\widehat{\mathbb{R}}$, and consequently it is everywhere tangent to the real line $\mathbb{R}$ (in a very strong sense!)

Proof. As in the proof of Proposition 10.14, let $K_1, K_2, \ldots, K_n,$ $\ldots$ be the circles of $\mathcal{C}$ that surround $\lambda(x)$, with each K_n surrounding the next circle K_{n+1}. Because x is irrational, $\lambda(x)$ belongs to no circle K_n.

To page 279

We will first consider the simpler case where, for infinitely many $n_1, n_2, \ldots, n_k, \ldots$, the circle K_{n_k+1} is disjoint from K_{n_k}. By definition of $\mathcal{C}$, the two circles K_{n_k} and K_{n_k+1} bound two faces of some tile $\gamma_k(X)$. In other words, there exists $\gamma_k \in \Gamma$ such that $\gamma_k^{-1}(K_{n_k})$ and

$\gamma_k^{-1}(K_{n_k})$ are two of the original circles C_1, C_2, C_3, C_4. By throwing away some of the n_k (but keeping infinitely many of them), we can assume that we always get the same two circles, namely, that there exists $i, j \in \{1, 2, 3, 4\}$ such that $\gamma_k^{-1}(K_{n_k}) = C_i$ and $\gamma_k^{-1}(K_{n_k}) = C_j$ for every $k \in \mathbb{N}$. Note that C_i and C_j are disjoint, and consequently bound two disjoint disks D_i and D_j in the Riemann sphere $\widehat{\mathbb{C}}$. (One of the two circles is actually a line, in which case the disk it bounds is a euclidean half-plane).

Suppose that the limit set Λ_Γ has a tangent line L at $\lambda(x)$. Then, for every $\eta > 0$, there exists an $\varepsilon > 0$ such that $\Lambda_\Gamma \cap B_{d_{\mathrm{euc}}}(\lambda(x), \varepsilon)$ is located between the lines L_η and $L_{-\eta}$ that make an angle of $\pm\eta$ with L at $\lambda(x)$, or more precisely it is contained in the two disk sectors of angle 2η and radius ε delimited by these two lines $L_{\pm\eta}$. By Lemma 10.15, the circles K_{n_k} and K_{n_k+1} are contained in the ball $B_{d_{\mathrm{euc}}}(\lambda(x), \varepsilon)$ for k large enough. We will use the corresponding γ_k to "zoom" over the region near x. Note that $\gamma_k(D_j)$ and $\gamma(\widehat{\mathbb{C}} - D_i)$ are contained in $B_{d_{\mathrm{euc}}}(\lambda(x), \varepsilon)$.

Consider the two circles $J_\eta^+ = \gamma_k^{-1}(L_\eta)$ and $J_\eta^- = \gamma_k^{-1}(L_{-\eta})$ in $\widehat{\mathbb{C}}$. These two circles have the following properties:

(1) They cross each other in one point $\gamma_k^{-1}\big(\lambda(x)\big) \in D_j$ and one point $\gamma_k^{-1}(\infty) \in D_i$, and make an angle of 2η with each other at these points. (Remember from Proposition 2.18 and Corollary 2.17 that linear fractional maps send circles to circles and respect angles).

(2) The part of the limit set Λ_Γ that is outside of D_i is located between J_η^+ and J_η^-. More precisely, $\Lambda_\Gamma - D_i$ is contained in the two "moons" that are delimited by the intersecting circles $J_\eta^\pm$ and whose angles at their vertices are equal to 2η.

We now vary η. Take a sequence $(\eta_m)_{m \in \mathbb{N}}$ of positive numbers converging to 0. For each of these η_m, the corresponding circle $J_{\eta_m}^+$ is completely determined by the four distinct points in which it intersects the two disjoint circles C_i and C_j. Replacing $(\eta_m)_{m \in \mathbb{N}}$ by a subsequence if necessary, we can assume by compactness of the circles C_i and C_j that these intersection points converge to four points in

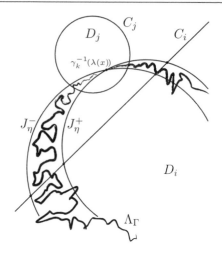

Figure 10.13

$C_i \cup C_j$. As a consequence, the circle $J_{\eta_m}^+$ converges to a circle J_0^+ as m tends to ∞; this limit circle J_0^+ may be tangent to the circle C_i or C_j, which will happen exactly when the two intersection points of $J_{\eta_m}^+$ with this circle converge to a single point. Further passing to a subsequence if necessary, we can similarly assume that in addition, the circle $J_{\eta_m}^-$ converges to a circle J_0^-.

The circles J_0^+ and J_0^- have one point in common in D_i and another one in D_j. Indeed, if $J_0^+ \cap J_0^- \cap D_i$ or $J_0^+ \cap J_0^- \cap D_j$ was empty, the same property would hold for $J_\eta^+ \cap J_\eta^- \cap D_i$ or $J_\eta^+ \cap J_\eta^- \cap D_j$ by continuity, contradicting the fact that J_η^+ and J_η^- meet at $\gamma_k^{-1}(\infty) \in D_i$ and $\gamma_k^{-1}(\lambda(x)) \in D_j$. Similarly, because intersection angles vary continuously, the circles $J_0^\pm$ make an angle of $\lim_{m\to\infty} \eta_m = 0$ at these two intersection points. These two properties imply that the circles actually coincide, namely, that $J_0^- = J_0^+$. (Here it is crucial that the circles C_i and C_j be disjoint, since otherwise J_0^+ and J_0^- could be two distinct circles passing through the point $C_i \cap C_j$ and tangent to each other.)

In particular, the two moons delimited by the circles $J_\eta^\pm$ and containing $\Lambda_\Gamma - D_i$ limit to the circle J_0^+, so that the part of the limit set that is outside D_i is completely contained in the circle J_0^+.

However, we saw that the limit set Λ_Γ contains all the vertices of the crooked Farey tessellation $\mathfrak{T}$, and so this is clearly impossible if s_1 (and $s_3 = -s_1 - s_5$) is not real. For instance, the circle J_0^+ must then contain the vertices 0, 1, ∞ and e^{-s_5} of $\mathfrak{T}$, and consequently must be $\widehat{\mathbb{R}}$. However, the vertices $\varphi_i(0)$, $\varphi_i(1)$ and $\varphi_i(\infty)$ are all outside of D_i, and cannot be all real by inspection of the formulas for the gluing maps φ_1, φ_2, φ_3, φ_4.

This concludes the proof of Proposition 10.19 under the additional assumption that the circles K_n and $K_{n+1} \in \mathcal{C}_x$ are disjoint for infinitely many n.

The proof is very similar in the remaining case where the circles K_n and K_{n+1} meet for every n large enough. Here it is convenient to look at the corresponding data for the standard Farey tessellation. There the circles K_n° and K_{n+1}° surround the point $x \in \widehat{\mathbb{R}}$ and touch each other at a rational point. This intersection point $K_n^\circ \cap K_{n+1}^\circ$ cannot be equal to x since x is irrational, so it will be either to the right or to the left of x. The intersection point $K_n^\circ \cap K_{n+1}^\circ$ cannot be systematically to the left or systematically to the right for n sufficiently large; indeed the point $K_n^\circ \cap K_{n+1}^\circ$ would then be independent of n, and K_n would eventually stop surrounding x since its diameter converges to 0. Therefore, the side changes for infinitely many values of n. Namely, we can find a sequence $(n_k)_{k \in \mathbb{N}}$ such that $K_{n_k}^\circ \cap K_{n_k+1}^\circ$ is to the left of x and $K_{n_k+1}^\circ \cap K_{n_k+2}^\circ$ is to its right, or conversely. A consequence of this is that K_{n_k} is now disjoint from K_{n_k+2} for every $k \in \mathbb{N}$.

Now, the circles K_{n_k} and K_{n_k+2} bound faces of two adjacent tiles of the tessellation of $\mathbb{H}^3$, which are of the form $\gamma_k(X)$ and $\gamma_k \circ \varphi_{i_k}(X)$ with $\gamma_k \in \Gamma$ and $i_k \in \{1, 2, 3, 4\}$. This again means that there are only finitely many possibilities for the circles $\gamma_k^{-1}(K_{n_k})$ and $\gamma_k^{-1}(K_{n_k+2})$. Therefore, replacing the subsequence $(K_{n_k})_{k \in \mathbb{N}}$ by a subsubsequence if necessary, we can assume without loss of generality that there exists fixed i, j and $l \in \{1, 2, 3, 4\}$ such that $\gamma_k^{-1}(K_{n_k}) = C_i$ and $\gamma_k^{-1}(K_{n_k+2}) = \varphi_l(C_j)$ for every $k \in \mathbb{N}$. The argument is then identical to that used in the first case, replacing C_j by $\varphi_l(C_j)$ everywhere. $\square$

From page 276

10.5.4. The parameter space. Our shearing and bending of the Farey tessellation depends on three complex parameters s_1, s_3, $s_5 \in \mathbb{C}$ such that $s_1 + s_3 + s_5 = 0$. Actually, by inspection of the formulas at the beginning of this chapter, these depend only on the exponentials e^{s_i}. (Remember that by definition of the complex exponential, $e^z = e^{z'}$ if and only if $z - z'$ is an integer multiple of $2\pi i$.)

We saw that for some values of these parameters, the tiling group Γ of the corresponding crooked Farey tessellation acts freely and discontinuously on the hyperbolic space $\mathbb{H}^3$. We might be interested in plotting all the values of the parameters for which this happens. However, even if we take into account the fact that $s_5 = -s_1 - s_3$, this leaves us with two free complex parameters e^{s_1} and e^{s_3}, so that a crooked Farey tessellation is determined by the point $(e^{s_1}, e^{s_3}) \in \mathbb{C}^2 = \mathbb{R}^4$. Most of us are not very comfortable with 4-dimensional pictures. We can try to decrease the dimension by imposing an additional condition. The simplest such condition is that $s_5 = 0$, so that $s_3 = -s_1$. The corresponding set of crooked Farey tessellation is known as the **Earle slice** in the set of all crooked Farey tessellations. An element of the Earle slice is completely determined by the complex number $v = e^{s_1} \in \mathbb{C}$, since $e^{s_3} = v^{-1}$ and $e^{s_5} = 1$.

Figure 10.14 represents the space of values of $v \in \mathbb{C}$ for which the tiling group Γ of the crooked Farey tessellation associated to v is a kleinian group acting freely on $\mathbb{H}^3$. More precisely, it uses a change a variables and, in the complex plane $\mathbb{C}$, the lighter shaded and cauliflower-shaped area of Figure 10.14 is the set of values of

$$u = \frac{2v - 1}{2v + 1} = \frac{2e^{s_1} - 1}{2e^{s_1} + 1}$$

for which the tiling group Γ acts discontinuously and freely on $\mathbb{H}^3$. Note that $v = e^{s_1}$ is easily recovered from u. This picture was drawn using the software *OPTi*.

The large circle in Figure 10.14, which is the circle of radius 1 centered at the origin, encloses the set D of values of u for which the hypotheses of Proposition 10.13 are satisfied. See Exercise 10.13.

The horizontal line segment corresponds to real values of u, for which the tiling group Γ is fuchsian. These fuchsian groups are exactly the tiling groups associated to the complete hyperbolic tori that we

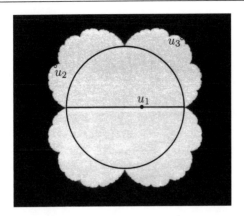

Figure 10.14. The Earle slice

considered in Section 6.7.2 (corresponding to the case $a = b \in (1, +\infty)$ in the notation of that section).

We have also indicated the values $u_1 = \frac{1}{3}$ corresponding to the standard Farey tessellation, and $u_2 = -1.205 + 0.714i$ associated to the example of Figure 10.8. Chapter 11 will be devoted to the tiling group corresponding to the point $u_3 = 1 + \frac{2}{\sqrt{3}}i$, located exactly on the boundary of the lighter shaded area. The point u_2 is near the boundary, but not quite on it.

In the general case, let $\Omega \subset \mathbb{C}^2$ be the set of values of the parameters (e^{s_1}, e^{s_3}) for which the tiling group of the associated crooked Farey tessellation acts discontinuously and freely on the hyperbolic space $\mathbb{H}^3$. It can be shown that the properties of the limit set Λ_Γ that we proved under the hypotheses of Proposition 10.13 hold for all parameters (e^{s_1}, e^{s_3}) located in the interior of Ω. More precisely, for these parameters, the limit set Λ_Γ is homeomorphic to the circle, and has no well-defined tangent line at most points unless the parameters all real.

For points (e^{s_1}, e^{s_3}) on the boundary of the parameter set Ω, the tiling group Γ of the corresponding crooked Farey tessellation is still a kleinian group acting freely and discontinuously on $\mathbb{H}^3$, but its geometry is much more complex. In particular, the limit set is not homeomorphic to the circle any more. We will investigate such a group in Chapter 11.

A result of Yair Minsky [**Minsky₁**] implies that the space Ω is path-connected, in the sense that any two points can be joined to each other by a continuous curve contained in Ω. However, Figure 10.14 suggests that its geometry is quite intricate near its boundary. It is actually much more intricate than what is visible on that picture. Ken Bromberg recently showed in [**Bromberg**] that there are points where Ω is not even locally connected. In particular, at such a point $P \in \Omega$, we can find a radius $\varepsilon > 0$ and points $Q, R \in \Omega$ which are arbitrarily close to P and such that any continuous curve from Q to R in Ω must leave the euclidean ball $B_{d_{\mathrm{euc}}}(P, \varepsilon) \subset \mathbb{C}^2 = \mathbb{R}^4$.

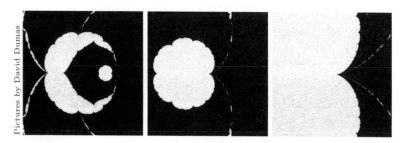

Figure 10.15

This is illustrated in Figure 10.15, which represents a different 2-dimensional cross-section of Ω. Each of the three squares is obtained by zooming in on a piece of the square immediately to the left. The little "islands" that converge toward the center P of the right-hand square can be connected to each other by a continuous curve in Ω (leaving the cross-section represented and using the additional two dimensions), but not by a curve which remains within ε of P. The pictures of Figure 10.15 were created by David Dumas, using his software *Bear* [**Dumas**].

What we observed for crooked Farey tessellations is part of a more general phenomenon. Any fuchsian group Γ° can be deformed to a family of kleinian groups depending on complex parameters belonging to a certain domain Ω in $\mathbb{C}^n$, where the dimension n depends on the topology of the quotient space $\mathbb{H}^2/\Gamma^\circ$. For any such kleinian group Γ

corresponding to an interior point of Ω, the limit set Λ_Γ is homeomorphic to the circle, and admits no tangent line at most points unless the group Γ is essentially fuchsian and the limit set is a circle in $\widehat{\mathbb{C}}$. If the quotient space $\mathbb{H}^2/\Gamma^\circ$ is compact, for instance for the tiling group Γ° of the tessellation of $\mathbb{H}^2$ by hyperbolic octagons that we constructed in Section 6.5.2, the limit set of the deformed kleinian group Γ even admits no tangent line at any point at all. The kleinian groups corresponding to points on the boundary of Ω are much more complex; it is only very recently that a reasonable understanding of their geometry has been reached [**Minsky$_2$, Brock & Canary & Minsky**].

10.6. Poincaré, Fuchs and Klein

The terminology of kleinian and fuchsian groups is due to Henri Poincaré (1854–1912), one of the mathematical giants of his time. Poincaré was led to hyperbolic geometry through an unexpected route. He was studying the linear differential equation

$$(10.1) \qquad x''(t) + p(t)x(t) = 0,$$

where the the function $p(t)$ is given, where the function $x(t)$ is the unknown, and where, more importantly, all quantities x, p, t are complex-valued. Compared to the setup usually taught in calculus or in a differential equations course, the fact that t is complex may be somewhat surprising, but the basic definitions and properties of differential equations immediately extend to this complex context.

When investigating a specific family of examples, where $p(t)$ was a rational function with real coefficients, Poincaré found that the space of local solutions to (10.1) gave rise to a certain group Γ of linear fractional maps with real coefficients. Using (and proving) the Poincaré Polygon Theorem 6.25, he then showed that Γ acts discontinuously on the upper half-space $\mathbb{H}^2$. In addition, the construction provided a preferred homeomorphism $f\colon X \to \mathbb{H}^2/\Gamma$ from the domain X of the given function $p(t)$ to the quotient space $\mathbb{H}^2/\Gamma$. The groups Γ considered by Poincaré in these examples are very closely related to the tiling groups associated to complete punctured tori that we investigated in Sections 5.5, 6.6 and 6.7.2.

In a subsequent work [**Poincaré₅**], Poincaré started perturbing equation 10.1 by allowing the coefficients of the rational fraction $p(t)$ to be complex, although close to real numbers. This process is very similar to the deformations of the standard Farey tessellation to the crooked Farey tessellations that we investigated in Sections 10.1 and 10.5. In particular, it yields a group Γ of linear fractional maps with complex coefficients, now acting on the Riemann sphere $\widehat{\mathbb{C}}$. Poincaré proved that these linear fractional maps extend to isometries of the hyperbolic space $\mathbb{H}^3$ and, using (and proving) the Poincaré Polyhedron Theorem 10.9, showed that the action of Γ on $\mathbb{H}^3$ is discontinuous. He then used this technique to produce a natural homeomorphism $f\colon X \to \Omega/\Gamma$ where, again, X is the domain of $p(t)$ and Ω is a subset of the Riemann sphere $\widehat{\mathbb{C}}$ invariant under the action of Γ. This subset Ω is not the upper half-plane $\mathbb{H}^2$ any more. Rather, the limit set Λ_Γ is homeomorphic to the circle, and Ω is one of two pieces of the complement $\widehat{\mathbb{C}} - \Lambda_\Gamma$. See Exercise 10.4 for a proof that Γ acts discontinuously on $\widehat{\mathbb{C}} - \Lambda_\Gamma$, so that the quotient space Ω/Γ makes sense.

In the real coefficient case, the homeomorphism $f\colon X \to \mathbb{H}^2/\Gamma$ and its derivatives are essentially equivalent to the data of certain functions which are well behaved with respect to the action of Γ. In the papers [**Poincaré₁, Poincaré₃, Poincaré₄**], Poincaré decided to call these functions *fuchsian functions* in honor of Lazarus Fuchs, whose articles [**Fuchs₁, Fuchs₂**] had inspired him. He also gave the name of *fuchsian groups* to the groups of isometries of $\mathbb{H}^2$ occurring in this way. Felix Klein then complained that he had already considered a notion equivalent to these fuchsian functions and groups acting discontinuously on $\mathbb{H}^2$, and consequently deserved more credit than Fuchs. Poincaré did not change his terminology but, when moving to the case with complex coefficients (which Klein had never considered at that time, although he did later on), Poincaré gave the name of *kleinian groups* to the discontinuous groups of isometries of $\mathbb{H}^3$ occurring in this context. See the note added by Klein, acting as editor of the journal, to Poincaré's announcement [**Poincaré₁**] and Poincaré's response in [**Poincaré₂**]. The upshot of the story is that fuchsian groups have almost no connection to Fuchs, and that Klein had little to do with kleinian groups.

There is another story about this. The late nineteenth century was a period of turmoil in mathematics, as the endeavor to set mathematics on a solid foundation would sometimes lead to rather unintuitive results. This included the emergence of pathological examples, such as functions which are continuous but nowhere differentiable. Poincaré had mixed feelings about this trend, and clearly thought that "real life" mathematical objects should be well behaved. The following quote from [**Poincaré₆**] (translated by the author) summarizes his thinking rather well.

> Logic sometimes generates monsters. The past half century has seen the emergence of a multitude of bizarre functions which try to resemble as little as possible the honest functions which have some applications. No more continuity, or else continuity but no derivatives, etc... Even more, from a logical point of view, it is these strange functions which are the most general, and those that one encounters without looking for them only appear as a special case. . . . In the old days, when one invented a new function, it was with the goal of some practical application; today, they are invented for the single purpose of exhibiting flaws in the reasonings of our fathers, and this is the only thing that one will ever extract out of them.

Little did he realize that these "monsters" already appeared in a natural way in his work. For instance, the article [**Poincaré₅**] contains many examples which are similar to the ones that we considered in this chapter. Poincaré was aware that in these examples, the partial tiling of the Riemann sphere that he was constructing would accumulate on a continuous curve Λ_Γ, and thought that this curve was unlikely to be very differentiable. However, it seems that the worst he envisioned was that this curve might not have second derivatives; see [**Poincaré₁**, page 559]. It is only the limited computing power available to him which prevented Poincaré from observing that this closed curve was nowhere differentiable, namely, it would produce one of these pathological curves that he thought would never occur in a natural way.

Exercises for Chapter 10

Exercise 10.1. Let φ be the isometry of $\mathbb{H}^3$ whose extension to the Riemann sphere $\widehat{\mathbb{C}}$ is the homothety $z \mapsto 2z$, and let $\Gamma = \{\varphi^n; n \in \mathbb{Z}\}$ be the transformation group of $\mathbb{H}^3$ generated by φ.

   a. Show that Γ is a kleinian group.

   b. Determine the limit set of Γ.

   c. Show that there exists a nonempty closed subset K of $\widehat{\mathbb{C}}$ which is invariant under Γ but which does not contain the limit set Λ_Γ. (Compare Proposition 10.3.)

Exercise 10.2. Show that the limit set of a kleinian group Γ is nonempty if and only if Γ is infinite.

Exercise 10.3. Consider a kleinian group Γ which fixes the point $\infty \in \widehat{\mathbb{C}}$.

   a. Show that every element of Γ is the isometric extension of a linear or antilinear fractional map of the form $\varphi_{a,b}(z) = az + b$ or $\psi_{a,b}(z) = a\bar{z} + b$, with $a \neq 0$.

   b. Show that if Γ contains an element γ_0 of the form $\varphi_{a,b}$ or $\psi_{a,b}$ with $|a| \neq 1$, then it also fixes a point of $\mathbb{C}$ in addition to ∞. Hint: For $\gamma \in \Gamma$, consider the elements $\gamma^{-1} \circ \gamma_0^{-n} \circ \gamma \circ \gamma_0^n \in \Gamma$ with $|n|$ large, and remember that Γ acts discontinuously on $\mathbb{H}^3$.

   c. Conclude that either Γ consists entirely of euclidean isometries of $(\mathbb{R}^3, d_{\mathrm{euc}})$ respecting $\mathbb{H}^3$ or it respects a unique complete geodesic g of $\mathbb{H}^3$.

Exercise 10.4 (Discontinuity domain). If a group Γ acts on a metric space (X, d), not necessarily by isometries, the action is **_discontinuous_** at $P \in X$ if there exists a ball $B_d(P, \varepsilon)$ which meets its images $\gamma(B_d(P, \varepsilon))$ for only finitely many $\gamma \in \Gamma$. When the action is by isometries, one readily checks that this definition is equivalent to the one given in Section 7.2.

   We want to show that the action of a kleinian group Γ on $\widehat{\mathbb{C}}$ is discontinuous at every point z of the complement $\mathbb{C} - \Lambda_\Gamma$ of the limit set Λ_Γ, for the euclidean metric d_{euc}. Since the limit set is closed, there exists a small euclidean ball $B_{d_{\mathrm{euc}}}(z, 2\varepsilon)$ which is disjoint from Λ_Γ. Suppose, in search of a contradiction, that there exists an infinite sequence $(\gamma_n)_{n \in \mathbb{N}}$ such that $\gamma_n(B_{d_{\mathrm{euc}}}(z, \varepsilon))$ meets $B_{d_{\mathrm{euc}}}(z, \varepsilon)$ for every $n \in \mathbb{N}$. By Proposition 9.10, each $\gamma_n(B_{d_{\mathrm{euc}}}(z, \varepsilon))$ is delimited in $\mathbb{R}^3$ by a euclidean sphere centered on $\widehat{\mathbb{C}}$; let r_n be the euclidean radius of this sphere. Pick a base point $P_0 \in \mathbb{H}^3$ on the euclidean sphere delimiting $B_{d_{\mathrm{euc}}}(z, \varepsilon)$.

a. Suppose, in addition, that $\lim_{n \to \mathbb{N}} r_n = 0$. Show that there exists a subsequence $\left(\gamma_{n_k}\right)_{k \in \mathbb{N}}$ such that $\left(\gamma_{n_k}(P_0)\right)_{k \in \mathbb{N}}$ converges to some point $z_\infty \in \Lambda_\Gamma$. Show that necessarily $d_{\text{euc}}(z, z_\infty) \leqslant \varepsilon$, which contradicts the choice of ε.

b. At the other extreme, suppose that the r_n are bounded from below, namely, that there exists $r_0 > 0$ such that $r_n \geqslant r_0$ for every $n \in \mathbb{N}$. Show that there exists a constant $C > 0$ such that for every $n \in \mathbb{N}$, there exists a point $P_n \in \gamma_n(B_{d_{\text{euc}}}(z, \varepsilon))$ with $d_{\text{hyp}}(P_0, P_n) \leqslant C$. Show that there exists a subsequence $\left(\gamma_{n_k}\right)_{k \in \mathbb{N}}$ for which $\left(\gamma_{n_k}^{-1}(P_0)\right)_{k \in \mathbb{N}}$ converges to some point $z_\infty \in \Lambda_\Gamma$. Use this property to show that necessarily $d_{\text{euc}}(z, z_\infty) \leqslant \varepsilon$, which contradicts the choice of ε. Hint: Compare the euclidean and hyperbolic distances from $\gamma_n^{-1}(P_0)$ to $\gamma_n^{-1}(P_n) \in B_{d_{\text{euc}}}(z, \varepsilon)$.

c. Combine parts a and b to reach a contradiction in all cases. Conclude that the action of Γ at every $z \in \mathbb{C} - \Lambda_\Gamma$ is discontinuous.

d. Show that for every $z \in \Lambda_\gamma \cap \mathbb{C}$, the action of Γ is not discontinuous at z.

A similar argument shows that for the obvious extension of definitions, the action of Γ is discontinuous at the point ∞ if and only if ∞ is not in the limit set Λ_Γ. For this reason, the complement $\widehat{\mathbb{C}} - \Lambda_\Gamma$ is called the *discontinuity domain* of the kleinian group Γ.

Exercise 10.5. Let $p \colon \mathbb{H}^3 \to \mathbb{H}^2$ be the orthogonal projection constructed in Section 10.3.

a. Let $\vec{v}$ be a vector based at $P \in \mathbb{H}^3$. Show that its image under the differential of p is such that $\|D_P p(\vec{v})\|_{\text{euc}} \leqslant \|\vec{v}\|_{\text{euc}}$ and $\|D_P p(\vec{v})\|_{\text{hyp}} \leqslant \|\vec{v}\|_{\text{hyp}} \cosh d$, where d is the hyperbolic distance from P to $p(P)$.

b. Let γ be a piecewise differentiable curve in $\mathbb{H}^3$ which stays at distance at least $D > 0$ from $\mathbb{H}^2$. Show that $\ell_{\text{hyp}}(p(\gamma)) \leqslant \ell_{\text{hyp}}(\gamma) \cosh D$.

In particular, when projecting from far away, the hyperbolic orthogonal projection $p \colon \mathbb{H}^3 \to \mathbb{H}^2$ decreases lengths much more that the euclidean orthogonal projection $\mathbb{R}^3 \to \mathbb{R}^2$.

Exercise 10.6. Let $p \colon \mathbb{H}^3 \to \mathbb{H}^2$ and $q \colon \mathbb{H}^3 \to \mathbb{R}$ be the orthogonal projection and the signed distance function introduced for Lemma 10.5. Endow the product $\mathbb{H}^2 \times \mathbb{R}$ with the product $d_{\text{hyp}} \times d_{\text{euc}}$ of the hyperbolic metric d_{hyp} of $\mathbb{H}^2$ and of the euclidean metric d_{euc} of $\mathbb{R}$, as defined in Exercise 1.6.

a. Show that the product function $p \times q \colon \mathbb{H}^3 \to \mathbb{H}^2 \times \mathbb{R}$, defined by $p \times q(P) = (p(P), q(P))$ for every $P \in \mathbb{H}^3$, is a homeomorphism.

b. Let Γ be a discontinuous group of isometries of $\mathbb{H}^2$, extended to a fuchsian group which we also denote by Γ. Endow the quotient spaces $\mathbb{H}^2/\Gamma$ and $\mathbb{H}^3/\Gamma$ with the quotient metrics $\bar{d}_{\text{hyp}}$ defined by the hyperbolic metrics d_{hyp} of $\mathbb{H}^2$ and $\mathbb{H}^3$ (and denoted by the same symbols). Finally, endow the product $\mathbb{H}^2/\Gamma \times \mathbb{R}$ with the product metric $\bar{d}_{\text{hyp}} \times d_{\text{euc}}$. Show that there is a homeomorphism

$$\bar{p} \times q \colon \mathbb{H}^3/\Gamma \to \mathbb{H}^2/\Gamma \times \mathbb{R}$$

defined by the property that $\bar{p} \times q(\bar{P}) = (\overline{p(P)}, q(P))$ for every $P \in \mathbb{H}^3$.

Exercise 10.7 (Twisted fuchsian groups). Let Γ be a fuchsian group, acting on $\mathbb{H}^3$, and suppose that we are given a map $\rho\colon \Gamma \to \{-1, +1\}$ such that $\rho(\gamma \circ \gamma') = \rho(\gamma)\rho(\gamma')$ (namely, if you know what this is, ρ is a group homomorphism from Γ to the group $\mathbb{Z}_2 = \{-1, +1\}$, where the group law is defined by multiplication). For every $\gamma \in \Gamma$, we define a new isometry γ^ρ of $\mathbb{H}^3$ as follows: If $\rho(\gamma) = 1$, then $\gamma^\rho = \gamma$; if $\rho(\gamma) = -1$, then $\gamma^\rho = \tau \circ \gamma$, where $\tau\colon \mathbb{H}^3 \to \mathbb{H}^3$ is the euclidean reflection across the vertical half-plane $\mathbb{H}^2 \subset \mathbb{H}^3$. Then consider the set

$$\Gamma^\rho = \{\gamma^\rho; \gamma \in \Gamma\}$$

of all γ^ρ obtained in this way.

a. Show that Γ^ρ is a group of isometries of $(\mathbb{H}^3, d_{\text{hyp}})$.

b. Show that the action of Γ^ρ on $\mathbb{H}^3$ is discontinuous. Hint: Compare Lemma 10.6.

A kleinian group Γ^ρ obtained in this way is called a ***twisted fuchsian group***.

c. Show that the limit set of a twisted fuchsian group is contained in $\widehat{\mathbb{R}} \subset \widehat{\mathbb{C}}$.

d. Let Γ' be a kleinian group whose limit set is contained in a euclidean circle and has at least three points. Show that there exists a hyperbolic isometry φ and a twisted fuchsian group Γ^ρ as above such that

$$\Gamma' = \{\varphi^{-1} \circ \gamma^\rho \circ \varphi; \gamma^\rho \in \Gamma^\rho\}.$$

Hint: Choose φ so that it sends the circle C to $\widehat{\mathbb{R}} \subset \widehat{\mathbb{C}}$.

Exercise 10.8. Let X be the hyperbolic half-space

$$X = \{(x, y, u) \in \mathbb{R}^3; y \geqslant 0, u > 0\}.$$

Consider X as a hyperbolic polyhedron by decomposing its boundary into one vertex $P = (0, 0, 1)$, two edges

$$E_1 = \{(x, y, u) \in \mathbb{R}^3; x = 0, y = 0, u \geqslant 1\}$$

$$\text{and } E_2 = \{(x, y, u) \in \mathbb{R}^3; x = 0, y = 0, 0 < u \leqslant 1\},$$

and two faces

$$F_1 = \{(x, y, u) \in \mathbb{R}^3; x \geqslant 0, y = 0, u > 0\}$$
$$\text{and } F_2 = \{(x, y, u) \in \mathbb{R}^3; x \leqslant 0, y = 0, u > 0\}.$$

Glue the faces F_1 and F_2 together by the restriction $\varphi_1 \colon F_1 \to F_2$ of the hyperbolic isometry $\varphi_1 \colon \mathbb{H}^3 \to \mathbb{H}^3$ extending the antilinear fractional map $z \mapsto -\frac{1}{\bar{z}}$.

a. Show that hypotheses (1–3) of Theorem 10.11 are satisfied, but that hypothesis (4) does not hold.

b. For any $r > 0$, endow the sphere

$$S = S_{\bar{d}_{\text{hyp}}}(\bar{P}_0, r) = \{\bar{P} \in \bar{X}; \bar{d}_{\text{hyp}}(\bar{P}, \bar{P}_0) = r\}$$

with the path metric d_S, where $d_S(\bar{P}, \bar{Q})$ is defined as the infimum of the hyperbolic lengths of all curves joining $\bar{P}$ to $\bar{Q}$ in S. Let $\frac{1}{\sinh r} d_S$ be the metric obtained by multiplying d_S by the factor $\frac{1}{\sinh r}$ (compare Exercise 9.4, and the proof of Lemma 10.10). Show that $(S_{\bar{d}_{\text{hyp}}}(\bar{P}_0, r), \frac{1}{\sinh r} d_S)$ is isometric to the projective plane $(\mathbb{RP}^2, \bar{d}_{\text{sph}})$ of Section 5.3.

c. Show that the quotient space $(\bar{X}, \bar{d}_X)$ is not locally isometric to $(\mathbb{H}^3, d_{\text{hyp}})$. It may be convenient to use the results of Exercise 5.11.

Exercise 10.9. Let $\mathcal{T}$ be a crooked Farey tessellation associated to parameters s_1, s_3 and s_5 with $s_1 + s_3 + s_5 = 0$, and let Γ be the corresponding tiling group. Consider the hyperbolic isometry defined by the linear fractional map $\rho(z) = -e^{-s_5} \frac{1}{z}$.

a. Show that ρ exchanges the two ideal triangles T^+ and T^-.

b. Show that $\rho \circ \varphi_1 = \varphi_1^{-1} \circ \rho$, $\rho \circ \varphi_1^{-1} = \varphi_1 \circ \rho$, $\rho \circ \varphi_2 = \varphi_2^{-1} \circ \rho$ and $\rho \circ \varphi_2^{-1} = \varphi_2 \circ \rho$.

c. Conclude that ρ respects the crooked Farey tessellation $\mathcal{T}$ and that when the tiling group Γ acts discontinuously on $\mathbb{H}^3$, ρ also respects the limit set Λ_Γ.

d. Let $\tau = \rho \circ \varphi_3 \circ \varphi_1$. Show that τ is a horizontal translation, whose translation vector is equal to half the translation vector of the translation $\varphi_4 \circ \varphi_2 \circ \varphi_3 \circ \varphi_1$.

e. Show that τ respects the crooked Farey tessellation $\mathcal{T}$ and, when the tiling group Γ acts discontinuously, that τ respects the limit set Λ_Γ.

Exercise 10.10 (Shear-bend parameters). We want to give a geometric interpretation of the parameters $s_i \in \mathbb{C}$ defining a crooked Farey tessellation $\mathcal{T}$. Do not let the long definitions intimidate you.

Let T_1 and T_2 be two adjacent triangles of $\mathcal{T}$, meeting along an edge E. Orient E to the left as seen from T_1. Namely, first orient, in the Farey tessellation $\mathcal{T}^\circ$, the edge E° corresponding to E to the left as seen from the face T_1° corresponding to T_1, and then transport this orientation to an orientation of E. (We need to do this because in dimension 3 the notion of right and left depends on which way we are standing on the triangle T_1.) Let P_1 and P_2 be the base points determined on E by the standard horocircles of T_1 and T_2, respectively, as defined in Lemma 8.4. The **shear parameter** between the triangles T_1 and T_2 of $\mathcal{T}$ is the signed distance from P_1 to P_2 in this oriented edge E. Compare the shear parameters of Section 8.4.

Let $\theta \in [0, 2\pi)$ be the angle by which one needs to rotate T_2 along E in order to bring it in the hyperbolic plane containing T_1, but on the side of E opposite T_1; here we measure the angles counterclockwise, as one looks in the direction of the orientation of E. This angle θ is the **external dihedral angle** θ between T_1 and T_2.

Suppose that E is the image of one of the standard edges E_i, with $i \in \{1, 3, 5\}$, under an element γ of the tiling group Γ. Show that the shear parameter t is equal to $\mathrm{Im}(s_i)$, and that the external dihedral angle θ is equal to $\mathrm{Re}(s_i)$ up to an integer multiple of 2π. Hint: First consider the case where γ is the identity map.

Exercise 10.11 (Schottky groups). Let B_1, B_2, B_3 and B_4 be four euclidean balls in $\mathbb{R}^3$, centered on the xy-plane $\mathbb{C}$, and far apart enough from each other that the corresponding closed balls are disjoint. Let X be the complement $\mathbb{H}^3 - B_1 \cup B_2 \cup B_3 \cup B_4$. In particular, X is a hyperbolic polyhedron in $\mathbb{H}^3$ delimited by four disjoint hyperbolic planes Π_1, Π_2, Π_3, Π_4, with no vertex at infinity. Choose isometries φ_1, φ_2, φ_3 and φ_4 of $(\mathbb{H}^3, d_{\mathrm{hyp}})$ such that the following holds: φ_1 sends Π_1 to Π_2, and sends X to the side of Π_2 that is opposite X; $\varphi_2 = \varphi_1^{-1}$; φ_3 sends Π_3 to Π_4, and sends X to the side of Π_4 that is opposite X; and $\varphi_4 = \varphi_2^{-1}$. Let Γ be the group generated by φ_1, φ_2, φ_3 and φ_4.

a. Show that Γ acts discontinuously on $\mathbb{H}^3$. Hint: Use Poincaré's Polyhedron Theorem 10.9.

b. Show that every element $\gamma \in \Gamma$ can be written in a unique way as

$$\gamma = \varphi_{i_1} \circ \varphi_{i_2} \circ \cdots \circ \varphi_{i_n},$$

where, for each j, i_j is an element of $\{1, 2, 3, 4\}$ and the set $\{i_j, i_{j+1}\}$ is different from $\{1, 2\}$ and $\{3, 4\}$ (which is a fancy way of saying that there is no obvious simplification); by convention, $n = 0$ when γ is the identity map. Hint: To prove the uniqueness, look at the tiles $\gamma'(X)$ that are crossed by an arbitrary geodesic going from a point of X to a point of $\gamma(X)$.

c. Let $(i_n)_{n \in \mathbb{N}}$ be a sequence valued in $\{1, 2, 3, 4\}$, such that the set $\{i_n, i_{n+1}\}$ is different from $\{1, 2\}$ and $\{3, 4\}$ for every $n \in \mathbb{N}$. Set

$$\gamma_n = \varphi_{i_1} \circ \varphi_{i_2} \circ \cdots \circ \varphi_{i_n}.$$

Show that the sequence $(\gamma_n(P_0))_{n \in \mathbb{N}}$ converges in $(\mathbb{R}^3, d_{\text{euc}})$ to some point z of the limit set Λ_Γ. Hint. Adapt the proof of Lemma 10.16.

d. Show that for every $z \in \Lambda_\Gamma$, there exists a unique sequence $(i_n)_{n \in \mathbb{N}}$ such that $z = \lim_{n \to \infty} \gamma_n(P_0)$, with the setup of part c. In particular, this proves that there is a one-to-one correspondence between points of the limit set Λ_Γ and sequences $(i_n)_{n \in \mathbb{N}}$ as in part c.

e. Choose a specific example and draw as many of the (euclidean circles delimiting the) hyperbolic planes $\varphi(\Pi_i)$, with $\varphi \in \Gamma$ and $i \in \{1, 2, 3, 4\}$ as you can. Similarly sketch the limit set Λ_Γ.

Exercise 10.12. Let Γ be one of the kleinian groups considered in Section 10.5, associated to a crooked Farey tessellation $\mathcal{T}$ whose parameters s_1, s_3, s_5 satisfy the hypotheses of Proposition 10.13. Let $\lambda \colon \widehat{\mathbb{R}} \to \Lambda_\Gamma$ be the homeomorphism constructed in Proposition 10.14.

a. Show that there is a strip in $\mathbb{C}$, delimited by two parallel lines, which contains all the limit set Λ_Γ (minus ∞, of course).

b. Show that for every rational point $\frac{p}{q} \in \mathbb{Q}$, there exists two disjoint open disks D_1 and D_2 in $\mathbb{C}$, delimited by two circles tangent to each other at $\lambda(\frac{p}{q})$, such that Λ_Γ is contained in $\widehat{\mathbb{C}} - D_1 \cup D_2$. Hint: Consider an element $\gamma \in \Gamma$ such that $\gamma(\infty) = \frac{p}{q}$, and use part a.

c. Conclude that the limit set Λ_Γ admits a tangent line at each $\lambda(\frac{p}{q})$ with $\frac{p}{q} \in \mathbb{Q}$.

Exercise 10.13. Let $\mathcal{T}$ be a crooked Farey tessellation associated to parameters s_1, s_3 and s_5 such that $s_5 = 0$ and $s_1 + s_3 = 0$. Let Γ be the corresponding tiling group, which is contained in the Earle slice of Section 10.5.4. Show that these parameters satisfy the conditions of Proposition 10.13, in the sense that $\cos(\text{Im}(s_1)) > 0$, if and only if $\left| \frac{e^{s_1} - 1}{e^{s_1} + 1} \right| < 1$. (Compare with the circle represented in Figure 10.14.)

Exercise 10.14 (Domino diagrams revisited). This is a continuation of Exercise 8.7. Let $\mathcal{T}$ be the crooked Farey tessellation determined by the parameters s_1, s_2, $s_3 \in \mathbb{C}$ with $s_1 + s_2 + s_3 = 0$. Let T be an ideal triangle of $\mathcal{T}$, associated to the Farey triangle T° in the standard Farey tessellation $\mathcal{T}^\circ$. We want to give an explicit formula for the vertices of T. We will restrict our attention to the case where T° is to the right of 0∞, namely, where all its vertices are nonnegative.

Let $S_1 S_2 \ldots S_n$, with each S_k equal to L of R, be the symbol sequence describing the Farey triangle T° as in Exercise 8.5, and consider the associated domino diagram. By construction, each domino is associated to one of the edges E of the Farey tessellation $\mathcal{T}^\circ$ that are traversed as one travels from the triangle 01∞ to the triangle T°, as in Exercise 8.5. The edge E is of the form $\gamma(E_i)$, where γ is an element of the tiling group Γ_0, and where E_i is one of the standard edges $E_1 = (-1)\infty$, $E_3 = 1\infty$ and $E_5 = 0\infty$. For such a domino associated to an edge $\gamma(E_i)$, label its upper right bullet by $e^{s_i/2}$ and its lower right bullet by $e^{-s_i/2}$. In particular, the two leftmost bullets of the domino diagram receive no labels.

For instance, the domino diagram (8.1) of Exercise 8.7 is labelled as follows.

For each path p in the domino diagram, let

$$w(p) = e^{\pm s_{i_1}/2} e^{\pm s_{i_2}/2} \ldots e^{\pm s_{i_n}/2}$$

be the product of the labels thus associated to the bullets traversed by p. Finally, for every $i, j \in \{1, 2\}$, let

$$s_{ij} = \sum_{\substack{p \text{ goes from } i \text{ to } j}} w(p).$$

a. Show that the triangle T of the crooked Farey tessellation $\mathcal{T}$ has vertices $\frac{s_{11}}{s_{21}}$, $\frac{s_{11}+s_{12}}{s_{21}+s_{22}}$, and $\frac{s_{12}}{s_{22}}$.

b. Devise a similar formula when T is associated to a standard Farey triangle T° which is on the left of 0∞, namely, with all vertices nonpositive.

We are not giving any hint, with the idea that this problem can be turned into a more challenging research project.

Chapter 11

The figure-eight knot complement

This chapter is devoted to the detailed analysis of one more example of a crooked Farey tessellation whose tiling group is a kleinian group. We will see that its features are quite different from those that we encountered in Chapter 10. This example also has an unexpected connection with the figure-eight knot that one can tie in a piece of string.

11.1. Another crooked Farey tessellation

Consider the crooked Farey tessellation corresponding to the shear-bend parameters $s_1 = \frac{2\pi}{3}\mathrm{i}$, $s_3 = -\frac{2\pi}{3}\mathrm{i}$ and $s_5 = 0$. Let us call this crooked tessellation $\mathcal{T}_8$, for reasons that we just hinted at.

In this case, the tiling group Γ_8 is generated by

$$\varphi_1(z) = \frac{z+1}{z+1+\mathrm{e}^{-\frac{2\pi}{3}\mathrm{i}}} = \frac{z+1}{z+\omega^{-1}}$$

and

$$\varphi_3(z) = \frac{z-1}{-z+1+\mathrm{e}^{-\frac{2\pi}{3}\mathrm{i}}} = \frac{z-1}{-z+\omega^{-1}},$$

if we set $\omega = \mathrm{e}^{\frac{\pi}{3}\mathrm{i}}$. Note that $\omega^3 = -1$, so that $\omega^2 - \omega + 1 = \frac{\omega^3+1}{\omega+1} = 0$ and $1 + \omega^{-2} = 1 - \omega = -\omega^2 = \omega^{-1}$.

This crooked tessellation is illustrated in Figure 11.1. However, this picture is necessarily imperfect, because one can only print a finite number of triangles (and because this finite number is further limited by the poor programming skills of the author).

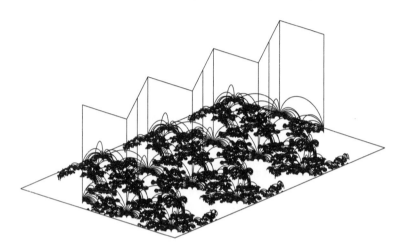

Figure 11.1. An approximation of the crooked Farey tessellation $\mathcal{T}_8$

Similarly, an approximation of the limit set Λ_Γ of Γ, namely, of the footprints of the crooked tessellation, is represented in Figure 11.2. Again, the picture cannot be completely accurate, since it includes only finitely many points.

In the examples of Section 10.1, the imperfections of the figures were unimportant because they were sufficiently close to the actual geometric object that the difference was essentially undetectable. In this case, however, the differences are very noticeable. Indeed, we will see in Section 11.3 that the limit set Λ_Γ is the whole Riemann sphere $\widehat{\mathbb{C}}$. In particular, Figure 11.2 should just be a (mathematically correct albeit aesthetically less pleasing) solid black rectangle.

11.2. Enlarging the group Γ_8

One can observe that although mathematically incomplete, Figure 11.2 exhibits some interesting symmetries under translation.

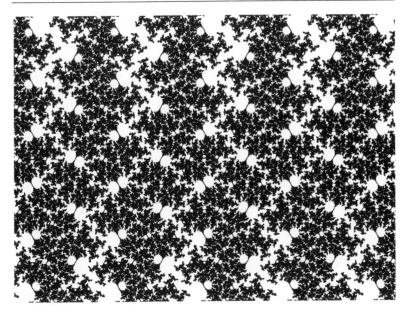

Figure 11.2. An approximation of the limit set of Γ_8

One of them, which is an actual symmetry of the picture, is by horizontal translation. You should remember that the figure has been rotated so as to make this symmetry horizontal; in reality, it corresponds to the element $\varphi_3^{-1} \circ \varphi_1^{-1} \circ \varphi_3 \circ \varphi_3$ which is a translation by $4 - 2\omega$ with, as before $\omega = e^{\frac{\pi}{3}i}$. Since this translation is an element of the tiling group Γ_8, it respects the crooked tessellation $\mathcal{T}_8$ in $\mathbb{H}^2$, and consequently respects its "footprints". As in Exercise 10.9, there actually is a translation symmetry by half this distance, namely, by $2 - \omega$.

However, we can also discern in the figure an imperfect symmetry by a vertical translation. The incompleteness of the computer output prevents this symmetry from being completely accurate on the picture, although the same incompleteness has the advantage of making the symmetry somewhat visible. When rotated back to its actual position, this symmetry corresponds to the translation τ defined by $\tau(z) = z + \omega$.

Let $\widehat{\Gamma}_8$ be the transformation group generated by φ_1, φ_3 and by this new translation τ. We will consider $\widehat{\Gamma}_8$ as a group of isometries of $\mathbb{H}^3$.

It turns out that discontinuity is easier to show for the action for $\widehat{\Gamma}_8$ than for Γ_8. Indeed, we will exhibit in the next section a relatively simple fundamental domain for $\widehat{\Gamma}_8$. It can be proven that Γ_8 admits no fundamental domain with finitely many faces. This is a special case of a theorem of Al Marden [**Marden₁**], but the proof of this property is far beyond the scope of this book.

Another direct proof that the actions of Γ_8 and $\widehat{\Gamma}_8$ on $\mathbb{H}^3$ are discontinuous is provided by Exercise 11.2.

11.2.1. A fundamental domain for the action of $\widehat{\Gamma}_8$.

In the hyperbolic space $\mathbb{H}^3$, let Δ_1 be the tetrahedron with vertices at infinity 0, 1, ∞ and $\omega = e^{\frac{\pi}{3}i}$, and let Δ_2 be the tetrahedron with vertices at infinity 0, 1, ∞ and $\omega^{-1} = 1 - \omega$. The union Δ of the two ideal tetrahedra Δ_1 and Δ_2 is a polyhedron with five ideal vertices, nine edges and six faces. We want to show that Δ is a fundamental domain for the action of $\widehat{\Gamma}_8$ on $\mathbb{H}^3$. In particular, this will prove that $\widehat{\Gamma}_8$ acts discontinuously on $\mathbb{H}^3$.

Figure 11.3 offers a top view of Δ_1 and Δ_2, namely, describes the vertical projection of these objects from the hyperbolic space $\mathbb{H}^3$ to the plane $\mathbb{C}$. In this projection, an ideal tetrahedron with one vertex equal to ∞ appears as a euclidean triangle in the plane. Similarly, an ideal triangle with a vertex at ∞ projects to a line segment, whereas any other ideal triangle projects to a euclidean triangle.

Note that Δ_1 and Δ_2 meet along the ideal triangle T^+ of the crooked tessellation $\mathfrak{T}_8$. Consequently, every time an element $\varphi \in \Gamma_8$ sends T^+ to an ideal triangle with one vertex equal to ∞, the ideal tetrahedra $\varphi(\Delta_1)$ and $\varphi(\Delta_2)$ have one vertex at ∞. Since the combinatorics of the crooked tessellation $\mathfrak{T}_8$ is controlled by that of the standard Farey tessellation $\mathfrak{T}^\circ$, we can find many examples of such φ. Some of these are illustrated in Figure 11.3. To save space, we have omitted the symbol $\circ$ when writing a composition of maps; you should also remember that $\varphi_2 = \varphi_1^{-1}$ and $\varphi_4 = \varphi_3^{-1}$.

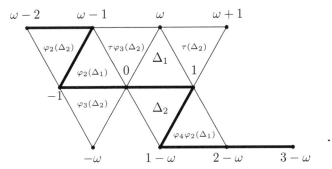

Figure 11.3. A fundamental domain for $\widehat{\Gamma}_8$

In particular, this analysis enables us to find several elements φ of the extended group $\widehat{\Gamma}_8$ for which Δ and $\varphi(\Delta)$ meet along along a face of Δ. More precisely, consider the elements $\psi_1 = \tau$, $\psi_3 = \varphi_4 \circ \varphi_2$ and $\psi_5 = \tau \circ \varphi_3$ of $\widehat{\Gamma}_8$. If $[a, b, c]$ denotes the ideal triangle with ideal vertices a, b, $c \in \widehat{\mathbb{C}}$, we see that ψ_1 sends the face $[0, 1 - \omega, \infty]$ of Δ to the face $[\omega, 1, \infty]$, that ψ_3 sends $[0, 1, \omega]$ to $[\infty, 1 - \omega, 1]$, and that ψ_5 sends $[0, 1, 1 - \omega]$ to $[0, \omega, \infty]$.

Proposition 11.1. *The group $\widehat{\Gamma}_8$ acts discontinuously and freely on the hyperbolic space $\mathbb{H}^3$, and the ideal polyhedron Δ is a fundamental domain for this action.*

Proof. We will apply Poincaré's Polyhedron Theorem 10.9. Glue the faces of the polyhedron Δ by the isometries ψ_i. More precisely, glue the face $F_1 = [0, 1 - \omega, \infty]$ to $F_2 = [\omega, 1, \infty]$ by ψ_1, $F_3 = [0, 1, \omega]$ to $F_4 = [\infty, 1 - \omega, 1]$ by ψ_3 and $F_5 = [0, 1, 1 - \omega]$ to $F_6 = [0, \omega, \infty]$ by ψ_5. As usual, set $\psi_2 = \psi_1^{-1}$, $\psi_4 = \psi_3^{-1}$ and $\psi_6 = \psi_5^{-1}$.

Let $(\bar{\Delta}, \bar{d}_{\mathrm{hyp}})$ be the quotient space so obtained.

The edges $[0, 1]$, $[\infty, 1 - \omega]$, $[\infty, 1]$ and $[0, \omega]$ are glued together to form an edge of $\bar{\Delta}$, as are the edges $[0, \infty]$, $[\omega, \infty]$, $[1, 1 - \omega]$ and $[\omega, 1]$. Also, if we orient each of the above edges $[a, b]$ by the direction from a to b, we see by inspection that these orientations are respected by the gluing maps φ_i. Therefore, the Edge Orientation Condition (2) of Theorem 10.9 holds.

The consideration of Figure 11.3 immediately shows that the dihedral angles of Δ along the edges $[0, \infty]$, $[1, \infty]$, $[\omega, \infty]$ and $[1 - \omega, \infty]$

are $\frac{2\pi}{3}$, $\frac{2\pi}{3}$, $\frac{\pi}{3}$, $\frac{\pi}{3}$, respectively. To determine the other dihedral angles, observe that the inversion across the sphere of radius 1 centered at 0 defines an isometry ρ of $\mathbb{H}^3$ that exchanges 0 and ∞, and fixes each of 1, ω and $1 - \omega$. It follows that ρ sends Δ to itself and consequently that for each $a = 1$, ω or $1 - \omega$, the dihedral angle of Δ along the edge $[0, a]$ is equal to the dihedral angle along $[\infty, a]$.

From these considerations, it follows that the sum of the dihedral angles of Δ along the edges that are glued to $[0, 1]$ is equal to $\frac{2\pi}{3} + \frac{\pi}{3} + \frac{2\pi}{3} + \frac{\pi}{3} = 2\pi$. Similarly, the dihedral angles along the edges that are glued to $[0, \infty]$ add up to $\frac{2\pi}{3} + \frac{\pi}{3} + \frac{\pi}{3} + \frac{2\pi}{3} = 2\pi$. Therefore, the Dihedral Angle Condition (1) in the hypotheses of Poincaré's Polyhedron Theorem 10.9 is satisfied.

For the Horosphere Condition (3), let S_0, S_1, S_ω and $S_{1-\omega}$ be the horosphere centered at 0, 1, ω and $1 - \omega$, respectively, that have euclidean diameter 1, and let S_∞ be the horizontal plane of equation $u = 1$. Note that for any face $[a, b, c]$ of Δ, the three horospheres S_a, S_b, S_c are tangent to each other. Therefore, if $[a, b, c]$ is glued to a face $[a', b', c']$ by a gluing map φ, this map φ sends S_a, S_b and S_c to horospheres centered at a', b' and c' and tangent to each other. By elementary euclidean geometry (compare Lemma 8.4), there are only three such horospheres, namely, $S_{a'}$, $S_{b'}$ and $S_{c'}$. Therefore, if the gluing map φ sends an ideal vertex a of Δ to $\varphi(a)$, it also sends the horosphere S_a to $S_{\varphi(a)}$. This shows that the Horosphere Condition (3) in the hypotheses of Poincaré's Polyhedron Theorem 10.9 is satisfied.

Poincaré's Polyhedron Theorem 10.9 then asserts that the group $\widehat{\Gamma}'_8$ generated by ψ_1, ψ_3, ψ_5 acts discontinuously on $\mathbb{H}^3$ and admits Δ as a fundamental domain.

Similarly, Theorem 10.11 shows that the action of $\widehat{\Gamma}'_8$ is free.

It remains to show that $\widehat{\Gamma}'_8 = \widehat{\Gamma}_8$. Every element of $\widehat{\Gamma}'_8$ can be expressed as a composition of terms $\psi_i^{\pm 1}$. Since $\psi_1 = \tau$, $\psi_3 = \varphi_4 \circ \varphi_2$ and $\psi_5 = \tau \circ \varphi_3$, it can be expressed as a composition of terms τ^{pm1}, $\varphi_1^{\pm 1}$ and $\varphi_3^{\pm 1}$. Therefore, every element of $\widehat{\Gamma}'_8$ is also an element of $\widehat{\Gamma}_8$. Conversely, we can manipulate the above equations and find that $\tau = \psi_1$, $\varphi_3 = \psi_1^{-1} \circ \psi_5$, $\varphi_1 = \psi_3^{-1} \circ \psi_5^{-1} \circ \psi_1$. Therefore, every element of $\widehat{\Gamma}_8$ can be written as a composition of terms $\psi_i^{\pm 1}$, and is consequently an element of $\widehat{\Gamma}'_8$. This proves that $\widehat{\Gamma}'_8 = \widehat{\Gamma}_8$. $\qquad\square$

Proposition 11.2. *The quotient space* $(\mathbb{H}^3/\widehat{\Gamma}_8, \bar{d}_{\text{hyp}})$ *is locally isometric to the hyperbolic space* $\mathbb{H}^3$.

Proof. This is an immediate consequence of Theorem 10.11. □

Since Γ_8 is contained in $\widehat{\Gamma}_8$, an immediate consequence is the following statement.

Corollary 11.3. *The tiling group* Γ_8 *is a kleinian group, namely, acts discontinuously on the hyperbolic space* $\mathbb{H}^3$. □

11.3. Limit sets

We now determine the limit sets of the kleinian groups Γ_8 and $\widehat{\Gamma}_8$.

Lemma 11.4. *The limit set of the enlarged group* $\widehat{\Gamma}_8$ *is the whole Riemann sphere* $\widehat{\mathbb{C}}$.

Proof. The proof is identical to that of Lemma 10.7. The key property is that if, for each ideal vertex ξ of Δ, S_ξ is the horosphere introduced in the proof of Proposition 11.1 and B_ξ is the ball bounded by S_ξ, then $X - (B_0 \cup B_1 \cup B_\omega \cup B_{1-\omega})$ is bounded in $\mathbb{H}^3$. □

It turns out that the two groups Γ_8 and $\widehat{\Gamma}_8$ have the same limit set. This follows from a relatively simple algebraic fact. We begin with a definition.

A **normal subgroup** of a transformation group Γ is a transformation group Γ' contained in Γ and such that for every $\gamma \in \Gamma$ and $\gamma' \in \Gamma'$, the composition $\gamma \circ \gamma' \circ \gamma^{-1}$ is also an element of Γ. Normal subgroups play an important role in algebra because they are well behaved with respect to quotient spaces; see Exercise 11.1.

In our case, we are more interested in the following property.

Proposition 11.5. *If* Γ' *is a normal subgroup of the kleinian group* Γ *and if the limit set of* Γ' *has at least two elements, then the limit sets* Λ_Γ *and* $\Lambda_{\Gamma'}$ *coincide.*

Proof. The key point in the proof is that $\Lambda_{\Gamma'}$ is invariant under the action of Γ. To prove this, consider $\xi \in \Lambda_{\Gamma'}$ and $\gamma \in \Gamma$. Fix a base point $P_0 \in \mathbb{H}^3$, and consider its orbit $\Gamma'(P_0)$ under the action of Γ'. Then there is a sequence of elements $\gamma_n' \in \Gamma'$ such that $\xi = \lim_{n\to\infty} \gamma_n'(P_0)$ in $\mathbb{R}^3 \cup \{\infty\}$ for the euclidean metric. Now,

$$\gamma(\xi) = \lim_{n\to\infty} \gamma \circ \gamma_n'(P_0) = \lim_{n\to\infty} \eta_n'(P_0')$$

if we set $P_0' = \gamma(P_0)$ and $\eta_n' = \gamma \circ \gamma_n' \circ \gamma^{-1} \in \Gamma'$. Since the limit set is independent of the choice of base point (Lemma 10.1), we conclude that $\gamma(\xi)$ is in $\Lambda_{\Gamma'}$ for every $\xi \in \Lambda_{\Gamma'}$ and $\gamma \in \Gamma$. In other words, the limit set $\Lambda_{\Gamma'}$ of the normal subgroup Γ' is invariant under the action of Γ.

Now, $\Lambda_{\Gamma'}$ is closed in $\widehat{\mathbb{C}}$ by Lemma 10.2, it is invariant under the action of Γ, and it has at least two points by hypothesis. By Proposition 10.3, it follows that $\Lambda_{\Gamma'}$ contains Λ_{Γ}. On the other hand, $\Lambda_{\Gamma'}$ is clearly contained in Λ_{Γ} since Γ' is contained in Γ. $\qquad\square$

We now prove that Γ_8 is a normal subgroup of $\widehat{\Gamma}_8$. This will follow from the following computation.

Lemma 11.6.

(11.1) $$\tau^{-1} \circ \varphi_1 \circ \tau = \varphi_3^{-1},$$

(11.2) $$\tau^{-1} \circ \varphi_3 \circ \tau = \varphi_3^2 \circ \varphi_1 \circ \varphi_3,$$

(11.3) $$\tau \circ \varphi_1 \circ \tau^{-1} = \varphi_1^2 \circ \varphi_3 \circ \varphi_1,$$

(11.4) $$\tau \circ \varphi_3 \circ \tau^{-1} = \varphi_1^{-1}.$$

Proof. We could use brute force computations of linear fractional maps but we prefer a more conceptual argument, which in addition should hint at the way these relations were discovered.

The gluing map ψ_1 sends the edge $[0, \infty]$ of Δ to the edge $[\omega, \infty]$, which is sent by $\psi_6 = \psi_5^{-1}$ to $[1, 1 - \omega]$, which is sent by $\psi_4 = \psi_3^{-1}$ to $[\omega, 1]$, which is sent by $\psi_2 = \psi_1^{-1}$ to $[0, 1 - \omega]$, which is sent back to $[0, \infty]$ by ψ_5.

Therefore, the element $\psi_5 \circ \psi_1^{-1} \circ \psi_3^{-1} \circ \psi_5^{-1} \circ \psi_1 \in \widehat{\Gamma}_8$ sends the oriented edge $[0, \infty]$ to itself, and in particular must fix the point $S_0 \cap [0, \infty]$ provided by the horosphere S_0. Since the action of $\widehat{\Gamma}_8$ is free by Proposition 11.1, this proves that

$$\psi_5 \circ \psi_1^{-1} \circ \psi_3^{-1} \circ \psi_5^{-1} \circ \psi_1 = \mathrm{Id}_{\mathbb{H}^3}.$$

Substituting $\psi_1 = \tau$, $\psi_3 = \varphi_4 \circ \varphi_2 = \varphi_3^{-1} \circ \varphi_1^{-1}$ and $\psi_5 = \tau \circ \varphi_3$, this gives

$$\tau \circ \varphi_3 \circ \tau^{-1} \circ \varphi_1 \circ \varphi_3 \circ \varphi_3^{-1} \circ \tau^{-1} \circ \tau = \mathrm{Id}_{\mathbb{H}^3},$$

and

$$\tau^{-1} \circ \varphi_1 \circ \tau = \varphi_3^{-1}$$

after simplification. This proves (11.1).

Considering a similar edge cycle beginning with the edge $[0,1]$, the same argument provides the relation

$$\psi_5^{-1} \circ \psi_3^{-1} \circ \psi_1 \circ \psi_3 = \mathrm{Id}_{\mathbb{H}^3}$$

and

$$\varphi_3^{-1} \circ \tau^{-1} \circ \varphi_1 \circ \varphi_3 \circ \tau \circ \varphi_3^{-1} \circ \varphi_1^{-1} = \mathrm{Id}_{\mathbb{H}^3},$$

which simplifies to

$$\tau^{-1} \circ \varphi_1 \circ \varphi_3 \circ \tau = \varphi_3 \circ \varphi_1 \circ \varphi_3.$$

Using (11.1) (which we just proved) on the left-hand side of the equation and simplifying gives

$$\tau^{-1} \circ \varphi_3 \circ \tau = \varphi_3^2 \circ \varphi_1 \circ \varphi_3,$$

which proves (11.2).

An algebraic manipulation of (11.2) gives

$$\varphi_3 = (\tau \circ \varphi_3 \circ \tau^{-1}) \circ (\tau \circ \varphi_3 \circ \tau^{-1}) \circ (\tau \circ \varphi_1 \circ \tau^{-1}) \circ (\tau \circ \varphi_3 \circ \tau^{-1})$$
$$= \varphi_1^{-2} \circ (\tau \circ \varphi_1 \circ \tau^{-1}) \circ \varphi_1^{-1},$$

using (11.1), from which (11.3) easily follows.

Finally, (11.4) is an immediate consequence of (11.1). □

Lemma 11.7. *The group Γ_8 is a normal subgroup of the enlarged group $\widehat{\Gamma}_8$.*

Proof. We need to show that for every $\gamma \in \widehat{\Gamma}_8$ and every $\gamma' \in \Gamma_8$, the element $\gamma \circ \gamma' \circ \gamma^{-1} \in \widehat{\Gamma}_8$ belongs to the subgroup Γ_8.

This is immediate for $\gamma = \varphi_1^{\pm 1}$ and $\gamma = \varphi_3^{\pm 1}$, since these elements belong to Γ_8, and for $\gamma = \tau^{\pm 1}$ by Lemma 11.6 since Γ_8 is generated by φ_1 and φ_3. The general result then follows, since every $\gamma \in \Gamma_8$ can be written as a composition of terms of the form $\varphi_1^{\pm 1}$, $\varphi_3^{\pm 1}$ or $\tau^{\pm 1}$. □

The combination of Lemma 11.4, Proposition 11.5, and Lemma 11.7 provides the following result.

Corollary 11.8. *The limit set of the kleinian group Γ_8 is equal to the whole Riemann sphere $\widehat{\mathbb{C}}$.*   □

As such, the tiling group Γ_8 of the crooked tessellation $\mathcal{T}_8$ is very different from the tiling groups Γ of the crooked tessellations that we analyzed in Section 10.5. Indeed, we had seen that the limit sets of those tiling groups were homeomorphic to the circle $\widehat{\mathbb{R}}$.

Actually, Jim Cannon and Bill Thurston proved a truly amazing property in [**Cannon & Thurston**].

For the kleinian groups Γ considered in Section 10.5, we constructed a homeomorphism $\lambda \colon \widehat{\mathbb{R}} \to \Lambda_\Gamma$ between $\widehat{\mathbb{R}} = \mathbb{R} \cup \{\infty\}$ and the limit set Λ_Γ of Γ. By continuity, the map λ is uniquely determined by the property that for every rational point $x \in \mathbb{Q} \cup \{\infty\}$ corresponding to a vertex of the standard Farey tessellation $\mathcal{T}^o$, its image $\lambda(x) \in \widehat{\mathbb{C}}$ is the corresponding vertex of the crooked Farey tessellation $\mathcal{T}$.

In the case of the crooked tessellation $\mathcal{T}_8$, we still have a well-defined map $\lambda \colon \mathbb{Q} \cup \{\infty\} \to \Lambda_{\Gamma_8}$, which to each vertex of $\mathcal{T}^o$ associates the corresponding vertex of $\mathcal{T}_8$. Cannon and Thurston prove:

Theorem 11.9 (Cannon-Thurston). *The above map λ has a unique extension $\lambda \colon \widehat{\mathbb{R}} \to \Lambda_{\Gamma_8} = \widehat{\mathbb{C}}$ which is both continuous and surjective.*

□

In particular, this natural map λ provides a continuous curve which passes through every point of the Riemann sphere $\widehat{\mathbb{C}}$. Poincaré would undoubtedly have been again very surprised to find out that his theory of kleinian groups would lead to such a mathematical "monster".

In contrast to what happened in Section 10.5, the map λ is very far from being injective.

The proof of Theorem 11.9 is widely beyond the scope of this text, and can be found in [**Cannon & Thurston**].

11.4. The figure-eight knot

The enlarged kleinian group $\widehat{\Gamma}_8$ of the previous section has an unexpected connection with the *figure-eight knot* familiar to sailors and rock climbers, and represented in Figure 11.4. Here, instead of a long rope, we should think of the knot as an infinite curve going from ∞ to ∞ in $\mathbb{R}^3$.

Figure 11.4. The figure-eight knot in $\mathbb{R}^3 \cup \{\infty\}$

If, in $\mathbb{R}^3 \cup \{\infty\}$, we move the knot away from ∞ and deform it to make the picture esthetically more pleasing, we obtain the closed curve K represented in Figure 11.5. We are particularly interested in the complement $X = \widehat{\mathbb{R}}^3 - K$ of this curve K in $\widehat{\mathbb{R}}^3 = \mathbb{R}^3 \cup \{\infty\}$. This is why the knot is drawn here as a "hollow" curve, to indicate that the knot is not really there.

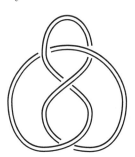

Figure 11.5. The figure-eight knot in $\mathbb{R}^3$

Theorem 11.10. *The complement $X = \widehat{\mathbb{R}}^3 - K$ of the figure-eight knot is homeomorphic to the quotient space $\mathbb{H}^3/\widehat{\Gamma}_8$.*

Proof. The proof of Theorem 11.10 will be a typical example of a "proof by pictures". This type of argument is standard fare for topologists, although sometimes regarded with a certain level of perplexity (or worse) by mathematicians from other fields. The general justification for such a proof is that it would be too time consuming to write (and read) it with all the details rigorously expounded. The sheer bulk of these details would even obscure the key ideas underlying the arguments. Consequently, the basic principle of the exposition of the proof is to explain these geometric ideas in such a way that, at each step, you should be able to convince yourself that you could rigorously write down the complete details if absolutely necessary. Of course, it takes some experience (and it can be a matter of appreciation) to decide which details are legitimate to skip and which ones must absolutely be included. However, with practice, one reaches a level of proficiency that makes this type of proof as mathematically valid as, say, a long algebraic computation.

Our proof of Theorem 11.10 will be a good introduction to this type of mathematical reasoning. As a warm-up, you can begin by trying to figure out a way to make rigorous the statement "let K be the closed curve in $\widehat{\mathbb{R}}^3$ represented in Figure 11.5". Next, you should convince yourself that one can prove the existence of a homeomorphism $\varphi\colon \widehat{\mathbb{R}}^3 \to \widehat{\mathbb{R}}^3$ sending the closed curve $K \subset \widehat{\mathbb{R}}^3$ of Figure 11.5 to the closed curve $K' \subset \widehat{\mathbb{R}}^3$ of Figure 11.4. For this, first consider the physical process of moving the string K to K' in $\widehat{\mathbb{R}}^3$; then mentally define $\varphi\colon \widehat{\mathbb{R}}^3 \to \widehat{\mathbb{R}}^3$ by the property that this process sends each particle $P \in \widehat{\mathbb{R}}^3$ of the universe to the particle $\varphi(P) \in \widehat{\mathbb{R}}^3$; and finally, convince yourself that this homeomorphism can be mimicked by a composition of homeomorphisms rigorously defined by explicit equations, although this exposition would be extraordinarily cumbersome.

We now begin the proof of Theorem 11.10. We will first show that the complement $X = \widehat{\mathbb{R}}^3 - K$ is homeomorphic to the space obtained by gluing two polyhedra X^+ and X^- along their faces. For this, we first arrange by a homeomorphism of $\widehat{\mathbb{R}}^3$ that most of the knot K lies in the xy-plane, except near the crossings, where one strand slightly rises above the plane while the other strand slightly dips under it. Roughly speaking, X^+ will consist of the part of X that lies above the xy-plane, while X^- will include those points which are below the plane. At least, this will be true away from the crossings.

More precisely, we cut X along a certain number of (curved) faces, which we now describe. Away from the crossings, these faces just coincide with the xy-plane. Understanding the arrangement of the faces near a crossing requires a little more thought. Figure 11.6 describes the situation near such a crossing.

We have here four faces meeting along an edge EF, joining a point E in the upper strand to a point F in the lower strand. The endpoints E and F belong to the knot K. In particular, they are not in its complement X, and consequently are not part of the edge EF.

Near the rectangle $PQRS$ drawn in the xy-plane, the first face coincides with a warped pentagon $AEFBQ$, where the edges AE and BF follow the knot and consequently are part of neither X nor the face. Similarly, the second face is a warped pentagon $BFECR$ with the edges BF and EC removed, the third face is $CEFDS$ minus the edges CE and FD, and the last face is $DFEAP$ minus DF and EA.

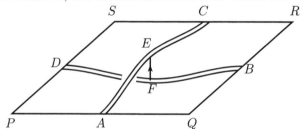

Figure 11.6. Splitting the knot complement near a crossing

Having described these four faces, we now need to understand how they decompose the knot complement X near the rectangle $PQRS$. The upper polyhedron X^+ of the decomposition touches the edge EF along two opposite dihedra, one delimited by $AEFBQ$ and $BFECR$, and the other delimited by $CEFDS$ and $DFEAP$. If we separate X^+ along the edge EF, this edge splits into two edges EF' and EF'', and X^+ can now be deformed to the part of $\mathbb{R}^3$ that lies *above* the diagram on the left-hand side of Figure 11.7, namely, in the picture, the part that lies in front of the sheet of paper.

Symmetrically, the lower polyhedron X^- touches the edge along the remaining two dihedra, namely, the one delimited by $AEFBQ$ and $DFEAP$, and the other one delimited by $BFECR$ and $CEFDS$.

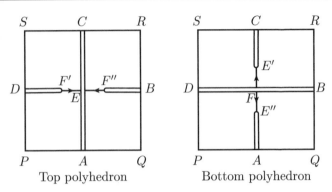

Figure 11.7

Separating X^- along the edge EF yields two new edges $E'F$ and $E''F$. Then X^+ can be deformed to the part of $\mathbb{R}^3$ that lies *below* the diagram on the right-hand side of Figure 11.7, namely, behind the sheet of paper.

Let the faces of the decomposition be defined in such a way that they coincide with the xy-plane away from the crossings, and that they are of the type described above near the crossings. If we split $X = \widehat{\mathbb{R}}^3 - K$ along these faces, we then obtain two pieces X^+ and X_-.

Before we describe these two pieces, it is convenient to introduce some notation. Let $\mathbb{R}^3_+ = \mathbb{R}^2 \times [0, +\infty)$ consist of those points of $\mathbb{R}^3$ that lie on or above the xy-plane $\mathbb{R}^2$, and let $\mathbb{R}^3_- = \mathbb{R}^2 \times (-\infty, 0]$ consist of those points that are located on or below $\mathbb{R}^2$. Then define $\widehat{\mathbb{R}}^3_\pm = \mathbb{R}^3_\pm \cup \{\infty\}$, considered as a subset of $\widehat{\mathbb{R}}^3 = \mathbb{R}^3 \cup \{\infty\}$.

As in Figure 11.7, the upper piece X^+ can be deformed (and is consequently homeomorphic) to the complement, in the upper part $\widehat{\mathbb{R}}^3_+$ of $\widehat{\mathbb{R}}^3$, of the four arcs that are indicated as hollow curves on the left-hand side of Figure 11.8. These four arcs correspond to the trace of the knot K on the boundary of X^+. In this picture, we also use the standard convention that $\mathbb{R}^3_+$ consists of those points of $\mathbb{R}^3$ which lie in front of the sheet of paper. The boundary of X^+ is decomposed into six faces and eight edges as indicated on the left-hand side of Figure 11.8, with two edges associated to each crossing of the picture in Figure 11.5.

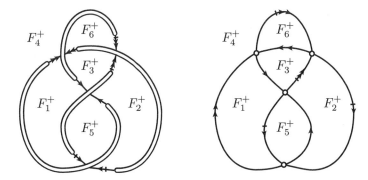

Figure 11.8. The top piece X^+ of the decomposition of the figure-eight knot complement

By shrinking each of the missing arcs to a point, X^+ can also be deformed to $\widehat{\mathbb{R}}^3_+$ minus the four points indicated on the right-hand side of Figure 11.8. The edges of the decomposition of its boundary are then as indicated in that picture. To clarify this deformation we should probably note that it does not provide a homeomorphism of $\mathbb{R}^3_+$, since it crushes each of the four arcs to a single point (and therefore is far from being injective). However, it can indeed be chosen so that it is injective outside of these arcs, and induces a homeomorphism between $\mathbb{R}^3_+$ minus the four arcs on the left and $\mathbb{R}^3_+$ minus the four points on the right.

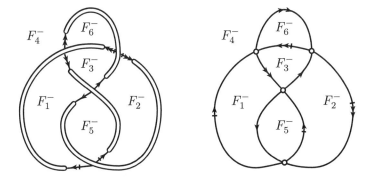

Figure 11.9. The bottom piece X^- of the decomposition of the figure-eight knot complement

Similarly, the lower piece X^- can be deformed to the lower half $\mathbb{R}^3_-$ of $\widehat{\mathbb{R}}^3$ minus the four arcs drawn on the left-hand side of Figure 11.9, or minus the four points indicated on the right-hand side. In particular, X^- now lies behind the sheet of paper in Figure 11.9. Again, the boundary of X^- is decomposed into six faces and eight edges, as drawn in these pictures.

The right-hand sides of Figures 11.8 and 11.9 begin to make the $X^\pm$ look like ideal polyhedra, namely, like polyhedra with their vertices deleted. However, X^+ and X^- each have two faces which are "digons", namely, faces with only two edges and two vertices (at infinity). This is clearly not possible for ideal polyhedra in the hyperbolic space $\mathbb{H}^3$, since two hyperbolic geodesics with the same endpoints at infinity must coincide. Consequently, we need to perform a last modification on the $X^\pm$.

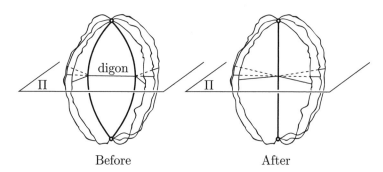

Before   After

Figure 11.10. Collapsing a digon to one edge

The two digons of X^+ and the two digons of X^- are glued together to give two digons in the decomposition of X. We will modify this decomposition by collapsing each of these two digons of X to a single edge. Under the following collapse, the two polyhedra $X^\pm$ are transformed into two other polyhedra $Y^\pm$.

This process is illustrated by Figure 11.10. We have indicated the cross section with a plane Π to clarify the picture.

In the collapse of the first digon corresponding to the faces $F_X^+ \subset X^+$ and $F_X^- \subset X^-$, the edges marked —▸— and —◂— get combined to form a single edge —▸— in the new decomposition. Similarly, in the second digon corresponding to $F_X^\pm \subset X^\pm$, the edges ⇢ and ⇠ are replaced by a single edge ⇢ .

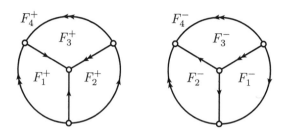

Figure 11.11. The tetrahedra $Y^\pm$

After collapsing these two digons, X^+ gets replaced by another region Y^+ of $\widehat{\mathbb{R}}^3$, homeomorphic to $\widehat{\mathbb{R}}_+^3$ minus the four points indicated on the left-hand side of Figure 11.11. In addition, Y^+ inherits from the nondigon faces of X^+ a decomposition of its boundary into four faces and six edges, as indicated on the left half of Figure 11.11. Compare Figures 11.8 and 11.11, and remember that $\widehat{\mathbb{R}}_+^3$ lies in front of the sheet of paper in these figures.

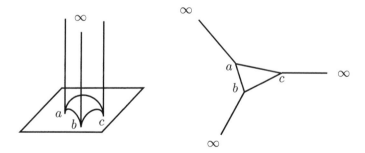

Figure 11.12. Unfolding an ideal tetrahedron

Similarly, after collapsing the digons and flipping the picture over, X^- is replaced by a region Y^- homeomorphic to $\widehat{\mathbb{R}}_+^3$ minus four points, and with four faces and six edges as indicated on the right-hand side of Figure 11.11. The flip is used to exchange front and half, since X^- was in the back of the sheet of paper in Figure 11.9.

With a little more thought, we can convince ourselves that Y^+ and Y^- are each homeomorphic to an ideal tetrahedron, by a homeomorphism sending edges to edges and faces to faces. It may be easier to go backward to see this, and start with an ideal tetrahedron in $\mathbb{H}^3$ with ideal vertices a, b, c, $\infty \in \widehat{\mathbb{R}}^2$. Unfolding its faces onto the xy-plane deforms the ideal tetrahedron to the upper part $\widehat{\mathbb{R}}^3_+$ of $\widehat{\mathbb{R}}^3$ minus $\{a, b, c, \infty\}$, and the four faces then are as indicated in the right-hand side of Figure 11.12. That picture is then easily deformed in $\widehat{\mathbb{R}}^2 = \mathbb{R}^2 \cup \{\infty\}$ to those of Figure 11.11, after moving the fourth vertex away from ∞.

The upshot of this discussion is that up to homeomorphism, the figure-eight knot complement $\widehat{\mathbb{R}}^3 - K$ is obtained from Y^+ and Y^- by gluing each face F_i^+ of Y^+ to the corresponding face F_i^- of Y^-, in a manner compatible with the edge identifications represented on Figure 11.11.

Let us now return to the fundamental domain Δ for Γ_8 that we considered in Section 11.2.1. This Δ was the union of the two tetrahedra $\Delta_1 = [0, 1, \omega, \infty]$ and $\Delta_2 = [0, 1, 1 - \omega, \infty]$, meeting along their common face $[0, 1, \infty]$. Namely, we can consider that Δ is abstractly obtained from the tetrahedra Δ_1 and Δ_2 by gluing the face $[0, 1, \infty]$ of Δ_1 to the face $[0, 1, \infty]$ of Δ_2 by the identity map. If we unfold these tetrahedra as in Figure 11.12, we then obtain the tetrahedra represented on Figure 11.13, where Δ_1 is on the left, Δ_2 is on the right, and both tetrahedra are in front of the sheet of paper.

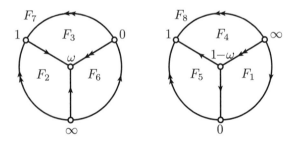

Figure 11.13. Gluing the tetrahedra Δ_1 and Δ_2.

The gluing data of the faces of Δ is such that the face $F_1 = [0, 1 - \omega, \infty]$ of Δ_2 is glued to the face $F_2 = [1, \omega, \infty]$ of Δ_1 by the gluing map φ_1, the face $F_3 = [0, 1, \omega]$ of Δ_1 is glued to the face $F_4 = [1, 1-\omega, \infty]$ of Δ_2 by φ_3, the face $F_5 = [0, 1, 1-\omega]$ of Δ_2 is glued to the face $F_6 = [-, \omega, \infty]$ of Δ_1 by φ_5, and the face $F_7 = [0, 1, \infty]$ of Δ_1 is glued to the face $F_8 = [0, 1, \omega]$ of Δ_2 by the identity map. In addition, the edges $[0, 1]$, $[\infty, 1-\omega]$, $[\infty, 1]$, $[0, \omega]$ are glued together, and marked on Figure 11.13 as ➤➤➤ ; similarly, the edges $[0, \infty]$, $[\omega, \infty]$, $[1, 1-\omega]$ and $[\omega, 1]$ are glued together and marked as ◄ .

Comparing Figures 11.11 and 11.13, it is now immediate that the gluing data are the same.

This shows that the figure-eight knot complement $\widehat{\mathbb{R}}^3 - K$ is homeomorphic to the quotient space $\bar{\Delta}$ obtained by gluing the ideal tetrahedra Δ_1 and Δ_2 according to the gluing maps described in Figure 11.11.

Theorem 10.9 shows that $\bar{\Delta}$ is homeomorphic to $\mathbb{H}^3/\widehat{\Gamma}_8$. Therefore, the quotient space $\mathbb{H}^3/\widehat{\Gamma}_8$ is homeomorphic to $\widehat{\mathbb{R}}^3 - K$, which concludes the proof of Theorem 11.10. □

We can rephrase Theorem 11.10 in a slightly different way. We say that two metrics d and d' on the same set X are **topologically equivalent**, or **induce the same topology**, if the identity map $\mathrm{Id}_X : (X, d) \to (X, d')$ is a homeomorphism. This is equivalent to the property that for every sequence $(P_n)_{n \in \mathbb{N}}$ in X, the sequence converges to the point P_∞ for the metric d if and only if it converges for to P_∞ for the metric d'; see Exercise 1.9.

In our case, the euclidean metric d_{euc} of $\mathbb{R}^3$ does not quite give a metric on $X = \widehat{\mathbb{R}}^3 - K$, because $d_{\mathrm{euc}}(P, Q)$ is undefined when $Q = \infty$. We will say that a metric d is topologically equivalent to, or induces the same topology as, the metric d_{euc} if a sequence $(P_n)_{n \in \mathbb{N}}$ in X converges to the point P_∞ (with possibly $P_\infty = \infty$) for the metric d if and only if it converges to P_∞ for the metric d_{euc}.

Theorem 11.11. *The complement $X = \widehat{\mathbb{R}}^3 - K$ of the figure-eight knot admits a metric d which is topologically equivalent to the euclidean metric and which satisfies the following two properties:*

(1) (X, d) *is complete;*

(2) (X, d) *is locally isometric to the hyperbolic plane $\mathbb{H}^3$.*

Proof. Let $\varphi \colon X \to \mathbb{H}^3/\widehat{\Gamma}_8$ be the homeomorphism provided by Theorem 11.10. If $\bar{d}_{\mathrm{hyp}}$ is the quotient metric induced on $\mathbb{H}^3/\widehat{\Gamma}_8$ by the hyperbolic metric d_{hyp} of $\mathbb{H}^3$, define a metric d on X by the property that $d(P,Q) = \bar{d}_{\mathrm{hyp}}(\varphi(P), \varphi(Q))$. Namely, d is uniquely determined by the property that φ is an isometry from (X, d) to $(\mathbb{H}^3/\widehat{\Gamma}_8, \bar{d}_{\mathrm{hyp}})$.

The metric $\bar{d}_{\mathrm{hyp}}$ is complete by Theorem 10.9, and locally isometric to $(\mathbb{H}^3, d_{\mathrm{hyp}})$ by Theorem 10.11. Since $\varphi \colon (X, d) \to (\mathbb{H}^3/\widehat{\Gamma}_8, \bar{d}_{\mathrm{hyp}})$ is an isometry, the same properties hold for d. $\qquad\square$

Theorem 11.10 was discovered and first proved by Bob Riley in [**Riley$_1$**]. This property originally appeared to set the figure-eight knot apart from the other knots until, a few years later, Bill Thurston showed that this is actually a manifestation of a more general result. Thurston's Geometrization Theorem completely revolutionized 3-dimensional topology. The next chapter is devoted to these results.

Exercises for Chapter 11

Exercise 11.1. Let Γ be a transformation group acting on the set X, and let Γ' be a normal subgroup of Γ. Let X/Γ' denote the quotient space of X under the action of Γ', and let Γ/Γ' be the quotient of Γ under the action of Γ' by left composition. Namely, $\gamma'(\gamma) = \gamma' \circ \gamma$ for every $\gamma' \in \Gamma'$ and $\gamma \in \Gamma$, and the image of $\gamma \in \Gamma$ in the quotient space Γ/Γ' is $\bar{\gamma} = \{\gamma' \circ \gamma; \gamma' \in \Gamma'\}$.

   a. For every $\bar{\gamma} \in \Gamma/\Gamma'$ and $\bar{P} \in X/\Gamma$, consider the element $\bar{\gamma}(\bar{P}) \in X/\Gamma$ represented by $\gamma(P) \in X$. Show that $\bar{\gamma}(\bar{P})$ is independent of the choice of $\gamma \in \bar{\gamma}$ and of $P \in \bar{P}$.

   b. Show that as $\bar{\gamma}$ ranges over all the elements of Γ/Γ', the corresponding maps $\bar{P} \mapsto \bar{\gamma}(\bar{P})$ form a transformation group Θ of X/γ'.

   c. Consider the map $\Gamma/\Gamma' \to \Theta$ which to $\bar{\gamma} \in \Gamma/\Gamma'$ associates the transformation $\bar{P} \mapsto \bar{\gamma}(\bar{P})$. Show that when $X = \Gamma$ and when the action of the group Γ on the set $X = \Gamma$ is by left multiplication, the above map $\Gamma/\Gamma' \to \Theta$ is a bijection, so that Θ has a natural identification with Γ/Γ'.

Exercise 11.2. For $\omega = e^{\frac{\pi}{3}i}$ as usual in this chapter, consider the subset $\mathbb{Z}[\omega] = \{m + n\omega; m, n \in \mathbb{Z}\}$ of $\mathbb{C}$, and let $\mathrm{PSL}_2(\mathbb{Z}[\omega])$ be the set of linear fractional maps $z \mapsto \frac{az+b}{cz+d}$ with a, b, c, $d \in \mathbb{Z}[\omega]$. Extend each of these linear fractional maps to an isometry of $(\mathbb{H}^3, d_{\mathrm{hyp}})$.

   a. Show that $\mathrm{PSL}_2(\mathbb{Z}[\omega])$ is a group of isometries of $\mathbb{H}^3$. Hint: Remember that $\omega^2 = \omega + 1$.

Our goal is to show that $\mathrm{PSL}_2\big(\mathbb{Z}[\omega]\big)$ acts discontinuously on $\mathbb{H}^3$. Assume, in search of a contradiction, that $\mathrm{PSL}_2\big(\mathbb{Z}[\omega]\big)$ does not act discontinuously at some $P_0 \in \mathbb{H}^3$. Namely assume that for every $\varepsilon > 0$, there are infinitely many $\gamma \in \mathrm{PSL}_2\big(\mathbb{Z}[\omega]\big)$ such that $\gamma(P_0) \in B_{d_{\mathrm{hyp}}}(P_0, \varepsilon)$.

b. Show that there is a sequence $(\gamma_n)_{n \in \mathbb{N}}$ of distinct elements of the group $\mathrm{PSL}_2\big(\mathbb{Z}[\omega]\big)$ such that $\lim_{n \to \infty} \gamma_n(P_0) = P_0$ in $(\mathbb{H}^3, d_{\mathrm{hyp}})$.

c. Show that there exists a subsequence $\big(\gamma_{n_k}\big)_{k \in \mathbb{N}}$ such that the sequences $\big(\gamma_{n_k}(0)\big)_{k \in \mathbb{N}}$, $\big(\gamma_{n_k}(1)\big)_{k \in \mathbb{N}}$, $\big(\gamma_{n_k}(\infty)\big)_{k \in \mathbb{N}}$ converge to points z_0, z_1, $z_\infty \in \widehat{\mathbb{C}}$, respectively, for the euclidean metric d_{euc}.

d. Show that the limits z_0, z_1 and z_∞ are pairwise distinct. Hint: Consider the complete geodesic g going from 0 to ∞, and the geodesic h joining the point $P_1 = (0, 0, 1) \in \mathbb{H}^3$ to the point $1 \in \mathbb{C}$; then observe that the images $\gamma_{n_k}(P_1)$ must remain at bounded hyperbolic distance from P_0.

e. Show that the coefficients a_{n_k}, b_{n_k}, c_{n_k}, $d_{n_k} \in \mathbb{Z}[\omega]$ of $\gamma_{n_k}(z) = \frac{a_{n_k} z + b_{n_k}}{c_{n_k} z + d_{n_k}}$ can be chosen so that they each converge in $\mathbb{C}$ as $k \to \infty$. Hint: If $\gamma(z) = \frac{az+b}{cz+d}$ with $ad - bc = 1$, express a, b, c, d in terms of $\gamma(0)$, $\gamma(1)$ and $\gamma(\infty)$, and note that these coefficients are uniquely determined up to multiplication by -1.

f. Show that for every $R > 0$, there are only finitely many $a \in \mathbb{Z}[\omega]$ such that $|a| \leqslant R$.

g. Conclude that we have reached the contradiction we were looking for.

As a consequence, $\mathrm{PSL}_2\big(\mathbb{Z}[\omega]\big)$ acts discontinuously on $\mathbb{H}^3$. In particular, so do the groups $\Gamma_8 \subset \widehat{\Gamma}_8 \subset \mathrm{PSL}_2\big(\mathbb{Z}[\omega]\big)$ considered earlier in this chapter.

Chapter 12

Geometrization theorems in dimension 3

This chapter generalizes to a wider framework the hyperbolic metric on the figure-eight knot complement that we encountered in Chapter 11. The goal is to show the dramatic impact of techniques of hyperbolic geometry on very classical problems in 3-dimensional topology. The first half of the discussion is focused on knot theory, where results are easier to state. After developing the necessary material, we then consider the general Geometrization Theorem for 3-dimensional manifolds, whose proof was completed very recently.

The proofs of many results in this chapter require a mathematical expertise which is much higher than what we have needed so far. We will not even attempt to explain the ideas behind these arguments. We will be content with providing the background necessary to understand the statements and illustrate these with a few applications.

12.1. Knots

Mathematical knot theory aims at analyzing the many different ways in which a piece of string can be tied into a knot. As for the figure-eight knot that we already encountered in Section 11.4, it is more convenient to consider strings where the two ends have been joined together.

Therefore, from a mathematical point of view, a **knot** in $\mathbb{R}^3$ is a regular simple closed curve K in $\mathbb{R}^3$. We need to make all these terms precise.

The fact that the curve is **regular** means that K can be parametrized by a differentiable function $\gamma\colon \mathbb{R} \to \mathbb{R}^3$ such that for each value $t \in \mathbb{R}$ of the parameter, the derivative $\gamma'(t)$ is different from the zero vector. Recall from multivariable calculus that this hypothesis guarantees that the curve K has a well-defined tangent line at each point.

The fact that the curve K is **closed** means that the above parametrization can be chosen so that it wraps onto itself, namely, so that there exists a $T > 0$ such that $\gamma(t + T) = \gamma(t)$ for every $t \in \mathbb{R}$.

Finally, the closed curve K is **simple** if it does not cross itself, namely, if $\gamma(t) \neq \gamma(t')$ for every $t, t' \in [0, T)$ with $t \neq t'$.

Given two knots K and K' in $\mathbb{R}^3$, a problem of interest to sailors and mathematicians alike is to determine whether it is possible to deform K to K' in $\mathbb{R}^3$ in such a way that the knot never crosses itself throughout the deformation. To express this in a mathematical way, we want to know if there is a family of homeomorphisms $\varphi_t\colon \mathbb{R}^3 \to \mathbb{R}^3$, depending continuously on a parameter $t \in [0, 1]$ such that the following holds: at the beginning, when $t = 0$, φ_0 is just the identity map; at the end, when $t = 1$, φ_1 sends the first knot K to the second one K'. To see how this is a reasonable model for the physical process that we are trying to analyze, think about what happens to each molecule of our world as one moves the string K to the string K'; for each value of the time parameter t, define $\varphi_t\colon \mathbb{R}^3 \to \mathbb{R}^3$ by the property that the molecule that was at the point P at the beginning is now at the point $\varphi_t(P)$ at time t.

When there is such a family of homeomorphisms φ_t, we will say that the knots K and K' are **isotopic**. The family of homeomorphisms φ_t is an **isotopy** from K to K'.

Knot theory is essentially concerned with the problem of classifying all possible knots up to isotopy. Ideally, one would like to have a complete catalogue of knots so that every possible knot is isotopic to one and only one entry of the catalogue. See the classic [**Ashley**] for a nonmathematical version of this catalogue; note that knot practitioners also care about additional properties, such as friction, so that isotopic knots appear several times in that book, under configurations that have different physical characteristics.

The figure-eight knot of Chapter 11 would be one entry in the catalogue. Another famous knot is the **unknot**, represented by a round circle in the plane $\mathbb{R}^2 \subset \mathbb{R}^3$. See Exercise 12.5 for a proof that the figure-eight knot and the unknot really represent distinct entries in the catalogue, namely that, in agreement with our own experience handling strings and ropes, it is not possible to unknot a string tied as a figure-eight knot.

12.1.1. Isotopy versus isomorphism. When K and K' are isotopic, we can just focus on the final homeomorphism $\varphi_1 \colon \mathbb{R}^3 \to \mathbb{R}^3$, which sends K to K'. Note that we can extend φ_1 to a homeomorphism of $\widehat{\mathbb{R}}^3 = \mathbb{R}^3 \cup \{\infty\}$ by setting $\varphi_1(\infty) = \infty$. We will say that two knots K and K' are **isomorphic** if there is a homeomorphism $\varphi \colon \widehat{\mathbb{R}}^3 \to \widehat{\mathbb{R}}^3$ which sends K to K' (but we do not necessarily require φ to send ∞ to ∞).

Clearly, isotopic knots are also isomorphic. However, the property of being isomorphic appears much weaker than being isotopic, since we only have to worry about finding one homeomorphism, as opposed to a whole family. In fact, it is only half as weak, as we will now explain.

A homeomorphism φ of $\widehat{\mathbb{R}}^3$ can be, either **orientation-preserving** or **orientation-reversing**. These properties are difficult to define for general homeomorphisms, but are easily explained when φ is differentiable at some point $P \in \mathbb{R}^3$. Recall that if the coordinate functions of φ are φ_1, φ_2, φ_3 (so that $\varphi(Q) = (\varphi_1(Q), \varphi_2(Q), \varphi_3(Q)) \in \mathbb{R}^3$), the **jacobian** of φ at P is the determinant

$$\operatorname{Jac}_P(\varphi) = \begin{vmatrix} \frac{\partial \varphi_1}{\partial x}(P) & \frac{\partial \varphi_2}{\partial x}(P) & \frac{\partial \varphi_3}{\partial x}(P) \\ \frac{\partial \varphi_1}{\partial y}(P) & \frac{\partial \varphi_2}{\partial y}(P) & \frac{\partial \varphi_3}{\partial y}(P) \\ \frac{\partial \varphi_1}{\partial z}(P) & \frac{\partial \varphi_2}{\partial z}(P) & \frac{\partial \varphi_3}{\partial z}(P) \end{vmatrix}$$

of the differential map $D_P\varphi \colon \mathbb{R}^3 \to \mathbb{R}^3$. If, additionally, $\operatorname{Jac}_P(\varphi) \neq 0$, then φ is orientation-preserving if $\operatorname{Jac}_P(\varphi) > 0$ and φ is orientation-reversing if $\mathrm{Jac}_P(\varphi) < 0$. Compare the 2-dimensional case that we encountered in Section 2.5.2. Using the branch of mathematics called algebraic topology, it can be shown that this property is independent of the point P with $\mathrm{Jac}_P(\varphi) \neq 0$ chosen, and it can even be defined when no such P exists. See for instance [**Hatcher**$_1$, Sect. 2.2], [**Massey**, Chap. VIII, §2], or any other textbook where the notion of degree of a continuous map is defined; a homeomorphism is orientation-preserving exactly when it has degree $+1$.

Suppose that we are given an isotopy $\varphi_t \colon \mathbb{R}^3 \to \mathbb{R}^3$, $0 \leqslant t \leqslant 1$, beginning with the identity map $\varphi_0 = \mathrm{Id}_{\mathbb{R}^3}$. The identity map is clearly orientation-preserving. The same algebraic topology techniques used to define the orientation-preserving property show that the final homeomorphism φ_1 is also orientation-preserving. As a consequence, if the two knots K and K' are isotopic, they are also isomorphic by an orientation-preserving homeomorphism. Namely, there exists an orientation-preserving homeomorphism $\varphi \colon \widehat{\mathbb{R}}^3 \to \widehat{\mathbb{R}}^3$ such that $\varphi(K) = K'$.

It turns out that by a deep result of topology, the converse is also true.

Theorem 12.1. *The knots K and K' are isotopic if and only if they are isomorphic by an orientation-preserving homeomorphism of $\widehat{\mathbb{R}}^3$.*

$\square$

We will not prove this result. The key to its traditional proof is a relatively simple construction of J. W. Alexander [**Alexander**], known as "Alexander's trick". However, this proof is somewhat unsatisfactory, because the isotopy that it provides is very badly behaved at at least one point. This would be considered very unrealistic by most nonmathematicians, who would require all maps involved to be differentiable. Fortunately, the result still holds in the differentiable context but its proof is much more difficult. This differentiable result is due to Jean Cerf [**Cerf**], and was later extended by Allen Hatcher [**Hatcher**$_1$].

In spite of Theorem 12.1, there is a difference between isotopy and isomorphism. For instance, the two trefoil knots represented in Figure 12.1 are isomorphic, since one can send one to the other by a reflection across a plane. However, it can be shown that there is no isotopy moving the left-handed trefoil knot to the right-handed trefoil. This fact is mathematically quite subtle, and cannot be easily proved within the framework of this book; see for instance [**Rolfsen**, §8.E.14].

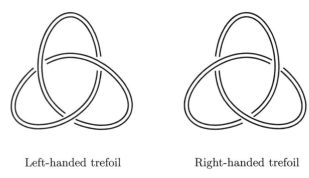

Left-handed trefoil   Right-handed trefoil

Figure 12.1. The two trefoil knots

Note that the jacobian of a reflection is equal to -1 at every point, so that the reflection exchanging the two trefoils is orientation-reversing.

12.2. The Geometrization Theorem for knot complements

This section is devoted to the statement of Thurston's Geometrization Theorem for knot complements. We first begin with a discussion of the two exceptions of this result, torus and satellite knots.

12.2.1. Torus knots. Among all knots, torus knots are among the simplest ones to describe. Given a rational number $\frac{p}{q} \in \mathbb{Q}$ with p, $q \in \mathbb{Z}$ coprime, the $\frac{p}{q}$-*torus knot* is the knot K formed by a curve contained in the standard torus T in $\mathbb{R}^3$, and which turns q times around the hole of the torus and p times counterclockwise around its core C. More precisely, for $R > r > 0$, thicken the horizontal circle C of radius R centered at the origin to a tube of width $2r$. Then, on the torus T bounding this tube, the $\frac{p}{q}$-torus knot K is the curve parametrized by the map $\gamma \colon \mathbb{R} \to \mathbb{R}^3$ defined by

$$\gamma(t) = \big((R + r \cos pt) \cos qt, (R + r \cos pt) \sin qt, r \sin pt\big).$$

Compare our parametrization of the torus in Section 5.1. It is immediate that up to isotopy, K is independent of the choice of the radii R and r.

Figure 12.2 describes the $\frac{-4}{5}$-torus knot as seen from above in $\mathbb{R}^3$. Namely, it represents the projection of this knot to the xy-plane.

Figure 12.2. The $\frac{-4}{5}$-torus knot

We have already encountered the $\frac{3}{2}$- and $\frac{-3}{2}$-torus knots. Indeed, it is relatively immediate that they are respectively isotopic to the left-handed and right-handed trefoil knots of Figure 12.1. Also, the $\frac{p}{\pm 1}$-torus knot is isotopic to the trivial knot, as is the $\frac{\pm 1}{q}$-torus knot.

It can be shown that when $|p|$, $|q|$, $|p'|$, $|q'|$ are all greater than 1, the $\frac{p}{q}$-torus knot is not isomorphic to the unknot, and it is isotopic to the $\frac{p'}{q'}$-torus knot if and only if $\frac{p'}{q'}$ is equal to $\frac{p}{q}$ or to $\frac{q}{p}$. See for instance [**Rolfsen**, §7D.10] for a proof, and compare Exercise 12.1.

12.2.2. Satellite knots. The satellite knot construction is a method for building complex knots from simpler ones.

Suppose that we are given two objects. The first one is a knot K in $\mathbb{R}^3$. The second one is a nontrivial knot L in the standard solid torus V. This requires some explanation.

The **standard solid torus** V is the inside of the standard torus in $\mathbb{R}^3$. Namely, supposing two radii R and r with $R > r > 0$ are given, V consists of those points of the form

$$\big((R + \rho \cos v)\cos u, (R + \rho \cos v)\sin u, \rho \sin v\big)$$

with $0 \leqslant \rho \leqslant r$ and u, $v \in \mathbb{R}$. In other words, V consists of those points which are at distance at most r from the horizontal circle C of radius R centered at the origin. A **nontrivial knot** in V is a regular simple closed curve L in the interior of V such that

(1) L is not isotopic to the central circle C in the solid torus V;

(2) L is not isotopic in V to a knot L' which is disjoint from one of the cross-section disks where the coordinate u is constant (namely, L is really spread around V).

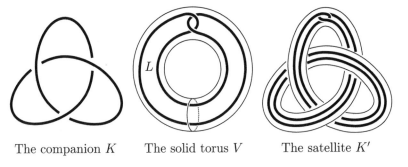

The companion K The solid torus V The satellite K'

Figure 12.3. The construction of a satellite knot

Given a knot K in $\mathbb{R}^3$ and a nontrivial knot L in the standard solid torus V, we can then tie V around the knot K, and consider the image of L. More precisely, choose an injective continuous map $\varphi \colon V \to \mathbb{R}^3$ which sends the central circle C to the knot K. Assume in addition that φ is differentiable and that its jacobian is everywhere different from 0, so that the image $K' = \varphi(L)$ is now a new knot in $\mathbb{R}^3$.

Any knot K' obtained in this way is said to be a **satellite** of the knot K. Conversely, K is a **companion** of K'.

Figure 12.3 offers an example.

A knot admits a unique factorization into companion knots, which is analogous to the factorization of an integer as a product of prime numbers. This factorization of knots was initiated by Horst Schubert [**Schubert₁, Schubert₂**] in the 1950s, generalized by Klaus Johannson [**Johannson**], Bus Jaco and Peter Shalen [**Jaco & Shalen**], and finally improved by Larry Siebenmann [**Bonahon & Siebenmann**, §2] in the 1970s. This unique factorization essentially reduces the analysis of knots to studying knots that are not satellites.[1]

12.2.3. The Geometrization Theorem for knot complements.
We are now ready to state Thurston's Geometrization Theorem for knot complements.

Theorem 12.2 (Geometrization Theorem for knot complements). *Let K be a knot in $\mathbb{R}^3$. Then exactly one of the following holds.*

(1) *K is a $\frac{p}{q}$-torus knot with $q \geqslant 2$;*

(2) *K is a satellite of a nontrivial knot;*

(3) *the complement $\widehat{\mathbb{R}}^3 - K$ admits a metric d which is:*
   (i) *complete;*
   (ii) *topologically equivalent to the euclidean metric;*
   (iii) *locally isometric to the hyperbolic metric d_{hyp} of the hyperbolic space $\mathbb{H}^3$.*

A knot K which satisfies conclusion (3) of Theorem 12.2 is said to be **hyperbolic**.

Theorem 12.2 was proved by Bill Thurston in the late 1970s. It completely revolutionized the world of knot theory (and 3-dimensional topology), for reasons we hope to make apparent in the next sections. This astounding breakthrough was one of the main reasons for which Thurston was awarded the Fields Medal (the mathematical equivalent of a Nobel Prize) in 1983. A key step in Thurston's proof was subsequently simplified by Curt McMullen [**McMullen**]. This part of McMullen's work was, again, one of the major contributions cited when he himself received the Fields Medal in 1998.

In addition to the announcement [**Thurston₁**], Thurston gave numerous lectures detailing key steps of the proof of his Geometrization Theorem. However, he never wrote a complete exposition beyond several influential preprints which never made it to final publication

---

[1]To be completely accurate, one needs to study nonsatellite *links*, consisting of finitely many disjoint knots in $\mathbb{R}^3$. The analysis of links in $\mathbb{R}^3$ is not much more complex than that of knots.

(with the exception of [**Thurston₃**]). Details are now available in [**Otal₁, Otal₂, Kapovich**], for instance.

Although stating this as a precise mathematical statement is not completely clear, it is an experimental fact that a knot randomly drawn in $\mathbb{R}^3$ is neither a torus knot nor a satellite knot. Therefore, in "most" cases, the Geometrization Theorem 12.2 provides a complete hyperbolic metric on the knot complement $X = \widehat{\mathbb{R}}^3 - K$, as in its conclusion (3).

Theorem 12.2 is an abstract existence theorem, whose proof does not provide a convenient method for finding the hyperbolic metric whose existence is asserted. The remarkable piece of software *Snap-Pea*, written by Jeff Weeks [**Weeks₂**] and improved over the years with various collaborators, usually succeeds in describing such a metric. Namely, given a knot K in $\mathbb{R}^3$, *SnapPea* attempts to find a group Γ acting isometrically, freely and discontinuously on the hyperbolic space $\mathbb{H}^3$, such that the quotient space $\mathbb{H}^3/\Gamma$, endowed with the quotient metric $\bar{d}_{\mathrm{hyp}}$ induced by the hyperbolic metric d_{hyp} of $\mathbb{H}^3$, is homeomorphic to $X = \widehat{\mathbb{R}}^3 - K$ by a homeomorphism $\varphi \colon X \to \mathbb{H}^3/\Gamma$. When it succeeds, the metric d defined on X by the property that $d(P, Q) = \bar{d}_{\mathrm{hyp}}\big(\varphi(P), \varphi(Q)\big)$ satisfies conclusion (3) of Theorem 12.2.

To find such a group Γ, *SnapPea* uses a method very similar to the one that we employed for the figure-eight knot in Chapter 11. Namely, it decomposes X into finitely many topological ideal tetrahedra, and then tries to identify them with geometric ideal tetrahedra in $\mathbb{H}^3$ in such a way that the corresponding gluing data satisfies the hypotheses of Poincaré's Polyhedron Theorem 10.9. This often requires changing the original decomposition of X into ideal tetrahedra, which the program does by (educated) trial and error. See [**Weeks₄**] for a more detailed description.

In practice and as long as the knot K is not too intricate (so that one does not encounter issues of computational complexity), *Snap-Pea* always finds the hyperbolic metric on $X = \widehat{\mathbb{R}}^3 - K$ that it is looking for, unless there is no such complete hyperbolic metric because the knot is a torus knot or a satellite knot. There is a curious mathematical situation here. The Geometrization Theorem 12.2 has been rigorously proved, but its abstract proof does not provide any method for explicitly finding the hyperbolic metric whose existence is usually asserted by this statement. Conversely, *SnapPea* almost always finds this hyperbolic metric, but at this point in time there is no rigorous proof that the algorithm used by *SnapPea* will always succeed in doing so.

Since it first became available, *SnapPea* has been a wonderful tool, widely used by researchers in topology. It is also an equally wonderful toy, and you are strongly encouraged to play with it and explore its many features.

We conclude this section with a complement, which sharpens Theorem 12.2.

Let (X, d) be a metric space which is locally isometric to $(\mathbb{H}^3, d_{\mathrm{hyp}})$. In particular, every point of X belongs to a ball $B_d(P, r)$ which is isometric to a ball $B_{d_{\mathrm{hyp}}}(P', r)$ in $\mathbb{H}^3$. We will say that (X, d) has **finite volume** if there exists a sequence of such balls $\{B_d(P_n, r_n)\}_{n \in \mathbb{N}}$, each isometric to a ball $B_{d_{\mathrm{hyp}}}(P'_n, r_n)$ in $\mathbb{H}^3$, such that

(1) the set X is the union of all balls $B_d(P_n, r_n)$. Namely, every $P \in X$ belongs to some ball $B_d(P_n, r_n)$;

(2) if $\mathrm{vol}_{\mathrm{hyp}}\big(B_{d_{\mathrm{hyp}}}(P'_n, r_n)\big) = \pi \sinh 2r_n - 2\pi r_n$ denotes the hyperbolic volume of the ball $B_{d_{\mathrm{hyp}}}(P'_n, r_n)$ in $\mathbb{H}^3$ (compare Exercise 9.14), the series

$$\sum_{n=1}^{\infty} \mathrm{vol}_{\mathrm{hyp}}\big(B_{d_{\mathrm{hyp}}}(P'_n, r_n)\big)$$

converges.

This definition is somewhat *ad hoc*, but will suffice for our purposes. See Exercise 12.6 for a more precise definition of the volume of (X, d).

Complement 12.3. *Whenever Theorem 12.2 provides a complete hyperbolic metric on the complement $\widehat{\mathbb{R}}^3 - K$ of a knot K, this hyperbolic metric has finite volume unless K is isotopic to the unknot.*

12.3. Mostow's Rigidity Theorem

What makes the Geometrization Theorem so powerful is that when it provides a complete metric locally isometric to the hyperbolic space, this metric is actually unique. This is a consequence of a fundamental result which, about 10 years earlier, had been proved by George Mostow [**Mostow₁**] and later improved by Gopal Prasad [**Prasad**] (see also [**Mostow₂, Benedetti & Petronio**] for a proof).

Theorem 12.4 (Mostow's Rigidity Theorem). *Let (X, d) and (X', d') be two complete metric spaces which are locally isometric to $\mathbb{H}^3$ and which have finite volume. If there exists a homeomorphism between (X, d) and (X, d'), then there exists an isometry $\varphi\colon (X, d) \to (X', d')$.* □

In the case that is currently of interest to us, where $X = \widehat{\mathbb{R}}^3 - K$ is the complement of a knot K which is neither a torus knot nor a satellite, the Geometrization Theorem 12.2 and Complement 12.3 provide a complete finite volume metric d which is locally isometric to the hyperbolic metric of the hyperbolic space $\mathbb{H}^3$. We can then rephrase Theorem 12.4 as follows.

Theorem 12.5. *Let K be a knot in $\mathbb{R}^3$ which is neither a torus knot nor a satellite knot, and let d and d' be two hyperbolic metrics on the knot complement $X = \widehat{\mathbb{R}}^3 - K$ as in conclusion (3) of the Geometrization Theorem 12.2. Namely, these two metrics are locally isometric to the hyperbolic metric of $\mathbb{H}^3$, are complete, have finite volume, and are topologically equivalent to the euclidean metric. Then the metric spaces (X, d) and (X, d') are isometric.*

Proof. The identity map Id_X defines a homeomorphism between (X, d) and (X, d') because d and d' are both topologically equivalent to the euclidean metric. It then suffices to apply Theorem 12.4 to these two spaces. □

In other words, the hyperbolic metric d provided by Theorem 12.2 is unique up to isometry. As a consequence, given two hyperbolic knots K and K', if we succeed in proving that the corresponding hyperbolic metrics are not isometric, then we are guaranteed that the two knots are not isomorphic. Here is a typical application.

Corollary 12.6. *Let K and K' be two knots for which Theorem 12.2 provides complete hyperbolic metrics d and d' on $X = \widehat{\mathbb{R}}^3 - K$ and*

$X' = \widehat{\mathbb{R}}^3 - K'$, *respectively. If (X, d) and (X, d') have different hyperbolic volumes (as defined in Exercise 12.6), then the two knots are not isomorphic.* □

In practice, this simple test is remarkably effective.

Mostow's Rigidity Theorem holds in a much more general framework than the one stated in Theorem 12.5. In particular, for any dimension n, one can define an n-dimensional hyperbolic space $\mathbb{H}^n$ by a somewhat straightforward extension of the formulas we used for $\mathbb{H}^2$ and $\mathbb{H}^3$. Mostow's Rigidity Theorem 12.5 extends to all dimensions $n \geqslant 3$ by systematically replacing $\mathbb{H}^3$ by $\mathbb{H}^n$ everywhere.

Surprisingly enough, the same statement is *false* for $n = 2$. Actually, we have already encountered this phenomenon in Section 8.4.2, when we associated to each triple of shear parameters s_1, s_3, $s_5 \in \mathbb{R}$ with $s_1 + s_3 + s_5 = 0$ a complete hyperbolic metric on the once punctured torus. The area of such a metric is twice the hyperbolic area of an ideal triangle in $\mathbb{H}^2$, namely, it is equal to 2π; in particular, it is finite. It can be shown that if we slightly change the shear parameters defining a hyperbolic metric d, the metric d' associated to the new shear parameters is not isometric to d.

12.4. Ford domains

A Ford domain is a variation of the Dirichlet domains that we encountered in Section 7.4. We first introduce some preparatory material.

12.4.1. Hyperbolic metrics and kleinian groups. In Section 7.2, we saw that if Γ acts by isometries, freely and discontinuously on the hyperbolic space $\mathbb{H}^3$, then the quotient space $\mathbb{H}^3/\Gamma$ is locally isometric to $\mathbb{H}^3$. In addition, Exercise 7.6 shows that $\mathbb{H}^3/\Gamma$ is complete, so that $\mathbb{H}^3/\Gamma$ is a complete hyperbolic 3-dimensional manifold.

Conversely, every connected complete hyperbolic 3-dimensional manifold can be obtained in this way, as a quotient of the hyperbolic space $\mathbb{H}^3$ by a kleinian group Γ acting freely. This is a general fact but, in the case we are interested in, we prefer to state this as a complement to Theorem 12.2, which will enable us to include a few properties that will be needed later on.

As in Exercise 9.10, an isometry of the hyperbolic space $\mathbb{H}^3$ is **parabolic** if, by definition, it fixes exactly one point of the sphere at infinity $\widehat{\mathbb{C}}$. A point $\xi \in \widehat{\mathbb{C}}$ is **parabolic** for a group Γ of isometries of $\mathbb{H}^3$ if ξ is fixed by some parabolic $\gamma \in \Gamma$.

Complement 12.7. *Whenever Theorem 12.2 and Complement 12.3 provide a finite volume complete hyperbolic metric d on the complement $X = \widehat{\mathbb{R}}^3 - K$ of a knot K, there exists a group Γ of isometries of $\mathbb{H}^3$ such that*

(1) *Γ acts freely and discontinuously on $\mathbb{H}^3$, and (X, d) is isometric to the quotient space $(\mathbb{H}^3/\Gamma, \bar{d}_{\mathrm{hyp}})$;*

(2) *the set Γ_∞ of those $\gamma \in \Gamma$ with $\gamma(\infty) = \infty$ is generated by two horizontal translations γ_1 and γ_2 along two linearly independent horizontal vectors $\vec{v}_1$ and $\vec{v}_2$, respectively; in particular, ∞ is a parabolic point;*

(3) *every parabolic point of γ is of the form $\gamma(\infty)$ for some $\gamma \in \Gamma$;*

(4) *there exists a euclidean half-space*

$$B_\infty = \{(x, y, u) \in \mathbb{H}^3; u \geqslant u_0\},$$

defined by some constant $u_0 > 0$, which is disjoint from all its images $\gamma(B_\infty)$ with $\gamma \in \Gamma - \Gamma_\infty$.

For (3), observe that conversely, every $\gamma(\infty) \in \widehat{\mathbb{C}}$ with $\gamma \in \Gamma$ is parabolic. In contrast to (4), note that (2) implies that $\gamma(B_\infty) = B_\infty$ for every $\gamma \in \Gamma_\infty$.

Proposition 12.8. *Under the hypotheses of Complement 12.7, let Γ and Γ' satisfy the conclusions of this statement. Then there exists an isometry φ of $(\mathbb{H}^3, d_{\mathrm{hyp}})$ such that*

$$\Gamma' = \{\varphi \circ \gamma \circ \varphi^{-1}; \gamma \in \Gamma\}.$$

In addition, one can arrange that $\varphi(\infty) = \infty$.

Proof. This is a relatively simple consequence of Mostow's Rigidity Theorem. Indeed, the version given in Theorem 12.5 provides an isometry $\psi \colon \mathbb{H}^3/\Gamma \to \mathbb{H}^3/\Gamma'$ between $(\mathbb{H}^3/\Gamma, \bar{d}_{\mathrm{hyp}})$ and $(\mathbb{H}^3/\Gamma', \bar{d}_{\mathrm{hyp}})$. Start with an arbitrary $P_0 \in \mathbb{H}^3$. Consider its image $Q_0 = \pi(P_0)$

under the quotient map $\pi\colon \mathbb{H}^3 \to \mathbb{H}^3/\Gamma$, the image $Q_0' = \varphi(Q_0) \in \mathbb{H}^3/\Gamma'$ of this point Q_0 under the isometry ψ, and finally a point $P_0' \in \pi'^{-1}(Q_0') \in \mathbb{H}^3$ of the preimage of Q_0' under the quotient map $\pi'\colon \mathbb{H}^3 \to \mathbb{H}^3/\Gamma'$, namely, such that $\pi'(P_0') = Q_0'$. The following diagram may be useful to keep track of all these maps and points.

$$
\begin{array}{ccc}
P_0 \in \mathbb{H}^3 & \stackrel{\varphi}{\longmapsto} & P_0' \in \mathbb{H}^3 \\
\pi \downarrow & & \downarrow \pi' \\
Q_0 \in \mathbb{H}^3/\Gamma & \stackrel{\psi}{\longmapsto} & Q_0' \in \mathbb{H}^3/\Gamma'
\end{array}
$$

By the proof of Theorem 7.8 and Corollary 7.9, the quotient map π is a local isometry. As a consequence, for $\varepsilon > 0$ sufficiently small, the restriction $\pi_{|B_{d_{\mathrm{hyp}}}(P_0,\varepsilon)}\colon B_{d_{\mathrm{hyp}}}(P_0,\varepsilon) \to B_{\bar{d}_{\mathrm{hyp}}}(Q_0,\varepsilon)$ is an isometry; in particular, it is bijective. Similarly, we can assume that $\pi'_{|B_{d_{\mathrm{hyp}}}(P_0',\varepsilon)}\colon B_{d_{\mathrm{hyp}}}(P_0',\varepsilon) \to B_{\bar{d}_{\mathrm{hyp}}}(Q_0',\varepsilon)$ is also an isometry, by taking ε small enough. Consider the composition

$$
\varphi = \left(\pi'_{|B_{d_{\mathrm{hyp}}}(P_0',\varepsilon)}\right)^{-1} \circ \psi \circ \pi_{|B_{d_{\mathrm{hyp}}}(P_0,\varepsilon)}\colon B_{d_{\mathrm{hyp}}}(P_0,\varepsilon) \to B_{d_{\mathrm{hyp}}}(P_0',\varepsilon).
$$

This map $\varphi\colon B_{d_{\mathrm{hyp}}}(P_0,\varepsilon) \to B_{d_{\mathrm{hyp}}}(P_0',\varepsilon)$ is an isometry between two hyperbolic balls of $\mathbb{H}^3$. Therefore, it extends to an isometry $\varphi\colon \mathbb{H}^3 \to \mathbb{H}^3$ by Lemma 9.9.

Consider an arbitrary $\gamma \in \Gamma$. By definition of the quotient map $\pi\colon \mathbb{H}^3 \to \mathbb{H}^3/\Gamma$, $\pi \circ \gamma = \pi$. Since $\pi' \circ \varphi(P) = \psi \circ \pi(P)$ for every $P \in B_{d_{\mathrm{hyp}}}(P_0,\varepsilon)$ by construction of φ, we conclude that

$$
\pi' \circ \varphi \circ \gamma(P) = \psi \circ \pi \circ \gamma(P) = \psi \circ \pi(P) = \pi' \circ \varphi(P)
$$

for every $P \in B_{d_{\mathrm{hyp}}}(P_0,\varepsilon)$. In the particular case where $P = P_0$, this implies that $\varphi \circ \gamma(P_0)$ and $\varphi(P_0)$ are glued together when one takes the quotient under the action of Γ', namely, that there exists an element $\gamma' \in \Gamma'$ such that $\gamma' \circ \varphi(P_0) = \varphi \circ \gamma(P_0)$.

Now, for every $P \in B_{d_{\mathrm{hyp}}}(P_0,\varepsilon)$,

$$
\pi' \circ (\gamma')^{-1} \circ \varphi \circ \gamma(P) = \pi' \circ \varphi \circ \gamma(P) = \pi' \circ \varphi(P).
$$

Note that $(\gamma')^{-1} \circ \varphi \circ \gamma(P_0) = \varphi(P_0) = P_0'$ by construction of γ' so that $(\gamma')^{-1} \circ \varphi \circ \gamma(P)$ and $\varphi(P)$ are both in the ball $B_{d_{\mathrm{hyp}}}(P_0',\varepsilon)$. Since the restriction of π' to this ball is injective, we conclude that $(\gamma')^{-1} \circ \varphi \circ \gamma(P) = \varphi(P)$.

This proves that the two isometries $(\gamma')^{-1} \circ \varphi \circ \gamma$ and φ of $(\mathbb{H}^3, d_{\mathrm{hyp}})$ coincide on the ball $B_{d_{\mathrm{hyp}}}(P_0, \varepsilon)$. Consequently, they coincide everywhere on $\mathbb{H}^3$ by the uniqueness statement in Lemma 9.9. Composing both of them with γ' on one side and φ^{-1} on the other side, we conclude that $\varphi \circ \gamma \circ \varphi^{-1} = \gamma' \in \Gamma'$.

This proves that $\{\varphi \circ \gamma \circ \varphi^{-1}; \gamma \in \Gamma\}$ is contained in Γ'.

Conversely, given $\gamma' \in \Gamma'$, the same argument but replacing φ by φ^{-1} provides an element $\gamma \in \Gamma$ such that $\varphi^{-1} \circ \gamma' \circ \varphi = \gamma$ or, equivalently, $\gamma' = \varphi \circ \gamma \circ \varphi^{-1}$. This proves that Γ' is contained in $\{\varphi \circ \gamma \circ \varphi^{-1}; \gamma \in \Gamma\}$. Since we just proved the reverse inclusion, these two sets are equal.

We need to prove the last statement of Proposition 12.8, namely that we can choose the isometry φ so that $\varphi(\infty) = \infty$.

For the isometry φ that we have constructed so far, we only know that $\varphi^{-1}(\infty)$ is a parabolic point of Γ. Indeed, if γ' is a parabolic element of Γ' fixing ∞, then $\varphi^{-1} \circ \gamma' \circ \varphi$ is a parabolic element of Γ which fixes $\varphi^{-1}(\infty)$.

Therefore, since Γ satisfies conclusion (3) of Complement 12.7, there exists an element $\gamma_0 \in \Gamma$ such that $\gamma_0(\infty) = \varphi^{-1}(\infty)$. Set $\varphi' = \varphi \circ \gamma_0$. Note that φ' is an isometry of $\mathbb{H}^3$ and sends ∞ to ∞. In addition

$$\{\varphi' \circ \gamma \circ (\varphi')^{-1}; \gamma \in \Gamma\} = \{\varphi \circ \gamma_0 \circ \gamma \circ \gamma_0^{-1} \circ \varphi^{-1}; \gamma \in \Gamma\}$$
$$= \{\varphi \circ \alpha \circ \varphi^{-1}; \alpha \in \Gamma\} = \Gamma'$$

as required. (To justify the second equality, note that $\alpha = \gamma_0 \circ \gamma \circ \gamma_0^{-1}$ belongs to the group Γ for every $\gamma \in \Gamma$ and that conversely, every $\alpha \in \Gamma$ can be written as $\alpha = \gamma_0 \circ \gamma \circ \gamma_0^{-1}$ for some $\gamma \in \Gamma$.) $\square$

12.4.2. Ford domains.

As indicated before, Ford domains are a variation of the Dirichlet domains of Section 7.4. For this reason, it may be useful to first repeat the definition of Dirichlet domains. Let Γ be a group of isometries of the metric space (X, d) whose action is discontinuous. The Dirichlet domain of Γ at $P \in X$ is the subset

$$\Delta_\Gamma(P) = \{Q \in X; d(P, Q) \leqslant d(\gamma(P), Q) \text{ for every } \gamma \in \Gamma\},$$

consisting of those points $Q \in X$ which are at least as close to P as to any other point of its orbit $\Gamma(P)$.

In Section 7.4, we considered the case where X is the hyperbolic plane $\mathbb{H}^2$ (as well as the euclidean plane, but this is irrelevant here). In particular, Theorem 7.13 showed that in this case the Dirichlet domain $\Delta_\Gamma(P)$ is a locally finite polygon in $\mathbb{H}^2$ and that, as γ ranges over all elements of γ, the polygons $\gamma(\Delta_\Gamma(P))$ form a tessellation of $\mathbb{H}^2$. In addition, the two tiles $\gamma(\Delta_\Gamma(P))$ and $\Delta_\Gamma(P)$ are distinct if and only if $\gamma(P) \neq P$.

The natural generalization of these results holds when X is the hyperbolic space $\mathbb{H}^3$. Namely, $\Delta_\Gamma(P)$ is then a locally finite polyhedron in $\mathbb{H}^3$. To prove this, one just needs to replace Lemma 7.14 by the statement that for any $P, Q \in \mathbb{H}^3$, the set of points R that are at the same hyperbolic distance from P and Q is a hyperbolic plane Π, while the set of R with $d_{\mathrm{hyp}}(P, R) \leqslant d_{\mathrm{hyp}}(Q, R)$ is a hyperbolic half-space H delimited by this perpendicular bisector plane Π. The proof of Theorem 7.13 then immediately generalizes to show that $\Delta_\Gamma(P)$ is a locally finite polyhedron and that the polyhedra $\gamma(\Delta_\Gamma(P))$ with $\gamma \in \Gamma$ form a tessellation of $\mathbb{H}^3$. Again, $\gamma(\Delta_\Gamma(P))$ is distinct from $\Delta_\Gamma(P)$ unless $\gamma(P) \neq P$.

For a hyperbolic knot K in $\mathbb{R}^3$, let Γ be a kleinian group satisfying the conclusions of Complement 12.7. In the definition of the Dirichlet domain $\Delta_\Gamma(P)$, we now replace the point $P \in \mathbb{H}^3$ by the point ∞. Namely, we want to consider the set $\Delta_\Gamma(\infty)$ of those points $Q \in \mathbb{H}^3$ which are at least as close to ∞ as to any other $\gamma(\infty)$ with $\gamma \in \Gamma$. There is of course a problem here, because a point $Q \in \mathbb{H}^3$ is at infinite hyperbolic distance from ∞, and from any $\gamma(\infty)$! However, some infinities are larger than others, and it turns out that we can really make sense of this statement.

For this, let us pursue the same formal reasoning. Since $\gamma \in \Gamma$ is a hyperbolic isometry, the distance from Q to $\gamma(\infty)$ should be the same as the distance from $\gamma^{-1}(Q)$ to ∞. Therefore, if we neglect the fact that we are talking about infinite distances, the statement that Q is closer to ∞ than to $\gamma(\infty)$ should be equivalent to the statement that ∞ is closer to Q than to $\gamma^{-1}(Q)$. If we remember our discussion of the Busemann function in Section 6.8 (see also Exercises 6.10–6.11), we can now make sense of this property. Indeed, if $Q_1 = (x_1, y_1, u_1)$ and $Q_2 = (x_2, y_2, u_2)$ are two points of $\mathbb{H}^3$,

$$\lim_{P \to \infty} d_{\text{hyp}}(P, Q_1) - d_{\text{hyp}}(P, Q_2) = \log \frac{u_1}{u_2},$$

where the limit is taken for the euclidean metric in $\widehat{\mathbb{R}}^3 = \mathbb{R}^3 \cup \{\infty\}$.

As a consequence, we can decide that $Q_1 = (x_1, y_1, u_1)$ is closer to ∞ than $Q_2 = (x_2, y_2, u_2)$ if $u_1 > u_2$. To avoid any ambiguity, we will use the euclidean terminology and say that in this case, Q_1 is **higher than** Q_2. Similarly, Q_1 is **at least as high as** Q_2 if $u_1 \geqslant u_2$.

This leads us to define the **Ford domain** of the kleinian group Γ as

$$\Delta_\Gamma(\infty) = \{Q \in \mathbb{H}^3; Q \text{ is at least as high as } \gamma^{-1}(Q) \text{ for every } \gamma \in \Gamma\}.$$

We will show that $\Delta_\Gamma(\infty)$ is a locally finite hyperbolic polyhedron.

Lemma 12.9. *Let γ be an isometry of $\mathbb{H}^3$ such that $\gamma(\infty) \neq \infty$. Then the set of those $Q \in \mathbb{H}^3$ which are at least as high as $\gamma^{-1}(Q)$ is a hyperbolic half-space H_γ, bounded by a hyperbolic plane Π_γ.*

Proof. For $u > 0$, let S_u be the horizontal plane passing through the point $(0, 0, u)$. In hyperbolic terms, $S(u)$ is a horosphere centered at ∞. By Proposition 9.11, its image $\gamma(S_u)$ is therefore a horosphere centered at the point $\gamma(\infty) \in \mathbb{C}$, namely, a euclidean sphere touching $\mathbb{C}$ at that point.

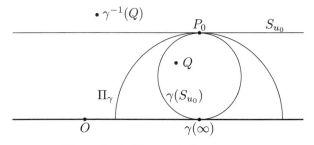

Figure 12.4. The proof of Lemma 12.9

When u is near $+\infty$, the euclidean radius of $\gamma(S_u)$ is very small and the horosphere $\gamma(S_u)$ is disjoint from S_u. On the other hand, when u is very close to 0, the euclidean radius of $\gamma(S_u)$ is very large and S_u is near the xy-plane, so that the two horospheres $\gamma(S_u)$ and S_u meet each other. (To check these two facts, note that $\gamma(S_u)$ contains the point $\gamma(0,0,u)$, which is on the hyperbolic geodesic g joining $\gamma(0,0,0)$ to $\gamma(\infty)$.) Therefore, there exists a value $u_0 > 0$ for which S_{u_0} and $\gamma(S_{u_0})$ are exactly tangent to each other, at a point P_0.

Let Π_γ be the hyperbolic plane that passes through P_0 and is orthogonal to the geodesic joining $\gamma(\infty)$ to ∞. Namely, Π_γ is the intersection with $\mathbb{H}^3$ of the euclidean sphere centered at $\gamma(\infty)$ and passing through P_0.

Let ρ be the hyperbolic reflection across Π_γ, which is also the restriction to $\mathbb{H}^3$ of the inversion across the euclidean sphere centered at $\gamma(\infty)$ and containing P_0. In particular, ρ exchanges ∞ and $\gamma(\infty)$, and fixes the point P_0. It follows that ρ also exchanges the horospheres S_{u_0} and $\gamma(S_{u_0})$. In particular, the hyperbolic isometry $\alpha = \rho \circ \gamma$ sends ∞ to ∞, and respects the horosphere S_{u_0}. As a consequence, α respects every horosphere centered at ∞, and it sends each point of $\mathbb{H}^3$ to a point which is at the same height.

If Q is a point of Π_γ, then $\gamma^{-1}(Q) = \alpha^{-1} \circ \rho^{-1}(Q) = \alpha^{-1}(Q)$ is at the same height as Q by the above observation on α.

If Q is a point of the open hyperbolic half-space delimited by Π_γ and adjacent to ∞, namely, if $Q \in \mathbb{H}^3$ is located outside of the closed euclidean ball centered at $\gamma(\infty)$ and containing P_0 in its boundary, then $\rho(Q)$ is lower than Q by elementary geometry. Noting that $\rho^{-1} = \rho$, the same property of α shows that $\gamma^{-1}(Q) = \alpha^{-1} \circ \rho^{-1}(Q)$ is lower than Q. In other words, Q is higher than $\gamma^{-1}(Q)$.

Similarly, if Q is in the other open hyperbolic half-space delimited by Π_γ, the same argument shows that Q is lower than $\gamma^{-1}(Q)$.

This completes the proof of Lemma 12.9. $\square$

By definition, the Ford domain $\Delta_\Gamma(\infty)$ is equal to the intersection of all these half-spaces H_γ as γ ranges over all elements of Γ such that $\gamma(\infty) \neq \infty$. Namely,

$$\Delta_\Gamma(\infty) = \bigcap_{\gamma \in \Gamma - \Gamma_\infty} H_\gamma$$

using in the notation the stabilizer $\Gamma_\infty = \{\gamma \in \Gamma; \gamma(\infty) = \infty\}$ of ∞. To show that this is a locally finite polyhedron, we need the following fact.

Lemma 12.10. *The family of hyperbolic planes* $\{\Pi_\gamma; \gamma \in \Gamma - \Gamma_\infty\}$ *defined by Lemma 12.9 is locally finite.*

Proof. We need to show that for every $P \in \mathbb{H}^3$, there exists an $\varepsilon > 0$ such that the ball $B_{d_{\mathrm{hyp}}}(P, \varepsilon)$ meets only finitely many Π_γ. Actually, any $\varepsilon > 0$ will do.

Note that we are not saying that there are only finitely many $\gamma \in \Gamma$ such that Π_γ meets $B_{d_{\mathrm{hyp}}}(P, \varepsilon)$. Indeed, if α is an element of the stabilizer Γ_∞ (namely, if $\alpha \in \Gamma$ and $\alpha(\infty) = \infty$), conclusion (2) of Complement 12.7 shows that α sends every point of $\mathbb{H}^3$ to a point at the same height. It then follows from our definitions that $\Pi_{\gamma \circ \alpha} = \Pi_\gamma$. In particular, there are infinitely many $\gamma \in \Gamma$ with the same associated plane Π_γ.

After these preliminary comments, let us begin the proof of Lemma 12.10. Suppose that the ball $B_{d_{\mathrm{hyp}}}(P, \varepsilon)$ meets infinitely many distinct Π_{γ_n}, with $n \in \mathbb{N}$. For each n, pick a point $Q_n \in B_{d_{\mathrm{hyp}}}(P, \varepsilon) \cap \Pi_{\gamma_n}$.

By Complement 12.7, the stabilizer Γ_∞ of ∞ is generated by two horizontal translations, with respective translation vectors $\vec{v}_1$ and $\vec{v}_2$. In particular, an arbitrary parallelogram $\mathcal{P}$ in the plane with sides respectively parallel to $\vec{v}_1$ and $\vec{v}_2$ is a fundamental domain for the action of Γ_n on the horizontal plane $\mathbb{R}^2$. As a consequence, there exists $\alpha_n \in \Gamma_\infty$ such that the first two coordinates (x_n, y_n) of $\alpha_n^{-1} \circ \gamma_n^{-1}(Q_n) = (x_n, y_n, u_n)$ are located in $\mathcal{P}$.

Note that the last coordinate u_n of $\alpha_n^{-1} \circ \gamma_n^{-1}(Q_n)$ is also the last coordinate of $\gamma_n^{-1}(Q_n)$, which is equal to the last coordinate of Q_n since $Q_n \in \Pi_{\gamma_n}$. As a consequence, since $d_{\mathrm{hyp}}(P, Q_n) < \varepsilon$, the usual comparison between euclidean and hyperbolic metrics (see Lemma 2.5) shows that $ue^{-\varepsilon} < u_n < ue^{\varepsilon}$, where u is the third coordinate of $P = (x, y, u)$.

It follows that all points $\alpha_n^{-1} \circ \gamma_n^{-1}(Q_n)$ are in the parallelepipedic box $\mathcal{P} \times [ue^{-\varepsilon}, ue^{\varepsilon}] \subset \mathbb{H}^3 \subset \mathbb{R}^3$. By compactness, or by another explicit comparison between hyperbolic and euclidean metrics, there is a large radius R such that this box $\mathcal{P} \times [ue^{-\varepsilon}, ue^{\varepsilon}]$ is contained in the ball $B_{d_{\text{hyp}}}(P, R)$. Then,

$$\begin{aligned}
d_{\text{hyp}}\big(P, \gamma_n \circ \alpha_n(P)\big) &\leqslant d_{\text{hyp}}(P, Q_n) + d_{\text{hyp}}\big(Q_n, \gamma_n \circ \alpha_n(P)\big) \\
&\leqslant d_{\text{hyp}}(P, Q_n) + d_{\text{hyp}}\big(\alpha_n^{-1} \circ \gamma_n^{-1}(Q_n), P\big) \\
&\leqslant \varepsilon + R.
\end{aligned}$$

Since the action of Γ is discontinuous, Lemma 7.15 shows that the ball $B_{d_{\text{hyp}}}(P, \varepsilon + R)$ contains only finitely many points of the orbit. In particular, there exists two indices $n_1 \neq n_2$ such that $\gamma_{n_1} \circ \alpha_{n_1}(P) = \gamma_{n_2} \circ \alpha_{n_2}(P)$. Additionally, because the action of Γ is free, $\gamma_{n_1} \circ \alpha_{n_1} = \gamma_{n_2} \circ \alpha_{n_2}$. However, by the observation that we made at the beginning of the proof,

$$\Pi_{\gamma_{n_1}} = \Pi_{\gamma_{n_1} \circ \alpha_{n_1}} = \Pi_{\gamma_{n_2} \circ \alpha_{n_2}} = \Pi_{\gamma_{n_2}},$$

contradicting our original assumption that the hyperbolic planes Π_{γ_n} are all distinct. $\qquad\square$

Proposition 12.11. *The Ford domain* $\Delta_\Gamma(\infty)$ *is a locally finite polyhedron in the hyperbolic space* $\mathbb{H}^3$.

Proof. We already saw that by definition, $\Delta_\Gamma(\infty)$ is equal to the intersection of all the half-spaces H_γ, defined in Lemma 12.9, as γ ranges over all elements of $\Gamma - \Gamma_\infty$.

By the local finiteness result of Lemma 12.10, every $P \in \Delta_\Gamma(\infty)$ is the center of a ball $B_{d_{\text{hyp}}}(P, \varepsilon)$ which meets only finitely many of the hyperbolic planes Π_γ bounding the H_γ. Let us write these planes as $\Pi_{\gamma_1}, \Pi_{\gamma_2}, \ldots, \Pi_{\gamma_n}$. As a consequence, the intersection of the ball $B_{d_{\text{hyp}}}(P, \varepsilon)$ with $\Delta_\Gamma(\infty)$ is also the intersection of $B_{d_{\text{hyp}}}(P, \varepsilon)$ with $\Pi_{\gamma_1} \cap \cdots \cap \Pi_{\gamma_n}$. It follows that the boundary of $\Delta_\Gamma(\infty)$ is the union of all sets $\Pi_\gamma \cap \Delta_\Gamma(\infty)$ with $\gamma \in \Gamma - \Gamma_\infty$.

By construction, the set $F_\gamma = \Pi_\gamma \cap \Delta_\Gamma(\infty)$ is the intersection of the hyperbolic plane Π_γ with all half-spaces $H_{\gamma'}$ with $\gamma' \in \Gamma - \Gamma_\infty$. Since the family of the planes $\Pi_{\gamma'}$ bounding these half-spaces is locally finite (Lemma 12.10), F_γ is either the empty set, a single point, a geodesic of $\mathbb{H}^3$, or a locally finite polygon in $\mathbb{H}^3$.

When F_γ is a polygon, we will say that it is a *face* of $\Delta_\Gamma(\infty)$. From the local picture of the Ford domain $\Delta_\Gamma(\infty)$ near each of its points P, it is immediate that two such faces can only meet along one edge or one vertex.

This concludes the proof that the Ford domain $\Delta_\Gamma(\infty)$ is a locally finite polyhedron. □

Remark 12.12. Here, let us collect a few simple properties of the Ford domain $\Delta_\Gamma(\infty)$.

(1) The intersection of $\Delta_\Gamma(\infty)$ with a vertical line is a half-line. Indeed, if $P = (x, y, u)$ is in the half-space H_γ of Lemma 12.9, every point $P' = (x, y, u')$ above it (namely, with $u' > u$) is also in H_γ.

(2) For the vertical projection of $\mathbb{R}^3$ to the xy-plane, every face of $\Delta_\Gamma(\infty)$ projects to a euclidean polygon in $\mathbb{R}^2$.

(3) The Ford domain $\Delta_\Gamma(\infty)$ is invariant under the stabilizer Γ_∞. Indeed, if $P \in H_\gamma$ and $\alpha \in \Gamma_\infty$, a simple algebraic manipulation shows that $\alpha(P)$ belongs to $H_{\alpha \circ \gamma}$.

(4) A consequence of (2) and (3) above is that the faces of $\Delta_\Gamma(\infty)$ are polygons with finitely many edges.

Figures 12.6 and 12.8 below provide a few examples of the top view of Ford domains, namely, of their projections to the xy-plane. Note the symmetry of these pictures under the action of the stabilizer Γ_∞. This stabilizer acts by translations on the plane, and its action is represented here by indicating a parallelogram which is a fundamental domain for the action; in particular, Γ_∞ is generated by the two translations along the sides of the parallelogram.

These pictures were drawn using the software *SnapPea* [**Weeks₂**]. Indeed, once we are given the kleinian group Γ, the definition of the Ford domain $\Delta_\Gamma(\infty)$ makes its computation quite amenable to computer implementation.

These examples are fairly typical, as projections of Ford domains provide (often quite intricate) tessellations of the euclidean plane $\mathbb{R}^2$ by euclidean polygons (usually triangles), which are invariant under the action of a group acting by translations.

12.4.3. Uniqueness and examples.

Proposition 12.13. *Let K be a knot in $\mathbb{R}^3$, to which Theorem 12.2, Complement 12.7 and Section 12.4.2 associate a kleinian group Γ and a Ford domain $\Delta_\Gamma(\infty)$. If Γ' and $\Delta_{\Gamma'}(\infty)$ are similarly associated to K, then there is an isometry $\varphi\colon \mathbb{H}^3 \to \mathbb{H}^3$ fixing ∞ and sending $\Delta_\Gamma(\infty)$ to $\Delta_{\Gamma'}(\infty)$.*

Proof. This is an immediate consequence of Proposition 12.8. □

To apply Proposition 12.13, note that an isometry φ of $\mathbb{H}^3$ that fixes ∞ is just the composition of a homothety and of a euclidean isometry of $\mathbb{R}^3$ respecting $\mathbb{H}^3$. (Look at the linear or antilinear fractional map induced on $\widehat{\mathbb{C}}$.) In addition, the possibilities for φ are further limited by the fact that Proposition 12.8 implies that it must send Γ_∞ to Γ'_∞, in the sense that $\Gamma'_\infty = \{\varphi \circ \gamma \circ \varphi^{-1}; \gamma \in \Gamma_\infty\}$.

Figure 12.5. Two very similar knots

Because of the wealth of information encoded in them, Proposition 12.13 makes Ford domains a very powerful tool to prove that two knots are not isomorphic. For instance, consider the two knots represented in Figure 12.5. Their respective Ford domains are represented in Figure 12.6. By inspection, there is no similitude (= composition of a homothety with a euclidean isometry) of the plane sending the projection of one of these Ford domains to the other. For instance, some vertices of the second tessellation are adjacent to 20 edges, whereas none have this property in the first tessellation. One can also look at the (finitely many) shapes of the tiles of each tessellation.

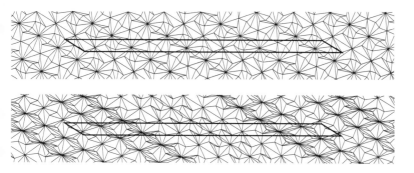

Figure 12.6. The Ford domains of the knots of Figure 12.5

It follows that the knots K and K' of Figure 12.5 are not isomorphic. These two knots were not completely chosen at random. The fact that they differ only by an exchange of two twisted parts (the technical term is "mutation") make them very hard to distinguish by the methods of algebraic topology which were standard prior to the introduction of hyperbolic geometry. This example illustrates the power of these new hyperbolic techniques.

As an aside, these two knots are still very close from the point of view of hyperbolic geometry. For instance, the hyperbolic metrics on their complements have exactly the same volume. It should also be clear from the parallelograms in each picture of Figure 12.6 (and it can be rigorously proved) that the stabilizers Γ_∞ and Γ_∞ coincide up to isometry of $\mathbb{H}^3$. Consequently, one needs the full force of Ford domains to distinguish these two knots.

As a final example, consider the two knots of Figure 12.7. *Snap-Pea* provides the same Ford domain for each of them, as indicated in Figure 12.8. In particular, we cannot use Ford domains to show that these two knots are not isomorphic.

There is a good reason for this, because the two knots represented are actually *isotopic...*! It took a long time to realize this fact. These two knots were listed as distinct in the early knot tables established by Charles Little [**Little**] in the nineteenth century, and remained so for over 90 years; see for instance [**Rolfsen**], where they appear as

Figure 12.7. Two more knots

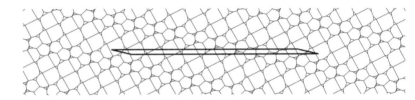

Figure 12.8. The Ford domain of the knots of Figure 12.7

the knots numbered 10_{161} and 10_{162}, respectively. It is only in 1974 that Kenneth Perko, an amateur mathematician, noticed that they were isotopic [**Perko**]. See Exercise 12.8.

We are mentioning this example because it illustrates very well the power of hyperbolic geometric methods in knot theory. Indeed, the isotopy which took 90 years to discover is now a simple consequence of the computation of Ford domains, provided we use a little more information than we have indicated so far.

More precisely, the faces of a Ford domain $\Delta_\Gamma(\infty)$ are paired together by the same principle that we already encountered for Dirichlet domains. Namely, using the notation of the proof of Proposition 12.11, if F_γ is a face of the Ford domain $\Delta_\Gamma(\infty)$, then $F_{\gamma^{-1}}$ is another face of $\Delta_\Gamma(\infty)$ and $\gamma \in \Gamma$ sends $F_{\gamma^{-1}}$ to F_γ. In particular, the faces F_γ and $F_{\gamma^{-1}}$ are glued together under the quotient map $\mathbb{H}^3 \to \mathbb{H}^3/\Gamma$.

As in the case of Dirichlet domains, we can therefore reconstruct the quotient space $\mathbb{H}^3/\Gamma$ from the following data: the Ford domain $\Delta_\Gamma(\infty)$; the action of the translation group Γ_∞ on $\Delta_\Gamma(\infty)$; the gluing data between the faces $\Delta_\Gamma(\infty)$. Indeed, $\mathbb{H}^3/\gamma$ is homeomorphic to the quotient space obtained by, first, taking the quotient $\Delta_\Gamma(\infty)/\Gamma_\infty$ and then gluing the faces of $\Delta_\Gamma(\infty)/\Gamma_\infty$ (namely, the images of the faces of $\Delta_\Gamma(\infty)$ under the quotient map) according to the gluing data.

If we are only interested in reconstructing $\mathbb{H}^3/\Gamma$ up to homeomorphism as opposed to up to isometry, we do not need to know exactly the gluing maps $F_\gamma \to F_{\gamma^{-1}}$, but only their combinatorics.

More precisely, let Γ and Γ' be two kleinian groups as in Complement 12.7, so that they have well-defined Ford domains $\Delta_\gamma(\infty)$ and $\Delta_{\Gamma'}(\infty)$. We will say that the Ford domains $\Delta_\Gamma(\infty)$ and $\Delta_{\Gamma'}(\infty)$ and their gluing data **have the same combinatorics** if there is a one-to-one correspondence between the faces, edges and vertices of $\Delta_\Gamma(\infty)$ and the faces, edges and vertices of $\Delta_{\Gamma'}(\infty)$ such that

(1) an edge of $\Delta_\Gamma(\infty)$ is contained in a given face if and only if the corresponding edge of $\Delta_{\Gamma'}(\infty)$ is contained in the corresponding face; similarly, a vertex of $\Delta_\Gamma(\infty)$ is contained in a given edge if and only if the corresponding vertex of $\Delta_{\Gamma'}(\infty)$ is contained in the corresponding edge of $\Delta_{\Gamma'}(\infty)$;

(2) two faces, edges or vertices of $\Delta_\Gamma(\infty)$ differ by the action of an element of Γ_∞ if and only if the corresponding faces, edges or vertices of $\Delta_{\Gamma'}(\infty)$ differ by the action of an element of Γ'_∞;

(3) two faces, edges or vertices of $\Delta_\Gamma(\infty)$ are glued together if and only if the corresponding faces, edges or vertices of $\Delta_{\Gamma'}(\infty)$ are glued together.

SnapPea easily computes the combinatorial data thus associated to the Ford domain of a hyperbolic knot.

Proposition 12.14. *Let K and K' be two knots, and let Γ and Γ' be two kleinian groups such that $\mathbb{H}^3/\Gamma$ and $\mathbb{H}^3/\Gamma'$ are homeomorphic to $\widehat{\mathbb{R}}^3 - K$ and $\widehat{\mathbb{R}}^3 - K'$, respectively, as in Complement 12.7. Suppose that the Ford domains $\Delta_\Gamma(\infty)$ and $\Delta_{\Gamma'}(\infty)$ and their gluing data are combinatorially equivalent.*

Then, the knots K and K' are isomorphic.

Proof. The definition of combinatorial equivalence is designed so that when it holds the quotient spaces $\mathbb{H}^3/\Gamma$ and $\mathbb{H}^3/\Gamma'$ are homeomorphic. Therefore, there exists a homeomorphism $\varphi\colon \widehat{\mathbb{R}}^3 - K \to \widehat{\mathbb{R}}^3 - K'$ between the corresponding knot complements. Then, a deep theorem of Cameron Gordon and John Luecke [**Gordon & Luecke**] shows that φ can be chosen so that it extends to a homeomorphism $\varphi\colon S^3 \to S^3$, such that $\varphi(K) = K'$. This proves that K and K' are isomorphic. $\square$

Ford domains were originally introduced, in the context of the hyperbolic plane $\mathbb{H}^2$, by Lester Ford in 1935 [**Ford$_1$**]. Bob Riley was probably the first one to realize that 3-dimensional Ford domains could be a powerful tool to distinguish knots, and one which was very amenable to computer implementation [**Riley$_2$, Riley$_3$, Riley$_4$**]. A conceptually different description of Ford domains is due to David Epstein and Bob Penner [**Epstein & Penner**].

12.5. The general Geometrization Theorem

The Geometrization Theorem 12.2 for knot complements is a special case of a more general result for 3-dimensional manifolds. To explain this statement, we need a few definitions about manifolds.

Consider the n-***dimensional euclidean space*** $\mathbb{R}^n$, consisting of all n-tuples $(x_1, x_2, \ldots, x_n)$ with each $x_i \in \mathbb{R}$. We endow $\mathbb{R}^n$ with the ***euclidean metric*** d_{euc} defined by

$$d_{\text{euc}}(P, P') = \sqrt{\sum_{i=1}^{n}(x_i - x'_i)^2}$$

when $P = (x_1, x_2, \ldots, x_n)$ and $P' = (x'_1, x'_2, \ldots, x'_n)$. This is of course the natural generalization of the cases where $n = 2$ and $n = 3$ that we have considered so far. When $p < n$, we identify the space $\mathbb{R}^p$ to the subset of $\mathbb{R}^n$ consisting of those points whose last $(n - p)$ coordinates are all equal to 0.

An n-***dimensional manifold*** is a metric space (X, d) which is locally homeomorphic to the euclidean space $(\mathbb{R}^n, d_{\text{euc}})$. Namely, for every $P \in X$, there exists a small ball $B_d(P, \varepsilon) \subset X$ and a homeomorphism $\varphi\colon B_d(P, \varepsilon) \to U$ between $B_d(P, \varepsilon)$ and a subset U of $\mathbb{R}^n$

which contains a small euclidean ball $B_{d_{\text{hyp}}}(\varphi(P), \varepsilon') \subset U$ centered at $\varphi(P)$.

A manifold X is **connected** if, for every P and $Q \in X$, there exists a continuous curve $\gamma \colon [a, b] \to X$ going from $P = \gamma(a)$ to $Q = \gamma(b)$ (where $[a, b]$ is an arbitrary closed interval in $\mathbb{R}$). Manifolds naturally appear as in the mathematical modelling of various physical phenomena and, consequently, they occur in many different branches of mathematics.

12.5.1. Geometrization of surfaces. A *surface* is the same thing as a 2-dimensional manifold. In Chapter 5, we constructed various euclidean, hyperbolic and spherical surfaces, namely, metric spaces locally isometric to the euclidean plane $(\mathbb{R}^2, d_{\text{euc}})$, the hyperbolic plane $(\mathbb{H}^2, d_{\text{hyp}})$ or the sphere $(\mathbb{S}^2, d_{\text{sph}})$.

This is part of a more general phenomenon. The following theorem is the culmination of work by a long line of mathematicians in the nineteenth century; see for instance [**Bonahon**, §1.1] for a discussion.

Theorem 12.15 (Geometrization Theorem for surfaces). *Let (X, d) be a connected surface. Then X can be endowed with another metric d' such that*

(1) *d' is topologically equivalent to the original metric d, in the sense that the identity map Id_X provides a homeomorphism between (X, d) and (X, d');*

(2) *the metric space (X, d') is complete;*

(3) *(X, d') is locally isometric to the euclidean plane $(\mathbb{R}^2, d_{\text{euc}})$, the hyperbolic plane $(\mathbb{H}^2, d_{\text{hyp}})$ or the sphere $(\mathbb{S}^2, d_{\text{sph}})$.*

In addition:

(1) *when (X, d') is locally isometric to the sphere, it is homeomorphic to the sphere or to the projective plane;*

(2) *when (X, d') is locally isometric to the euclidean plane, it is homeomorphic to the plane, the cylinder, the Möbius strip, the torus, or the Klein bottle;*

(3) *when (X, d') is locally isometric to the hyperbolic plane, it is* not *homeomorphic to the sphere, the projective plane, the torus, or the Klein bottle.* □

In particular, outside of four exceptions, every connected surface can be endowed with a complete hyperbolic metric. This includes the relatively nice surfaces that we encountered in Chapter 5, but also wilder examples such as the surface of infinite genus represented in Figure 12.9 (also known as the "infinite Loch Ness Monster"), or the complement of a Cantor set in the plane (if you know what this is).

Figure 12.9. The infinite Loch Ness Monster

As indicated earlier, Mostow's Rigidity Theorem 12.4 is not valid in dimension 2, so that this hyperbolic metric is not unique up to isometry. However, there are still relatively few such metrics. Indeed, for a given surface (X, d), we can consider the set of all complete euclidean, hyperbolic or spherical metrics d' as in Theorem 12.15, and identify two such metrics when they are isometric. There is a way to turn the corresponding quotient space into a metric space[2] called the **moduli space** $\mathcal{M}_{\text{euc}}(X)$, $\mathcal{M}_{\text{hyp}}(X)$ or $\mathcal{M}_{\text{sph}}(X)$, of euclidean, hyperbolic or spherical metrics on the surface X. When X is compact, or more generally when it is obtained from a compact surface by removing finitely many points, this moduli space is almost a manifold in the sense that it is locally isometric to the quotient of a manifold by the action of a finite group of isometries; such an object is called an **orbifold**. In particular, it is locally a manifold at most of its points, but can have singularities similar to the cone singularities of Exercise 7.14 at those points that are fixed by nontrivial elements of the finite group.

For instance, when X is homeomorphic to the sphere or to the projective plane, the moduli space $\mathcal{M}_{\text{sph}}(X)$ consists of a single point

---

[2]with several classical choices possible for the metric

(and consequently it is a manifold of dimension 0), which is another way to say that any two spherical metrics on X are isometric.

For euclidean metrics, the moduli space $\mathcal{M}_{\text{euc}}(X)$ is a 3-dimensional orbifold when X is homeomorphic to the torus or the Klein bottle, and is a 1-dimensional orbifold (homeomorphic to a semi-open interval) when X is a cylinder or a Möbius strip.

When X is obtained by removing p points from the compact orientable surface of genus g, then $\mathcal{M}_{\text{hyp}}(X)$ is an orbifold of dimension $3(2g+p-2)$. The fact that this dimension is finite can be interpreted as the property that a complete hyperbolic metric on X is essentially controlled by finitely many parameters (such as the shear parameters of Section 6.7.2) or, as a physicist would say, that there are only finitely many degrees of freedom in the choice of such a metric. For infinite surfaces such as the infinite Loch Ness Monster, the moduli space $\mathcal{M}_{\text{hyp}}(X)$ is infinite-dimensional.

12.5.2. Essential surfaces in 3-dimensional manifolds. Let $(\mathbb{S}^3, d_{\text{sph}})$ be the **3-*dimensional sphere*** defined by straightforward generalization of the 2-dimensional sphere $(\mathbb{S}^2, d_{\text{sph}})$ to three dimensions. Namely,

$$\mathbb{S}^3 = \{(x_1, x_2, x_3, x_4) \in \mathbb{R}^4; x_1^2 + x_2^2 + x_3^2 + x_4^2 = 1\},$$

and the ***spherical distance*** $d_{\text{sph}}(P, q)$ is defined as the infimum of the euclidean lengths of all piecewise differentiable curves joining P to Q in $\mathbb{S}^3$.

One might optimistically hope that a version of Theorem 12.15 also holds for 3-dimensional manifolds, and that every 3-dimensional manifold admits a metric which is locally isometric to the euclidean space $(\mathbb{R}^3, d_{\text{euc}})$, the hyperbolic space $(\mathbb{H}^3, d_{\text{hyp}})$ or the 3-dimensional sphere $(\mathbb{S}^3, d_{\text{sph}})$. This is not quite the case.

Let X be a 3-dimensional manifold.

A typical example of 3-dimensional manifold is the complement $X = \widehat{\mathbb{R}}^3 - K$ of a knot K in $\mathbb{R}^3$. In that case, Theorem 12.2 asserted the existence of a complete hyperbolic metric on X provided K is neither a satellite nor a torus knot. For a general 3-dimensional manifold X, the nonsatellite condition is replaced by a condition on a certain type of surfaces in X.

A **two-sided surface** in the 3-dimensional manifold X is a subset $S \subset X$ for which there exists a 2-dimensional manifold Y and a homeomorphism $\varphi \colon Y \times (-\varepsilon, \varepsilon) \to U$ to a subset $U \subset X$ such that $\varphi(Y \times \{0\}) = S$. Here, we are endowing the product $Y \times (-\varepsilon, \varepsilon)$ with the product of the metric of Y and of the euclidean metric of the interval $(-\varepsilon, \varepsilon)$, as defined in Exercise 1.6. The same convention will hold for all products in this section.

One can always find two-sided surfaces in a 3-dimensional manifold X by using relatively trivial constructions. For instance, by definition of manifolds, X contains a subset U which is homeomorphic to an open ball V in $\mathbb{R}^3$; the preimages of small euclidean spheres in V then provide many two-sided spheres in X. If we start from a simple closed curve K in X, we can also try to thicken it to a small tube, and then take the boundary of this tube; in most cases, the boundary of the tube will be a two-sided torus, but it can also be a two-sided Klein bottle.

Such constructions are clearly too simple for the corresponding surfaces to have much significance. We will focus attention on surfaces which are not obtained in this way. Such surfaces will be called *essential*, where this word is defined on a case-by-case basis.

An **essential sphere** in the 3-dimensional manifold X is a two-sided surface S which is homeomorphic to the sphere, and which is contained in no subset $B \subset X$ where B is homeomorphic to a ball in $\mathbb{R}^3$.

An **essential projective plane** in X is simply a two-sided surface which is homeomorphic to the projective plane, with no further condition.

The definition of an essential torus involves more properties. Namely, an **essential torus** in the 3-dimensional manifold X is a two-sided surface T such that

(1) T is homeomorphic to the torus $\mathbb{T}^2$;

(2) T is not contained in a subset $B \subset X$ which is homeomorphic to a ball in $\mathbb{R}^3$;

(3) T is not the boundary of a solid torus $V \subset X$; namely, there is no homeomorphism $\varphi \colon \mathbb{S}^1 \times \mathbb{D}^2 \to V \subset X$ such that $T = \varphi(\mathbb{S}^1 \times \mathbb{S}^1)$, where $\mathbb{S}^1$ and $\mathbb{D}^2$ denote the unit circle and the closed unit disk in $\mathbb{R}^2$, respectively;

(4) T is not the boundary of a collar W going to infinity in X; namely, there is no homeomorphism $\varphi \colon \mathbb{T}^2 \times [0, \infty) \to W \subset X$ such that $T = \varphi(\mathbb{T}^2 \times \{0\})$ and such that $\varphi(\mathbb{T}^2 \times \{t\})$ goes to infinity in X as t tends to $+\infty$, in the sense that

$$\lim_{t \to +\infty} \inf \left\{ d(P_0, P); P \in \varphi(\mathbb{T}^2 \times \{t\}) \right\} = +\infty$$

for an arbitrary base point $P_0 \in X$.

The definition of essential Klein bottles will involve twisted (also called nonorientable) solid tori. Recall that the circle $\mathbb{S}^1$ is homeomorphic to the quotient space obtained from the interval $[0, 1]$ by gluing the point 0 to the point 1 (see Exercise 4.5). As a consequence, the solid torus $\mathbb{S}^1 \times \mathbb{D}^2$ is homeomorphic to the quotient space obtained from $[0, 1] \times \mathbb{D}^2$ by gluing each point $(0, P)$ with $P \in \mathbb{D}^2$ to the point $(1, P)$.

Let the **twisted solid torus** $\mathbb{S}^1 \widetilde{\times} \mathbb{D}^2$ be the quotient space obtained from $[0, 1] \times \mathbb{D}^2$ by gluing each point $(0, P)$ to $(1, \bar{P})$, where $\bar{P}$ denotes the complex conjugate of $P \in \mathbb{D}^2 \subset \mathbb{R}^2 = \mathbb{C}$. The **boundary Klein bottle** of this twisted solid torus is the image $\mathbb{S}^1 \widetilde{\times} \mathbb{S}^1$ of $[0, 1] \times \mathbb{S}^1$ under the quotient map.

An **essential Klein bottle** in the 3-dimensional manifold X is a two-sided surface K such that

(1) K is homeomorphic to the Klein bottle $\mathbb{K}^2$;

(2) K is not the boundary of a twisted solid torus $V \subset X$; namely, there is no homeomorphism $\varphi \colon \mathbb{S}^1 \widetilde{\times} \mathbb{D}^2 \to V \subset X$ such that $K = \varphi(\mathbb{S}^1 \widetilde{\times} \mathbb{S}^1)$;

(3) K is not the boundary of a collar W going to infinity in X; namely, there is no homeomorphism $\varphi \colon \mathbb{K}^2 \times [0, \infty) \to W \subset X$ such that $K = \varphi(\mathbb{T}^2 \times \{0\})$ and such that

$$\lim_{t \to +\infty} \inf \{ d(P_0, P); P \in \varphi(\mathbb{K}^2 \times \{t\}) \} = +\infty.$$

Proposition 12.16. *Let* (X, d) *be a* 3-*dimensional manifold. If* d *is topologically equivalent to a complete hyperbolic metric* d', *then* X *contains no essential sphere, projective plane, torus or Klein bottle.*
□

See [**Scott, Bonahon**] for a proof and further references.

12.5.3. Seifert fibrations. Recall that the standard solid torus is the subset V of $\mathbb{R}^3$ consisting of all points of the form

$$\big((R + \rho \cos v) \cos u, (R + \rho \cos v) \sin u, \rho \sin v\big)$$

with $0 \leqslant \rho \leqslant r$ and $u, v \in \mathbb{R}$, where the two radii R and r with $R > r > 0$ are given. For any rational $\frac{p}{q} \in \mathbb{Q}$, we can consider in V all the torus knots parametrized by

$$t \mapsto \big((R + \rho \cos qt) \cos(pt + t_0), (R + \rho \cos qt) \sin(pt + t_0), -\rho \sin qt\big)$$

with $0 \leqslant \rho \leqslant R$ and $t_0 \in \mathbb{R}$. When $\rho = 0$, the corresponding curve is the core circle of V. When $\rho > 0$, the corresponding curves are disjoint $\frac{p}{q}$-torus knots wrapping around this core.

There is another possible presentation of this partition of the solid torus. Indeed, we already saw that V is homeomorphic to the quotient space $\mathbb{S}^1 \times \mathbb{D}^2$ obtained from $[0, 1] \times \mathbb{D}^2$ by gluing each point $(0, P)$ to $(1, P)$. However, if $\sigma \colon \mathbb{D}^2 \to \mathbb{D}^2$ denotes the rotation of angle $-2\pi \frac{p}{q}$, V is also homeomorphic to the other quotient space obtained from $[0, 1] \times \mathbb{D}^2$ by gluing each $(0, P)$ to $(1, \sigma(P))$. (To check this, first use your personal experience with playdough, and then try to build a more rigorous argument.) Then the partition of $[0, 1] \times \mathbb{D}^2$ into line segments $[0, 1] \times \{P\}$ projects to a partition of this second quotient space into disjoint simple closed curves, which is homeomorphic to the partition of V by $\frac{p}{q}$-torus knots as above.

This second presentation immediately extends to the twisted solid torus $W = \mathbb{S}^1 \widetilde{\times} \mathbb{D}^2$. Recall that W is the quotient space obtained from $[0, 1] \times \mathbb{D}^2$ by gluing $(0, P)$ to $(1, \bar{P})$, where $\bar{P}$ is the complex conjugate image of $P \in \mathbb{D}^2 \subset \mathbb{R}^2 = \mathbb{C}$ under the reflection across the x-axis. Again, the partition of $[0, 1] \times \mathbb{D}^2$ into line segments $[0, 1] \times \{P\}$ projects to a partition of W into disjoint simple closed curves. Note that the curves corresponding to P in the x-axis (so that $\bar{P} = P$) wrap once around $\mathbb{S}^1 \,\widetilde{\times}\, \mathbb{D}^2$, but that all other curves wrap twice around this twisted solid torus.

A **Seifert fibration** of a 3-dimensional manifold X is a partition of X into disjoint simple closed curves, called the **fibers** of the fibration, which locally looks like the above decompositions of the twisted and untwisted solid torus. More precisely, for every fiber K, there is a subset $U \subset X$ containing K, a rational number $\frac{p}{q} \in \mathbb{Q}$ and a homeomorphism $\varphi \colon U \to \mathbb{S}^1 \times \mathbb{D}^2$ or $\mathbb{S}^1 \,\widetilde{\times}\, \mathbb{D}^2$ between U and the solid torus $\mathbb{S}^1 \times \mathbb{D}^2$ or the twisted solid torus $\mathbb{S}^1 \,\widetilde{\times}\, \mathbb{D}^2$, such that

(1) U is a union of fibers;

(2) φ sends K to the core circle of the twisted or untwisted solid torus;

(3) φ sends any other fiber contained in U to a $\frac{p}{q}$-torus knot in $\mathbb{S}^1 \times \mathbb{D}^2$ or to one of the curves of the decomposition of $\mathbb{S}^1 \,\widetilde{\times}\, \mathbb{D}^2$, as above.

For most fibers, the solid torus U is untwisted and we can take $\frac{p}{q} = \frac{0}{1}$; all other fibers are called **exceptional**.

A 3-dimensional manifold that admits a Seifert fibration is a **Seifert manifold**.

Since a Seifert fibration of X is a partition, it gives rise to a quotient space $\bar{X}$. It easily follows from the local description of the Seifert fibration that this quotient space is a surface with boundary. To a large extent, X behaves very much like the product $\bar{X} \times \mathbb{S}^1$ of $\bar{X}$ with the circle $\mathbb{S}^1$, and we should think of X as a certain thickening of the surface $\bar{X}$.

This intuition is expressed in the early work [**Seifert**] of Herbert Seifert, who classified Seifert fibrations up to homeomorphism respecting the fibration, in terms of the quotient surface $\bar{X}$ and of the local type of the exceptional fibers. A little over thirty years later, Friedhelm Waldhausen [**Waldhausen**] and Peter Orlik, Elmar Vogt and Heiner Zieschang [**Orlik, Vogt & Zieschang**] showed that the Seifert fibration of a Seifert manifold is unique up to homeomorphism, outside of a few well-understood exceptions, so that the classification of Seifert manifolds up to homeomorphism is equivalent to Seifert's classification up to homeomorphism respecting the fibration. The main consequence of these results is that Seifert manifolds form a class of 3-dimensional manifolds which is very well understood.

12.5.4. The general Geometrization Theorem. We are now almost ready state the Geometrization Theorem for 3-dimensional manifolds. However, we need a last couple of definitions.

Let $\mathbb{R}_+^n$ denote the closed half-space $\mathbb{R}^{n-1} \times [0, \infty) \subset \mathbb{R}^n$ consisting of those points of the euclidean space $\mathbb{R}^n$ whose last coordinate is nonnegative. An n-dimensional **manifold-with-boundary** (in one word) is a metric space (X, d) which is locally homeomorphic to $(\mathbb{R}_+^n, d_{\mathrm{euc}})$. Namely, for every $P \in X$, there exists a small ball $B_d(P, \varepsilon) \subset X$ and a homeomorphism $\varphi : B_d(P, \varepsilon) \to U$ betwen $B_d(P, \varepsilon)$ and a subset $U \subset \mathbb{R}_+^n$ which contains a small ball $\mathbb{R}_+^n \cap B_{d_{\mathrm{euc}}}(P, \varepsilon')$ of the metric space $(\mathbb{R}_+^n, d_{\mathrm{euc}})$. The **boundary** ∂X of X consists of those P as above which are sent to a point of $\mathbb{R}^{n-1} \times \{0\}$ by the homeomorphism. Its **interior** is the complement $X - \partial X$.

A manifold is said to have **finite topological type** if it is homeomorphic to the interior of a compact manifold-with-boundary. For instance, a compact manifold has finite topological type. The infinite Loch Ness Monster surface of Figure 12.9 is a typical example of a manifold of infinite topological type.

Theorem 12.17 (The Geometrization Theorem for 3-dimensional manifolds). *Let (X, d) be a connected 3-dimensional manifold with finite topological type. Then at least one of the following holds:*

(1) *X admits a complete hyperbolic metric d' which is topologically equivalent to d;*

(2) *X contains an essential sphere, projective plane, torus, or Klein bottle;*

(3) *X is a Seifert manifold.* □

In addition, there is relatively little overlap with these three possibilities. We already saw in Proposition 12.16 that (1) and (2) are incompatible. With respect to (1) and (3) there are, up to homeomorphism, only six exceptional connected Seifert manifolds that admit complete hyperbolic metrics, such as the interior of the solid torus, or the product of the torus $\mathbb{T}^2$ with the line $\mathbb{R}$. Similarly, only two connected Seifert manifolds contain an essential sphere or projective plane. However, most Seifert manifolds contain many essential tori or Klein bottles.

Complement 12.18. *Under the hypothesis of Theorem 12.17, suppose in addition that X is the union of a compact subset C and of finitely many subsets homeomorphic to $\mathbb{T}^2 \times [0, \infty)$ and $\mathbb{K}^2 \times [0, \infty)$, where $\mathbb{T}^2$ and $\mathbb{K}^2$ denote the torus and the Klein bottle, respectively. Then, we can additionally require in conclusion (1) of Theorem 12.17 that the hyperbolic metric d' has finite volume.* □

It then follows from Mostow's Rigidity Theorem 12.4 that the hyperbolic metric d' is unique up to isometry under the hypotheses and conclusions of Complement 12.18.

Theorem 12.17 may not quite look like the Geometrization Theorem 12.15 for surfaces because it involves only one geometry, the hyperbolic geometry. However, Seifert manifolds can be endowed with other ***geometric structures***, namely, complete metrics which are locally isometric to some homogeneous model space.

Among homogeneous spaces of dimension 3, we have already encountered the hyperbolic space $(\mathbb{H}^3, d_{\mathrm{hyp}})$, the euclidean space $(\mathbb{R}^3, d_{\mathrm{euc}})$ and the 3-dimensional sphere $(\mathbb{S}^3, d_{\mathrm{sph}})$. These three spaces are also isotropic. However, in dimension 3, there also exist homogeneous spaces which are not isotropic, such as the products $\mathbb{H}^2 \times \mathbb{R}$ and $\mathbb{S}^2 \times \mathbb{R}$. There also exist twisted versions of these products. Altogether, if we identify two geometric models when they only differ by rescaling, and if we require each of them to appear for at least one 3-dimensional manifold with finite volume, there are only eight possible geometries in dimension 3. This observation is due to Bill Thurston, and is explained in detail in [**Scott, Bonahon**], for instance.

It turns out that every Seifert manifold can be endowed with a geometric structure modeled after one of six of these eight geometries. Therefore, only essential spheres, projective planes, tori and Klein bottles can create problems for endowing a 3-dimensional manifold with a geometric structure.

It turns out that earlier theories provide a kind of unique factorization of 3-dimensional manifolds of finite topological type into pieces which either contain no such essential surfaces or can be Seifert fibered. To some extent, these factorizations are analogous to the unique factorization of an integer as a product of primes. The first such factorization is the theory of connected sums initiated by Helmuth Kneser [**Kneser**] and completed by Wolfgang Haken [**Haken**] and John Milnor [**Milnor**] (see also the graduate textbook [**Hempel**]), which essentially reduces the analysis of 3-dimensional manifolds to that of manifolds containing no essential sphere or projective planes. The second such factorization is provided by the Characteristic Toric Decomposition, originally suggested by Friedhelm Waldhausen and fully developed by Klaus Johannson [**Johannson**] and Bus Jaco and Peter Shalen [**Jaco & Shalen**].

Combined with these factorizations, Theorem 12.17 then asserts that every connected 3-dimensional manifold of finite topological type has a natural splitting into pieces, each of which admits a geometric structure. In particular, this explains why this statement is called a Geometrization Theorem, and not just a Hyperbolization Theorem.

Theorem 12.17 and Complement 12.18 were proved by Bill Thurston in the late 1970s in the case of unbounded manifolds, or for bounded manifolds which contain essential two-sided surfaces (in a sense which we will not define here). See [**Thurston$_1$**], and expositions of the proof in [**Otal$_1$, Otal$_2$, Kapovich**]. The Geometrization Theorem for knot complements that we presented as Theorem 12.2 is a special case of this result.

The final cases of the proof of Theorem 12.17 were completed by Grisha Perelman shortly after 2000, following a program started in the early 1980s by Richard Hamilton. Perelman was offered the Fields Medal for this work in 2006, but declined the award. Following a definite pattern in this area of mathematics, the complete details of the proof have not been written down by Perelman, who only released a few unpublished preprints [**Perelman$_1$, Perelman$_2$, Perelman$_3$**]. However, detailed expositions are now beginning to become available [**Chow & Lu & Ni, Chow & al., Cao & Zhu, Kleiner & Lott, Morgan & Tian$_1$, Morgan & Tian$_2$**].

An important corollary of Theorem 12.17 is that it solves a century-old problem, the Poincaré Conjecture.

We saw that a manifold X is connected if any two points P, $Q \in X$ can be joined by a continuous curve $\gamma \colon [0,1] \to X$ with $\gamma(0) = P$ and $\gamma(1) = Q$. The manifold X is **simply connected** if, in addition, the space of curves going from P to Q is connected; namely, if, for any two curves $\gamma_0 \colon [0,1] \to X$ and $\gamma_1 \colon [0,1] \to X$ with endpoints $\gamma_0(0) = \gamma_1(0) = P$ to $\gamma_0(1) = \gamma_1(1) = Q$, there exists a family of curves $\gamma_s \colon t \mapsto \gamma_s(t)$ depending continuously on a parameter $s \in [0,1]$, all going from $\gamma_s(0) = P$ to $\gamma_s(1) = Q$ and coinciding with the original curves γ_0 and γ_1 when $s = 0,\ 1$. The fact that the curves γ_s depend continuously on s means that the map $H \colon [0,1] \times [0,1] \to X$ defined by $H(s,t) = \gamma_s(t)$ is continuous.

Examples of simply connected 3-dimensional manifolds include the euclidean space $\mathbb{R}^3$ and the 3-dimensional sphere

$$\mathbb{S}^3 = \{(x,y,z,t) \in \mathbb{R}^4; x^2 + y^2 + z^2 + t^2 = 1\}.$$

In 1904, Henri Poincaré proposed the following conjecture, which became a theorem 100 years later after many unsuccessful attempts in the intervening years.

Theorem 12.19 (The Poincaré Conjecture). *Every simply connected bounded 3-dimensional manifold is homeomorphic to the 3-dimensional sphere $\mathbb{S}^3$.*   □

The Poincaré Conjecture comes as a corollary of the Geometrization Theorem 12.17 by using relatively simple arguments on hyperbolic manifolds, by exploiting the Kneser-Haken-Milnor theory of connected sums to reduce the problem to manifolds without essential spheres or projective planes, and by applying an earlier result of Herbert Seifert [**Seifert**] that $\mathbb{S}^3$ is the only simply connected Seifert manifold.

Exercises for Chapter 12

Exercise 12.1.

a. Show by a series of pictures that the $\frac{2}{3}$-torus knot is isotopic to the $\frac{3}{2}$-torus knot.

b. More generally, show that the $\frac{p}{q}$-torus knot is isotopic to the $\frac{q}{p}$-torus knot for every $\frac{p}{q} \in \mathbb{Q}$.

Exercise 12.2. Let K be the figure-eight knot represented in Figure 11.5, and let K' be its mirror image, namely, let K' be the image of K under a reflection across a plane of $\mathbb{R}^3$. Show by a series of pictures that K and K' are isotopic.

Exercise 12.3. Consider the unknot, formed by the unit circle $\mathbb{S}^1$ in $\mathbb{R}^2 \subset \mathbb{R}^3$. Show that there exists a kleinian group Γ_0, generated by a single element and acting freely on $\mathbb{H}^3$, such that the quotient space $\mathbb{H}^3/\Gamma$ is homeomorphic to the complement $\widehat{\mathbb{R}}^3 - \mathbb{S}^1$. Possible hints: First show that $\widehat{\mathbb{R}}^3 - \mathbb{S}^1$ is homeomorphic to $\mathbb{R}^3 - \mathbb{R} \times \{(0,0)\}$, perhaps by using a suitable Möbius transformation; then consider the map $\mathbb{H}^3 \to \mathbb{R}^3 - \mathbb{R} \times \{(0,0)\}$ defined by $(x, y, u) \mapsto (x, u \cos y, u \sin y)$.

Exercise 12.4. Let Γ and Γ' be two groups acting freely and discontinuously on the hyperbolic space $\mathbb{H}^3$. Suppose that the corresponding quotient spaces are homeomorphic by a homeomorphism $\varphi \colon \mathbb{H}^3/\Gamma' \to \mathbb{H}^3/\Gamma$.

a. Let $\alpha \colon [a, b] \to \mathbb{H}^3/\Gamma$ be a parametrized continuous curve in $(\mathbb{H}^3/\Gamma, \bar{d}_{\mathrm{hyp}})$, beginning at $\bar{P}_0 = \alpha(a)$ with $P_0 \in \mathbb{H}^3$. Show that there exists a parametrized continuous curve $\widetilde{\alpha} \colon [a, b] \to \mathbb{H}^3$ such that $\widetilde{\alpha}(a) = P_0$ and $\alpha = \pi \circ \widetilde{\alpha}$, where $\pi \colon \mathbb{H}^3 \to \mathbb{H}^3/\Gamma$ is the quotient map. Hint: Consider the supremum of the set of those $t \in [a, b]$ for which such a map $\widetilde{\alpha}$ can be defined on $[a, t]$, and remember that π is a local isometry by (the proof of) Theorem 7.8.

b. Show that in part a, the map $\widetilde{\alpha}$ is unique. Hint: Given another such map $\widetilde{\alpha}'$, consider the infimum of those $t \in [a, b]$ such that $\widetilde{\alpha}'(t) \neq \widetilde{\alpha}(t)$.

c. Suppose that we are now given a family of continuous curves $\alpha_u \colon [a, b] \to \mathbb{H}^3/\Gamma$, depending continuously on a parameter $u \in [c, d]$ in the sense that the map $[a, b] \times [c, d] \to \mathbb{H}^3/\Gamma$ defined by $(t, u) \mapsto \alpha_u(t)$ is continuous. Assume in addition that $\alpha_u(a) = \bar{P}_0$ for every $u \in [c, d]$, and let $\widetilde{\alpha}_u \colon [a, b] \to \mathbb{H}^3$ be associated to α_u as in parts a and b. Show that the map $[a, b] \times [c, d] \to \mathbb{H}^3$ defined by $(t, u) \mapsto \widetilde{\alpha}_u(t)$ is continuous. Hint: If the map was not continuous at some (t_0, u_0), consider the infimum of those $t \in [a, b]$ such that the map $(t, u) \mapsto \widetilde{\alpha}_u(t)$ is not continuous at (t, u_0).

d. Under the hypotheses and conclusions of part c, suppose in addition that the map $u \mapsto \alpha_u(b) \in \mathbb{H}^3/\Gamma$ is constant. Show that the map $u \mapsto \widetilde{\alpha}_u(b) \in \mathbb{H}^3$ is also constant.

e. Fix two base points P_0, $P_0' \in \mathbb{H}^3$ such that $\varphi(\bar{P}_0) = \bar{P}_0'$. For $P \in \mathbb{H}^3$, choose a continuous curve $\widetilde{\alpha} \colon [a, b] \to \mathbb{H}^3$ with $\widetilde{\alpha}(a) = P_0$ and $\widetilde{\alpha}(b) = P$, and lift the curve $\alpha' = \varphi \circ \pi \circ \alpha \colon [a, b] \to \mathbb{H}^3/\Gamma'$ to a continuous curve

$\widetilde{\alpha}' \colon [a, b] \to \mathbb{H}^3$ with $\widetilde{\alpha}'(a) = P_0'$ as in part a. Use parts a–d to show that its endpoint $P' = \widetilde{\alpha}'(b)$ depends only on P, and not on the choice of the curve $\widetilde{\alpha}$. In particular, there is a well-defined map $\widetilde{\varphi} \colon \mathbb{H}^3 \to \mathbb{H}^3$ associating to each $P \in \mathbb{H}^3$ the point $\widetilde{\varphi}(P) = P'$ as above.

f. Show that the above map $\widetilde{\varphi} \colon \mathbb{H}^3 \to \mathbb{H}^3$ is a homeomorphism.

g. Show that for every $\gamma \in \Gamma$, the composition $\widetilde{\varphi} \circ \gamma \circ \widetilde{\varphi}^{-1}$ is an element of Γ'.

h. Show that the map $\rho \colon \Gamma \to \Gamma'$ defined by $\rho(\gamma) = \widetilde{\varphi} \circ \gamma \circ \widetilde{\varphi}^{-1}$ is a bijection, and that $\rho(\gamma_1 \circ \gamma_2) = \rho(\gamma_1) \circ \rho(\gamma_2)$ for every γ_1, $\gamma_2 \in \Gamma$. (Namely, ρ is a group isomorphism between the groups Γ and Γ', if you know what this is.)

Exercise 12.5. Let $\widehat{\Gamma}_8$ be the kleinian group of Section 11.2, such that $\mathbb{H}^3 / \widehat{\Gamma}_8$ is homeomorphic to the complement $\widehat{\mathbb{R}}^3 - K$ of the figure-eight knot, and let Γ_0 be a kleinian group as in Exercise 12.3, generated by a single element and for which $\mathbb{H}^3 / \Gamma_0$ is homeomorphic to the complement $\widehat{\mathbb{R}}^3 - \mathbb{S}^1$ of the unknot.

a. Show that $\widehat{\Gamma}_8$ contains elements γ_1, γ_2 such that $\gamma_1 \circ \gamma_2 \neq \gamma_2 \circ \gamma_1$.

b. Show that $\gamma_1 \circ \gamma_2 = \gamma_2 \circ \gamma_1$ for every γ_1, $\gamma_2 \in \Gamma_0$.

c. Use part h of Exercise 12.4 to conclude that $\widehat{\mathbb{R}}^3 - K$ and $\widehat{\mathbb{R}}^3 - \mathbb{S}^1$ are not homeomorphic, and therefore that the figure-eight knot is not isomorphic to the unknot.

Exercise 12.6. Let the kleinian group Γ act freely on $\mathbb{H}^3$, so that the quotient space $\mathbb{H}^3 / \Gamma$ is locally isometric to $\mathbb{H}^3$ by Theorem 7.8. We want to define the hyperbolic volume of $\mathbb{H}^3 / \Gamma$.

a. Let A be a subset of $\mathbb{H}^3 / \Gamma$ which is contained in a ball $B_{\bar{d}_{\mathrm{hyp}}}(\bar{P}, \varepsilon) \subset \mathbb{H}^3 / \Gamma$ isometric to a ball of $\mathbb{H}^3$ by an isometry $\varphi \colon B_{\bar{d}_{\mathrm{hyp}}}(\bar{P}, \varepsilon) \to B_{d_{\mathrm{hyp}}}(Q, \varepsilon) \subset \mathbb{H}^3$. Define the hyperbolic volume $\mathrm{vol}_{\mathrm{hyp}}(A)$ to be the hyperbolic volume of $\varphi(A) \subset \mathbb{H}^3$, as introduced in Exercise 9.12. (Here, we are implicitly assuming that the subset A is sufficiently "nice" that the triple integral involved in the definition of the hyperbolic volume of $\varphi(A)$ makes sense; all subsets involved in this exercise will satisfy this property, so that we do not have to worry about the deeper mathematical issues that could arise in the general case.) Show that this hyperbolic volume $\mathrm{vol}_{\mathrm{hyp}}(A)$ is independent of the choice of the ball $B_{\bar{d}_{\mathrm{hyp}}}(\bar{P}, \varepsilon)$ and of the isometry φ. Hint: Use Exercise 9.12.

b. Now consider a subset A which is contained in the union of finitely many balls B_1, B_2, ..., B_n in $\mathbb{H}^3 / \Gamma$, each of which is isometric to a ball in $\mathbb{H}^3$. For every subset I of $\{1, 2, \ldots, n\}$, let B_I be the subset of those $\bar{P} \in \mathbb{H}^3/\Gamma$ such that $\{i; \bar{P} \in B_i\} = I$. Define

$$\text{vol}_{\text{hyp}}(A) = \sum_{I \subset \{1, \ldots, n\}} \text{vol}_{\text{hyp}}(A \cap B_I),$$

where each $\text{vol}_{\text{hyp}}(A \cap B_I)$ is defined by part a. Show that $\text{vol}_{\text{hyp}}(A)$ is independent of the choice of the balls B_i.

c. Fix a base point $P_0 \in \mathbb{H}^3$. Show that for every $R > 0$, there exists $\varepsilon_R > 0$ such that $d_{\text{hyp}}(P, \gamma(P)) \geqslant \varepsilon_R$ for every $P \in B_{d_{\text{hyp}}}(P_0, R)$ and every $\gamma \in \Gamma - \{\text{Id}_\Gamma\}$. Conclude that the ball $B_{\bar{d}_{\text{hyp}}}(\bar{P}_0, R)$ is contained in the union of finitely many balls B_1, B_2, ..., B_n in $\mathbb{H}^3/\Gamma$, each of which is isometric to a ball in $\mathbb{H}^3$.

As a consequence, the hyperbolic volume $\text{vol}_{\text{hyp}}\big(B_{\bar{d}_{\text{hyp}}}(\bar{P}_0, R)\big)$ is well defined.

d. Show that the limit

$$\text{vol}_{\text{hyp}}(\mathbb{H}^3/\Gamma) = \lim_{R \to +\infty} \text{vol}_{\text{hyp}}\big(B_{\bar{d}_{\text{hyp}}}(\bar{P}_0, R)\big)$$

exists (possibly infinite), and is independent of the choice of the base point P_0.

Exercise 12.7.

a. Consider the **perpendicular bisector** of two points P and $Q \in \mathbb{H}^3$, defined as

$$\Pi_{PQ} = \{R \in \mathbb{H}^3; d_{\text{hyp}}(R, P) = d_{\text{hyp}}(R, Q)\}.$$

Show that Π_{PQ} is a hyperbolic plane. Hint: Compare Exercise 2.4.

b. Let γ be an isometry of $\mathbb{H}^3$ such that $\gamma(\infty) \neq \infty$, and let $(P_n)_{n \in \mathbb{N}}$ be a sequence in $\mathbb{H}^3$ converging to the point ∞ for the euclidean metric. Show that the perpendicular bisector plane $\Pi_{P_n \gamma(P_n)}$ converges to the hyperbolic plane Π_γ defined in Lemma 12.9, in the sense that the center and radius of $\Pi_{P_n \gamma(P_n)}$, considered as a euclidean hemisphere, converges to the center and radius of Π_γ as n tends to infinity.

Exercise 12.8. Show by a series a pictures that the two knots of Figure 12.7 are indeed isotopic. Tying the knots in shoe laces or another type of string may (or may not) be helpful.

Exercise 12.9. Let K be the $\frac{p}{q}$-torus knot in $\mathbb{R}^3$, drawn on a torus T consisting of those points that are at distance r from a central circle C of radius $R > r$. Show that the complement $X = \widehat{\mathbb{R}}^3 - K$ of K in $\widehat{\mathbb{R}}^3 = \mathbb{R}^3 \cup \{\infty\}$ admits a Seifert fibration where one fiber is the union of the z-axis and of the point ∞, where another fiber is the central circle C, and where all other fibers are isotopic to the $\frac{p}{q}$-torus knot.

Tool Kit

This appendix, referred to as the TOOL KIT in the text, is a quick list of notation and basic mathematical definitions used in the book.

T.1. Elementary set theory

For us, a **set** X is a collection of objects called its **elements**. In theory, we need to be more careful in the definition of sets in order to avoid logical inconsistencies. However, these deep and subtle issues do not arise at the level of the mathematics described in this book. Consequently, we will be content with the above intuitive definition.

When the object x is an element of the set X, we say that x **belongs to** X and we write $x \in X$.

In practice, a set can be described by listing all of its elements between curly brackets or by describing a property that characterizes the elements of the set. For instance, the set of all even integers that are strictly between -3 and 8 is denoted by $\{-2, 0, 2, 4, 6\}$, or by

$\{x;$ there exists an integer n such that $x = 2n$ and $-3 < x < 8\}.$

A **subset** of a set X is a set Y such that every element of Y is also an element of X. We then write $Y \subset X$.

A particularly useful set is the **empty set** $\varnothing = \{\ \}$, which contains no element.

A diagonal bar across a symbol indicates that the corresponding property does not hold. For instance, $x \notin X$ means that x does not belong to the set X.

Here is a list of classical sets of numbers, with the notation used in this book:

$\mathbb{N} = \{1, 2, 3, \dots\}$ is the set of all positive integers; in particular, 0 is not an element of $\mathbb{N}$, a convention which is not universal.

$\mathbb{Z} = \{\dots, -3, -2, -1, 0, 1, 2, 3, \dots\}$ is the set of all integers.

$\mathbb{Q}$ is the set of all rational numbers, namely, of all numbers that can be written as a quotient $\frac{p}{q}$ where p, $q \in \mathbb{Z}$ are integers with $q \neq 0$.

$\mathbb{R}$ is the set of all real numbers.

$\mathbb{C}$ is the set of all complex numbers. (See Section T.4 later in this TOOL KIT.)

Given two sets X and Y, their **intersection** $X \cap Y$ consists of all elements that are in both X and Y. Their **union** $X \cup Y$ consists of those objects that are in X or in Y (or in both). The **complement** $X - Y$ consists of those elements of X which do not belong to Y. For instance, if $X = \{1, 2, 3, 4\}$ and $Y = \{3, 4, 5\}$, then

$$X \cap Y = \{3, 4\},$$
$$X \cup Y = \{1, 2, 3, 4, 5\}$$
$$\text{and } X - Y = \{1, 2\}.$$

Two sets are **disjoint** when their intersection is empty.

More generally, is $\mathcal{X}$ is a set of sets, namely, a set whose elements are themselves sets, the **union** of these $X \in \mathcal{X}$ is the set

$$\bigcup_{X \in \mathcal{X}} X = \{x; x \in X \text{ for some } X \in \mathcal{X}\}$$

of those elements x that belong to at least one $X \in \mathcal{X}$. Similarly, their **intersection** is the set

$$\bigcap_{X \in \mathcal{X}} X = \{x; x \in X \text{ for all } X \in \mathcal{X}\}$$

of those elements x that belong to all $X \in \mathcal{X}$.

We can also consider the **product** of X and Y, which is the set $X \times Y$ consisting of all ordered pairs (x, y) where $x \in X$ and $y \in Y$. More generally, the product $X_1 \times X_2 \times \cdots \times X_n$ of n sets X_1, X_2, ..., X_n consists of all ordered n-tuples $(x_1, x_2, \ldots, x_n)$ where each coordinate x_i is an element of X_i.

In particular,

$$\mathbb{R}^2 = \mathbb{R} \times \mathbb{R} = \{(x, y); x \in \mathbb{R}, y \in \mathbb{R}\}$$

is naturally identified to the plane through cartesian coordinates. The same holds for the 3-dimensional space

$$\mathbb{R}^3 = \mathbb{R} \times \mathbb{R} \times \mathbb{R} = \{(x, y, z); x \in \mathbb{R}, y \in \mathbb{R}, z \in \mathbb{R}\}.$$

A **map** or **function** $\varphi \colon X \to Y$ is a rule φ which to each $x \in X$ associates an element $\varphi(x) \in Y$. We also express this by saying that the map φ is defined by $x \mapsto \varphi(x)$. Note the slightly different arrow shape.

When $X = Y$, there is a special map, called the **identity map** $\mathrm{Id}_X \colon X \to X$ which to $x \in X$ associates itself, namely, such that $\mathrm{Id}_X(x) = x$ for every $x \in X$.

The **composition** of two maps $\varphi \colon X \to Y$ and $\psi \colon Y \to Z$ is the map $\psi \circ \varphi \colon X \to Z$ defined by the property that $\psi \circ \varphi(x) = \psi(\varphi(x))$ for every $x \in X$. In particular, $\varphi = \mathrm{Id}_Y \circ \varphi = \varphi \circ \mathrm{Id}_X$ for any function $\varphi \colon X \to Y$.

The map φ is **injective** or **one-to-one** if $\varphi(x) \neq \varphi(x')$ for every x, $x' \in X$ with $x \neq x'$. It is **surjective** or **onto** if every $y \in Y$ is the image $y = \varphi(x)$ of some $x \in X$. The map φ is **bijective** if it is both injective and surjective, namely, if every $y \in Y$ is the image $y = \varphi(x)$ of a unique $x \in X$. In this case, there is a well-defined **inverse map** $\varphi^{-1} \colon Y \to X$, for which $\varphi^{-1}(y)$ is the unique $x \in X$ such that $y = \varphi(x)$. In particular, $\varphi \circ \varphi^{-1} = \mathrm{Id}_Y$ and $\varphi^{-1} \circ \varphi = \mathrm{Id}_X$. When $\varphi \colon X \to Y$ is bijective, we also say that φ is a **bijection**, or that it defines a **one-to-one correspondence** between elements of X and elements of Y.

The **image** of a subset $A \subset X$ under the map $\varphi \colon X \to Y$ is the subset

$$\varphi(A) = \{y \in Y; y = \varphi(x) \text{ for some } x \in X\}$$

of Y. The **preimage** of $B \subset Y$ under $\varphi\colon X \to Y$ is the subset

$$\varphi^{-1}(B) = \{x \in X; \varphi(x) \in B\}$$

of X. Note that the preimage φ^{-1} is defined even when φ is not bijective, in which case the inverse map φ^{-1} may not be defined and the preimage of a point may be empty or consist of many points.

The map $\varphi\colon X \to X$ **preserves** or **respects** a subset $A \subset X$ if $\varphi(A)$ is contained in A. A **fixed point** for φ is an element $x \in X$ such that $\varphi(x) = x$; equivalently, we then say that φ **fixes** x.

If we have a map $\varphi\colon X \to Y$ and a subset $A \subset X$, the **restriction** of φ to A is the function $\varphi_{|A}\colon A \to X$ defined by restricting attention to elements of A, namely, defined by the property that $\varphi_{|A}(a) = \varphi(a)$ for every $a \in A$.

When the map $\varphi\colon \mathbb{N} \to X$ is defined on the set $\mathbb{N}$ of all positive integers, it is called a **sequence**. In this case, it is customary to write $\varphi(n) = P_n$ (with the integer n as a subscript) and to denote the sequence by a list $P_1, P_2, \ldots, P_n, \ldots$, or by $(P_n)_{n \in \mathbb{N}}$ for short.

T.2. Maximum, minimum, supremum, and infimum

If a set $A = \{x_1, x_2, \ldots, x_n\}$ consists of finitely many real numbers, there is always one of these numbers which is larger than all the other ones and another one which is smaller than the other ones. These are the **maximum** $\max A$ and the **minimum** $\min A$ of A, respectively.

However, the same does not hold for infinite subsets of $\mathbb{R}$. For instance, the set $A = \{2^n; n \in \mathbb{Z}\}$ does not have a maximum, because it contains elements that are arbitrarily large. It has no minimum either because there is no $a \in A$ such that $a \leqslant 2^n$ for every $n \in \mathbb{Z}$.

We can fix this problem by doing two things. First, we introduce a point $\pm\infty$ at each end of the number line $\mathbb{R} = (-\infty, +\infty)$, so as to get a new set $[-\infty, +\infty] = \mathbb{R} \cup \{-\infty, +\infty\}$. Then, we will say that an element $M \in [-\infty, +\infty]$ is a **supremum** for the subset $A \subset \mathbb{R}$ if:

(1) $a \leqslant M$ for every $a \in A$;

(2) M is the smallest number with this property, in the sense that there is no $M' < M$ such that $a \leqslant M'$ for every $a \in A$.

The second condition is equivalent to the property that we can find elements of A that are arbitrarily close to M.

Similarly, an **infimum** for $A \subset \mathbb{R}$ is an element $m \in [-\infty, +\infty]$ such that

(1) $a \geqslant m$ for every $a \in A$;

(2) m is the largest number with this property, in the sense that there is no $m' > m$ such that $a \geqslant m'$ for every $a \in A$.

It is a deep result of real analysis that any subset $A \subset \mathbb{R}$ admits a *unique supremum* $M = \sup A$, and a *unique infimum* $m = \inf A$. The proof of this statement requires a deep understanding of the nature of real numbers. To a large extent, real numbers were precisely introduced for this property to hold true, and some people even use it as an axiom in the construction of real numbers. We refer to any undergraduate textbook on real analysis for a discussion of this statement.

For instance,

$$\sup\{2^n; n \in \mathbb{Z}\} = +\infty$$
$$\text{and } \inf\{2^n; n \in \mathbb{Z}\} = 0.$$

It may happen that $\sup A$ is an element of A, in which case we say that the supremum is also a **maximum** and we write $\sup A = \max A$; otherwise, the maximum of A does not exist. Similarly, the **minimum** $\min A$ of A is equal to $\inf A$ if this infimum belongs to A, and does not exist otherwise. In particular, the maximum and the minimum are elements of A when they exist. The supremum and infimum always exist, but are not necessarily in A.

For instance, $\min\{2^n; n \in \mathbb{N}\} = 2$, but $\min\{2^n; n \in \mathbb{Z}\}$ does not exist since $\inf\{2^n; n \in \mathbb{Z}\} = 0 \notin \{2^n; n \in \mathbb{Z}\}$.

Be aware of the behavior of suprema and infima under arithmetic operations. For instance, if we are given two sequences $(x_n)_{n \in \mathbb{N}}$ and $(y_n)_{n \in \mathbb{N}}$ of real numbers, it is relatively easy to check that

$$\sup\{x_n + y_n; n \in \mathbb{N}\} \leqslant \sup\{x_n; n \in \mathbb{N}\} + \sup\{y_n; n \in \mathbb{N}\}$$
$$\text{and } \inf\{x_n + y_n; n \in \mathbb{N}\} \geqslant \inf\{x_n; n \in \mathbb{N}\} + \inf\{y_n; n \in \mathbb{N}\}.$$

However, these inequalities will be strict in most cases. Similarly,

$$\sup\{-x_n; n \in \mathbb{N}\} = -\inf\{x_n; n \in \mathbb{N}\}$$
$$\text{and } \inf\{-x_n; n \in \mathbb{N}\} = -\sup\{x_n; n \in \mathbb{N}\}.$$

Finally, it may be enjoyable to consider the case of the empty set $\varnothing$, and to justify the fact that $\sup \varnothing = -\infty$ and $\inf \varnothing = +\infty$.

T.3. Limits and continuity. Limits involving infinity

In Section 1.3, we define limits and continuity in metric spaces by analogy with the corresponding notions that one encounters in calculus. It may be useful to review these calculus definitions.

Let $f \colon \mathcal{D} \to \mathbb{R}$ be a function with domain $\mathcal{D} \subset \mathbb{R}$. The function f is ***continuous*** at $x_0 \in \mathcal{D}$ if $f(x)$ is arbitrary close to $f(x_0)$ when $x \in \mathcal{D}$ is sufficiently close x_0. This intuitive statement is made rigorous by quantifying the adverbs "arbitrarily" and "sufficiently" with appropriate numbers ε and δ. In this precise definition of continuity, the function f is continuous at x_0 if, for every $\varepsilon > 0$, there exists a number $\delta > 0$ such that $|f(x) - f(x_0)| < \varepsilon$ for every $x \in \mathcal{D}$ with $|x - x_0| < \delta$. This property is more relevant when ε and δ are both small, and this is the situation that we should keep in mind to better understand the meaning of the definition.

We can reinforce the analogy with the metric space definition given in Section 1.3 by using the notation $d(x, y) = |x - y|$, namely, by considering the usual metric d of the real line $\mathbb{R}$. The above definition can then be rephrased by saying that f is continuous at $x_0 \in \mathcal{D}$ if, for every $\varepsilon > 0$, there exists a $\delta > 0$ such that $d\big(f(x), f(x_0)\big) < \varepsilon$ for every $x \in \mathcal{D}$ with $d(x, x_0) < \delta$.

Also, a sequence of real numbers $x_1, x_2, \ldots, x_n, \ldots$ ***converges*** to $x_\infty \in \mathbb{R}$ if x_n is arbitrarily close to x_∞ when the index n is sufficiently large. More precisely, the sequence $(x_n)_{n \in \mathbb{N}}$ converges to x_∞ if, for every $\varepsilon > 0$, there exists an n_0 such that $|x_n - x_\infty| < \varepsilon$ for every $n \geqslant n_0$. Again, if we replace the statement $|x_n - x_\infty| < \varepsilon$ by $d(x_n, x_\infty) < \varepsilon$, we recognize here the definition of limits in metric spaces that is given in Section 1.3.

In calculus, we also encounter infinite limits, and limits as the variable tends to $\pm\infty$. Recall that $f(x)$ has a limit $L \in \mathbb{R}$ as x tends to $+\infty$ if, for every $\varepsilon > 0$, there exists a number $\eta > 0$ such that $|f(x) - L| < \varepsilon$ for every x with $x > \eta$. Similarly, $f(x)$ converges to L as x tends to $-\infty$ if, for every $\varepsilon > 0$, there exists a number $\eta > 0$ such that $|f(x) - L| < \varepsilon$ for every x with $x < -\eta$. In both cases, the more relevant situation is that where ε is small and η is large.

In the book, we combine $+\infty$ and $-\infty$ into a single infinity ∞. Then, by definition, $f(x)$ **converges** to L as x tends to ∞ if, for every $\varepsilon > 0$, there exists a number $\eta > 0$ such that $|f(x) - L| < \varepsilon$ for every x with $|x| > \eta$.

Beware that the symbols ∞ and $+\infty$ represent different mathematical objects in these statements. In particular, $\lim_{x \to \infty} f(x) = L$ exactly when the properties that $\lim_{x \to +\infty} f(x) = L$ and $\lim_{x \to -\infty} f(x) = L$ *both* hold.

Similarly, $f(x)$ **converges** to ∞ as x tends to x_0 if, for every number $\eta > 0$, there exists a $\delta > 0$ such that $|f(x)| > \eta$ for every x with $0 < |x - x_0| < \delta$. In particular, $\lim_{x \to x_0} f(x) = \infty$ if either $\lim_{x \to x_0} f(x) = +\infty$ or $\lim_{x \to x_0} f(x) = -\infty$. However, the converse is not necessarily true, as illustrated by the fact that $\lim_{x \to 0} \frac{1}{x} = \infty$ but that neither $\lim_{x \to 0} \frac{1}{x} = +\infty$ nor $\lim_{x \to 0} \frac{1}{x} = -\infty$ hold.

Immediate generalizations of these limits involving infinity occur in the book when we consider the Riemann sphere $\widehat{\mathbb{C}} = \mathbb{C} \cup \{\infty\} = \mathbb{R}^2 \cup \{\infty\}$ in Chapter 9, $\mathbb{H}^3 \cup \widehat{\mathbb{C}} = \mathbb{H}^3 \cup \mathbb{C} \cup \{\infty\}$ in Chapter 9, or $\widehat{\mathbb{R}}^3 = \mathbb{R}^3 \cup \{\infty\}$ in Chapter 11.

T.4. Complex numbers

In the plane $\mathbb{R}^2$, we can consider the x-axis $\mathbb{R} \times \{0\}$ as a copy of the real line $\mathbb{R}$, by identifying the point $(x, 0) \in \mathbb{R} \times \{0\}$ to the number $x \in \mathbb{R}$. If we set $\mathrm{i} = (0, 1)$, then every point of the plane can be written as a linear combination $(x, y) = x + \mathrm{i}y$. When using this notation, we will consider $x + \mathrm{i}y$ as a generalized number, called a ***complex number***.

It is most likely that you already have some familiarity with complex numbers, and we will just review a few of their properties.

Complex numbers can be added in the obvious manner

$$(x + \mathrm{i}y) + (x' + \mathrm{i}y') = (x + x') + \mathrm{i}(y + y'),$$

and multiplied according to the rule that $\mathrm{i}^2 = -1$, namely,

$$(x + \mathrm{i}y)(x' + \mathrm{i}y') = (xx' - yy') + \mathrm{i}(x'y + xy').$$

These additions and multiplications behave according to the standard rules of algebra. For instance, given three complex numbers $z = x + \mathrm{i}y$, $z' = x' + \mathrm{i}y'$ and $z'' = x'' + \mathrm{i}y''$, we have that $z(z' + z'') = zz' + zz''$ and $z(z'z'') = (zz')z''$.

For a complex number $z = x + \mathrm{i}y$, the x-coordinate is called the **real part** $\mathrm{Re}(z) = x$ of z, and the y-coordinate is its **imaginary part** $\mathrm{Im}(z) = y$. The **complex conjugate** of z is the complex number

$$\bar{z} = x - \mathrm{i}y$$

and the **modulus**, or **absolute value** of z is

$$|z| = \sqrt{x^2 + y^2} = \sqrt{z\bar{z}}.$$

In particular,

$$\frac{1}{x + \mathrm{i}y} = \frac{1}{z} = \frac{\bar{z}}{z\bar{z}} = \frac{\bar{z}}{|z|^2} = \frac{x}{x^2 + y^2} - \mathrm{i}\frac{y}{x^2 + y^2}.$$

Also,

$$\overline{zz'} = (xx' - yy') - \mathrm{i}(xy' + yx') = (x - \mathrm{i}y)(x' - \mathrm{i}y') = \bar{z}\bar{z}'$$

and

$$|zz'| = \sqrt{zz'\bar{z}\bar{z}'} = \sqrt{z\bar{z}}\sqrt{z'\bar{z}'} = |z||z'|$$

for every $z = x + \mathrm{i}y$ and $z' = x' + \mathrm{i}y' \in \mathbb{C}$.

In the book, we make extensive use of Euler's **exponential notation**, where

$$\cos\theta + \mathrm{i}\sin\theta = e^{\mathrm{i}\theta}$$

for every $\theta \in \mathbb{R}$. In particular, any complex number $z = x + \mathrm{i}y$ can be written as $z = re^{\mathrm{i}\theta}$, where $[r, \theta]$ are polar coordinates describing the same point z as the cartesian coordinates (x, y) in the plane $\mathbb{R}^2$.

There are many ways to justify this exponential notation. For instance, we can remember the Taylor expansions

$$\sin\theta = \sum_{k=0}^{\infty}(-1)^k\frac{\theta^{2k+1}}{(2k+1)!} = \theta - \frac{\theta^3}{3!} + \frac{\theta^5}{5!} - \frac{\theta^7}{7!} + \cdots$$

$$\cos\theta = \sum_{k=0}^{\infty}(-1)^k\frac{\theta^{2k}}{(2k)!} = 1 - \frac{\theta^2}{2!} + \frac{\theta^4}{4!} - \frac{\theta^6}{6!} + \cdots$$

$$e^\theta = \sum_{n=0}^{\infty}\frac{\theta^n}{n!} = 1 + \theta + \frac{\theta^2}{2!} + \frac{\theta^3}{3!} + \frac{\theta^4}{4!} + \frac{\theta^5}{5!} + \frac{\theta^6}{6!} + \frac{\theta^7}{7!} + \cdots$$

valid for every $\theta \in \mathbb{R}$. If, symbolically, we replace θ by $i\theta$ in the last equation and remember that $i^2 = -1$,

$$e^{i\theta} = 1 + (i\theta) + \frac{(i\theta)^2}{2!} + \frac{(i\theta)^3}{3!} + \frac{(i\theta)^4}{4!} + \frac{(i\theta)^5}{5!} + \frac{(i\theta)^6}{6!} + \frac{(i\theta)^7}{7!} + \cdots$$

$$= 1 + i\theta - \frac{\theta^2}{2!} - i\frac{\theta^3}{3!} + \frac{\theta^4}{4!} + i\frac{\theta^5}{5!} - \frac{\theta^6}{6!} - i\frac{\theta^7}{7!} + \cdots$$

$$= \left(1 - \frac{\theta^2}{2!} + \frac{\theta^4}{4!} - \frac{\theta^6}{6!} + \cdots\right) + i\left(\theta - \frac{\theta^3}{3!} + \frac{\theta^5}{5!} - \frac{\theta^7}{7!} + \cdots\right)$$

$$= \cos\theta + i\sin\theta.$$

There is actually a way to justify this symbolic manipulation by proving the absolute convergence of this infinite series of complex numbers.

In the same vein, using the addition formulas for trigonometric functions,

$$e^{i\theta}e^{i\theta'} = (\cos\theta + i\sin\theta)(\cos\theta' + i\sin\theta')$$

$$= (\cos\theta\cos\theta' - \sin\theta\sin\theta') + i(\cos\theta\sin\theta' + \sin\theta\cos\theta')$$

$$= \cos(\theta + \theta') + i\sin(\theta + \theta')$$

$$= e^{i(\theta+\theta')},$$

which is again consistent with the exponential notation.

Note the special case

$$e^{i\pi} = -1,$$

known as Euler's Formula, which combines two of the most famous mathematical constants (three if one includes the number 1 among famous constants).

More generally, for an arbitrary complex number $z = x + \mathrm{i}y \in \mathbb{C}$, we can define

$$\mathrm{e}^z = \mathrm{e}^{x+\mathrm{i}y} = \mathrm{e}^x \mathrm{e}^{\mathrm{i}y} = \mathrm{e}^x(\cos y + \mathrm{i}\sin y).$$

It immediately follows from the above observations that this ***complex exponential*** satisfies many of the standard properties of real exponentials, and in particular that $\mathrm{e}^{z+z'} = \mathrm{e}^z \mathrm{e}^{z'}$ for every $z, z' \in \mathbb{C}$.

Supplemental bibliography and references

In this section, we list the references that were mentioned in the text, but begin with a bibliography suggesting additional material for further reading. Some entries appear in both lists.

Supplemental bibliography

This supplemental material is rated with bullets, according to its difficulty. One bullet • indicates a textbook which is roughly at the same mathematical level as the present monograph or easier. Two bullets •• are used for more advanced textbooks, at the graduate level. Three bullets •• denote research-level material.

We have included software [**Heath, Wada, Weeks$_2$, Weeks$_3$**] that currently is freely available on the Internet, and which the reader is strongly encouraged to explore. Of course, these electronic references are quite likely to become unstable with time. The beautiful (and mathematically challenging) movie [**NotKnot**] is also highly recommended.

Some of this suggested reading, such as the excellent [**Stillwell₁**] and [**Mumford, Series & Wright**], has a significant overlap with the current text but offers a different emphasis and viewpoint.

More elementary background on hyperbolic geometry can be found in [**Anderson**], [**Greenberg**], [**Henle**] or [**Stillwell₁**]. Advanced textbooks include the very influential [**Thurston₄**], as well as [**Benedetti & Petronio**], [**Marden₂**], [**Maskit**], [**Ratcliffe**], [**Mumford, Series & Wright**]. See also [**Stillwell₂**] for easy access and an introduction to some historical references.

The topological classification of surfaces, which is not discussed in this book, is nicely presented in [**Massey**, Chapter 1].

With respect to the topics discussed towards the end of the book, [**Adams**], [**Lickorish**] and [**Livingston**] provide nice introductions to knot theory, and more advanced material can be found in [**Rolfsen**] and [**Burde & Zieschang**]. See also [**Flapan**] for applications of knot theory.

A topological approach to 3-dimensional manifolds can be found in [**Hempel**], [**Jaco**] or [**Rolfsen**]. The book [**Weeks₁**] is a lively introduction to a more geometric approach to 3-dimensional manifolds, while the articles [**Scott**] and [**Bonahon**] survey more advanced topics.

You should also consider learning some topology and algebraic topology. There are numerous textbooks in these areas, such as [**Armstrong**], [**Gamelin & Greene**], [**Hatcher₂**], [**Massey**] and [**McCleary**].

[**Adams**]• Colin C. Adams, *The knot book. An elementary introduction to the mathematical theory of knots*, American Mathematical Society, Providence, RI, 2004.

[**Anderson**]• James W. Anderson, *Hyperbolic geometry*, Springer Undergraduate Mathematics Series, Springer-Verlag London Ltd., London, 1999.

[**Armstrong**]• M. Anthony Armstrong, *Basic Topology*, Undergraduate Texts in Mathematics, Springer-Verlag, New York-Berlin, 1983.

[**Benedetti & Petronio**] •• Riccardo Benedetti, Carlo Petronio, *Lectures on hyperbolic geometry*, Universitext, Springer-Verlag, Berlin, 1992.

[**Bonahon**] ♣ Francis Bonahon, "Geometric structures on 3-manifolds", in: *Handbook of geometric topology* (R. J. Daverman and R. B. Sher, eds.), 93-164, North-Holland, Amsterdam, 2002.

[**Burde & Zieschang**] ♣ Gerhardt Burde, Heiner Zieschang, *Knots*, (Second edition), de Gruyter Studies in Mathematics 5, Walter de Gruyter GmbH, Berlin, Germany, 2003.

[**Flapan**] •• Erica Flapan, *When topology meets chemistry. A topological look at molecular chirality*, Cambridge University Press, Cambridge, United Kingdom; Mathematical Association of America, Washington, DC, 2000.

[**Gamelin & Greene**] • Theodore W. Gamelin, Robert E. Greene, *Introduction to topology*, (Second Edition), Dover Publications, Mineola, NY, 1999.

[**Greenberg**] • Marvin J. Greenberg, *Euclidean and non-Euclidean geometries: development and history*, W. H. Freeman and Co., San Francisco, Calif., 1973.

[**Hatcher$_2$**] •• Allen E. Hatcher, *Algebraic topology*, Cambridge University Press, 2001.

[**Heath**] Daniel J. Heath, *Geometry playground*, mathematical software, freely available at www.plu.edu/~heathdj/java/.

[**Hempel**] •• John Hempel, *3-Manifolds*, Annals of Mathematics Studies 86, Princeton University Press, Princeton, NJ, 1976.

[**Henle**] • Michael Henle, *Modern geometries. The analytic approach*, Prentice Hall, Inc., Upper Saddle River, NJ, 1997.

[**Jaco**] •• William Jaco, *Lectures on three-manifold topology*, CBMS. Regional Conference Series in Mathematics 43, American Matheamtical Society, Providence, RI, 1980.

[**Lickorish**] •• W. B. Raymond Lickorish, *An introduction to knot theory*, Graduate Texts in Mathematics 175, Springer-Verlag, New York, NY, 1997

[**Livingston**] • Charles Livinston, *Knot theory*, Mathematical Association of America, Washington, DC, 1993.

[**Marden₂**] •• Albert Marden, *Outer circles: an introduction to hyperbolic 3-manifolds*, Cambridge University Press, Cambridge, United Kingdom, 2007.

[**Massey**] •• William S. Massey, *A basic course in algebraic topology*, Graduate Texts in Mathematics 127, Springer-Verlag, Berlin, 1997.

[**Maskit**] •• Bernard Maskit, *Kleinian groups*, Grundlehren der Mathematischen Wissenschaften 287, Springer-Verlag, Berlin, 1988.

[**McCleary**] • A first course in topology, Student Mathematical Library 31, American Mathematical Society, Providence, RI, 2006.

[**Mumford, Series & Wright**] • David Mumford, Caroline Series, David Wright, *Indra's pearls. The vision of Felix Klein*, Cambridge University Press, New York, 2002.

[**NotKnot**] • *Not Knot*, video by the Geometry Center of the University of Minnesota, A.K. Peters Ltd, Wellesley, MA, 1994.

[**Ratcliffe**] •• John Ratcliffe, *Foundations of hyperbolic manifolds*, Graduate Texts in Mathematics 149, Springer-Verlag, New York, 1994.

[**Rolfsen**] •• Dale Rolfsen, *Knots and links*, Publish or Perish, Berkeley, 1976.

[**Scott**] ⁂ G. Peter Scott, "The geometries of 3-manifolds", *Bulletin of the London Mathematical Society* 15 (1983), 401–487.

[**Stillwell₁**] • John C. Stillwell, *Geometry of surfaces*, Universitext, Springer-Verlag, New York, 1992.

[**Stillwell₂**] •• John C. Stillwell, *Sources of hyperbolic geometry*, History of Mathematics 10, American Mathematical Society, Providence, RI, 1996.

[**Thurston₄**] •• William P. Thurston, *Three-dimensional geometry and topology, Vol. 1* (Edited by Silvio Levy), Princeton Mathematical Series 35, Princeton University Press, Princeton, NJ, 1997.

[**Wada**] Masaaki Wada, *OPTi*, mathematical software, freely available at `http://vivaldi.ics.nara-wu.ac.jp/~wada/OPTi`.

[**Weeks₁**] • Jeffrey Weeks, *The shape of space* (Second edition), Monographs and Textbooks in Pure and Applied Mathematics 249, Marcel Dekker, Inc., New York, 2002.

[**Weeks₂**] Jeffrey Weeks, *SnapPea*, mathematical software, freely available at `www.geometrygames.org`.

[**Weeks₃**] Jeffrey Weeks, *Curved Spaces*, mathematical software, freely available at `www.geometrygames.org`.

References

Note that some of the references below correspond to books, while others point to articles in academic journals.

[**Alexander**] James W. Alexander, "On the deformation of an *n*-cell", *Proceedings of the National Academy of Sciences of the United States of America* 9 (1923), 406–407.

[**Ashley**] Clifford Ashley, *The Ashley book of knots*, Doubleday, New York, 1944.

[**Bonahon & Siebenmann**] Francis Bonahon, Laurence C. Siebenmann, *New geometric splittings of classical knots, and the classification and symmetries of arborescent knots*, to appear, Geometry & Topology Monographs, Coventry, UK.

[**Brock & Canary & Minsky**] Jeffrey F. Brock, Richard D. Canary, Yair N. Minsky, "The classification of Kleinian surface groups, II: The Ending Lamination Conjecture", preprint, available at `http://arxiv.org/abs/math/0412006`.

[**Bromberg**] Kenneth W. Bromberg, "The space of Kleinian punctured torus groups is not locally connected", preprint, available at `http://arxiv.org/abs/0901.4306`.

[**Cannon & Thurston**] James W. Cannon, William P. Thurston, "Group invariant Peano curves", *Geometry & Topology* 11 (2007), 1315–1355.

[**Cao & Zhu**] Huai-Dong Cao, Xi-Ping Zhu, "A complete proof of the Poincaré and geometrization conjectures—application of the

Hamilton-Perelman theory of the Ricci flow", *Asian Journal of Mathematics* 10 (2006), 165–492; "Erratum", *Asian Journal of Mathematics* 10 (2006), 663.

[**Cerf**] Jean Cerf, *Sur les difféomorphismes de la sphère de dimension trois* ($\Gamma_4 = 0$), Lecture Notes in Mathematics 53, Springer-Verlag, Berlin-New York, 1968.

[**Chow & Knopf**] Bennett Chow, Dan Knopf, *The Ricci flow: an introduction*, Mathematical Surveys and Monographs 110, American Mathematical Society, Providence, RI, 2005.

[**Chow & Lu & Ni**] Bennett Chow, Peng Lu, Lei Ni, *Hamilton's Ricci flow*, Graduate Studies in Mathematics 77, American Mathematical Society, Providence, RI; Science Press, New York, 2006.

[**Chow & al.**] Bennett Chow, Sun-Chin Chu, David Glickenstein, Christine Guenther, James Isenberg, Tom Ivey, Dan Knopf, Peng Lu, Feng Luo, Lei Ni, *The Ricci flow: techniques and applications, Part I. Geometric aspects*, Mathematical Surveys and Monographs 135, American Mathematical Society, Providence, RI, 2007; *Part II. Analytic aspects*, Mathematical Surveys and Monographs 144, American Mathematical Society, Providence, RI, 2008.

[**Dirichlet**] Gustav Lejeune Dirichlet, "Über die Reduktion der positiven quadratischen Formen mit drei unbestimmten ganzen Zahlen", *Journal für die Reine und Angewandte Mathematik* 40 (1850), 209–227.

[**Dumas**] David Dumas, *Bear, a tool for studying Bers slices of punctured tori*, mathematical software, freely available at `bear.sourceforge.net`.

[**Epstein & Penner**] David B. A. Epstein, Robert C. Penner, "Euclidean decompositions of noncompact hyperbolic manifolds", *Journal of Differential Geometry* 27 (1988), 67–80.

[**Farey**] John Farey, "On a curious property of vulgar fractions", *Philosophical Magazine* 47 (1816), 385-386.

[**Ford₁**] Lester R. Ford, "The fundamental region for a Fuchsian group", *Bulletin of the American Mathematical Society* 31 (1935), 531–539.

[**Ford₂**] Lester R. Ford, "Fractions", *American Mathematical Monthly* 45 (1938), 586-601.

[**Fuchs₁**] Lazarus Fuchs, "Sur quelques propriétés des intégrales des équations différentielles, auxquelles satisfont les modules de périodicité des intégrales elliptiques des deux premières espèces", *Journal für die Reine und Angewandte Mathematik* 83 (1877), 13-57.

[**Fuchs₂**] Lazarus Fuchs, "Über eine Klasse von Functionen mehrerer Variabeln, welche durch Umkehrung der Integrale von Lösungen des linearen Differentialgleichungen mit rationalen Coefficienten entstehen", *Journal für die Reine und Angewandte Mathematik* 89 (1880), 151–169.

[**Gordon & Luecke**] Cameron McA. Gordon, John Luecke, "Knots are determined by their complements", *Journal of the American Mathematical Society* 2 (1989), 371–415.

[**Haken**] Wolfgang Haken, "Ein Verfahren zur Aufspaltung einer 3-Mannigfaltigkeit in irreduzible 3-Mannigfaltigkeiten", *Mathematische Zeitschrift* 76 (1961), 427–467.

[**Hardy & Wright**] Godfrey H. Hardy, Edward M. Wright, *An introduction to the theory of numbers*, Fifth edition, The Clarendon Press, Oxford University Press, New York, 1979.

[**Hatcher₁**] Allen E. Hatcher, "A proof of a Smale conjecture $\mathrm{Diff}(S^3) \simeq O(4)$", *Annals of Mathematics* 117 (1983), 553-607.

[**Hatcher₂**] Allen E. Hatcher, *Algebraic topology*, Cambridge University Press, 2001.

[**Jaco & Shalen**] William H. Jaco, Peter B. Shalen, *Seifert fibered spaces in 3-manifolds*, Memoirs of the American Mathematical Society 220, American Mathematical Society, Providence, 1979.

[**Johannson**] Klaus Johannson, *Homotopy equivalences of 3-manifolds with boundary*, Lecture Notes in Mathematics 761, Springer-Verlag, Berlin-New York, 1979.

[**Kapovich**] Michael Kapovich, *Hyperbolic manifolds and discrete groups*, Progress in Mathematics 183, Birkhäuser Boston, Boston, MA, 2001.

[**Kneser**] Hellmuth Kneser, "Geschlossene Flächen in dreidimensionalen Mannigfaltigkeiten", *Jahresbericht der Deutschen Mathematiker-Vereinigung* 38 (1929), 248-260.

[**Kleiner & Lott**] Bruce Kleiner, John Lott, "Notes on Perelman's papers", *Geometry & Topology* 12 (2008), 2587–2855.

[**Little**] Charles N. Little, "Non-alternate ± knots", *Transactions of the Royal Society of Edinburgh* 39 (1898-9), 771-778.

[**Marden$_1$**] Albert Marden, "The geometry of finitely generated Kleinian groups", *Annals of Mathematics* 99 (1974), 383-462.

[**McMullen**] Curtis T. McMullen, "Iteration on Teichmüller space", *Inventiones Mathematicæ* 99 (1990), 425–454.

[**Milnor**] John W. Milnor, "A unique decomposition theorem for 3-manifolds", *American Journal of Mathematics* 84 (1962), 1–7.

[**Minsky$_1$**] Yair N. Minsky, "The classification of punctured-torus groups", *Annals of Mathematics* 149 (1999), 559–626.

[**Minsky$_2$**] Yair N. Minsky, "The classification of Kleinian surface groups, I: Models and bounds", to appear in *Annals of Mathematics*.

[**Morgan & Tian$_1$**] John W. Morgan, Gang Tian, *Ricci flow and the Poincaré conjecture*, Clay Mathematics Monographs 3, American Mathematical Society, Providence; Clay Mathematics Institute, Cambridge, MA, 2007.

[**Morgan & Tian$_2$**] John W. Morgan, Gang Tian, *Completion of the proof of the geometrization conjecture*, preprint, 2008, available at http://arxiv.org/abs/0809.4040.

[**Mostow$_1$**] George D. Mostow, "Quasi-conformal mappings in n-space and the rigidity of hyperbolic space forms", *Publications Mathématiques, Institut des Hautes Études Scientifiques* 34 (1968), 53-104.

[**Mostow₂**] George D. Mostow, *Strong rigidity of locally symmetric spaces*, Annals of Mathematics Studies 78, Princeton University Press, Princeton, N.J., 1973.

[**Orlik, Vogt & Zieschang**] Peter Orlik, Elmar Vogt, Heiner Zieschang, "Zur Topologie gefaserter dreidimensionaler Mannigfaltigkeiten", *Topology* 6 (1967), 49–64.

[**Otal₁**] Jean-Pierre Otal, *Le théorème d'hyperbolisation pour les variétés fibrées de dimension* 3, Astérisque 235, Société Mathématique de France, Paris, France, 1996.

[**Otal₂**] Jean-Pierre Otal, "Thurston's hyperbolization of Haken manifolds", in: *Surveys in differential geometry, Vol. III (Cambridge, MA, 1996)*, 77–194, International Press, Boston, MA, 1998.

[**Perelman₁**] Grigori Y. Perelman, "The entropy formula for the Ricci flow and its geometric applications", unpublished preprint, 2002, available at http://arxiv.org/abs/math/0211159.

[**Perelman₂**] Grigori Y. Perelman, "Ricci flow with surgery on three-manifolds", unpublished preprint, 2003, available at http://arxiv.org/abs/math/0303109.

[**Perelman₃**] Grigori Y. Perelman, "Finite extinction time for the solutions to the Ricci flow on certain three-manifolds", unpublished preprint, 2003, available at http://arxiv.org/abs/math/0307245.

[**Perko**] Kenneth A. Perko, "On the classification of knots", *Proceedings of the American Mathematical Society* 45 (1974), 262-266.

[**Poincaré₁**] Henri Poincaré, "Sur les fonctions uniformes qui se reproduisent par des substitutions linéaires", *Mathematische Annalen* 19 (1882), 553-564.

[**Poincaré₂**] Henri Poincaré, "Sur les fonctions uniformes qui se reproduisent par des substitutions linéaires (Extrait d'une lettre adressée à Mr. F. Klein)", *Mathematische Annalen* 20 (1882), 52-53.

[**Poincaré₃**] Henri Poincaré, "Théorie des groupes fuchsiens", *Acta Mathematica* 1 (1882), 1–62.

[**Poincaré₄**] Henri Poincaré, "Mémoire sur les fonctions fuchsiennes", *Acta Mathematica* 1 (1882), 193-294.

[**Poincaré₅**] Henri Poincaré, "Mémoire sur les groupes kleinéens", *Acta Mathematica* 3 (1983), 49–92.

[**Poincaré₆**] Henri Poincaré, *Science et Méthode*, Flammarion, Paris, 1908.

[**Prasad**] Gopal Prasad, "Strong rigidity of $\mathbb{Q}$-rank 1 lattices", *Inventiones Mathematicæ* 21 (1973), 255–286.

[**Riley₁**] Robert F. Riley, "A quadratic parabolic group", *Mathematical Proceedings of the Cambridge Philosophical Society* 77 (1975), 281–288.

[**Riley₂**] Robert F. Riley, "Applications of a computer implementation of Poincaré's theorem on fundamental polyhedra", *Mathematics of Computation* 40 (1983), 607–632.

[**Riley₃**] Robert F. Riley, "An elliptical path from parabolic representations to hyperbolic structures", in: *Topology of low-dimensional manifolds (Proceedings of the Second Sussex Conference, Chelwood Gate, 1977)*, 99–133, Lecture Notes in Mathematics 722, Springer, Berlin, 1979.

[**Riley₄**] Robert F. Riley, "Seven excellent knots", in: *Low-dimensional topology (Bangor, 1979)*, 81–151, London Mathematical Society Lecture Note Series 48, Cambridge University Press, Cambridge-New York, 1982.

[**Rolfsen**] Dale Rolfsen, *Knots and links*, Publish or Perish, Berkeley, 1976.

[**Schubert₁**] Horst Schubert, "Die eindeutige Zerlegbarkeit eines Knotens in Primknoten", *Sitzungsberichte der Heidelberger Akademie der Wissenschaften. Mathematisch-Naturwissenschaftliche Klasse* 1949 (1949), 57–104.

[**Schubert₂**] Horst Schubert, "Knotten und Vollringe", *Acta Mathematica* 90 (1953), 131–286.

[**Seifert**] Herbert Seifert, "Topologie dreidimensionaler gefaserte Räume", *Acta Mathematica* 60 (1933), 147–238.

[**Scott**] G. Peter Scott, "The geometries of 3-manifolds", *Bulletin of the London Mathematical Society* 15 (1983), 401–487.

[**Thurston$_1$**] William P. Thurston, "Hyperbolic geometry and 3-manifolds", in: *Low-dimensional topology (Bangor, 1979)*, 9–25, London Mathematical Society Lecture Note Series 48, Cambridge University Press, Cambridge–New York, 1982.

[**Thurston$_2$**] William P. Thurston, "Three-dimensional manifolds, Kleinian groups and hyperbolic geometry", *Bulletin of the American Mathematical Society* 6 (1982), 357–381.

[**Thurston$_3$**] William P. Thurston, "Hyperbolic Structures on 3-manifolds, I: Deformation of acylindrical manifolds", *Annals of Mathematics* 124 (1986), 203-246.

[**Thurston$_4$**] William P. Thurston, *Three-dimensional geometry and topology, Vol. 1* (Edited by Silvio Levy), Princeton Mathematical Series 35, Princeton University Press, Princeton, NJ, 1997.

[**Voronoi$_1$**] Georgy Voronoi, "Nouvelles applications des paramètres continus à la théorie des formes quadratiques", *Journal für die Reine und Angewandte Mathematik* 133 (1908), 97–178.

[**Voronoi$_2$**] Georgy Voronoi, "Recherches sur les paralléloèdres primitives", *Journal für die Reine und Angewandte Mathematik* 134 (1908), 198-287.

[**Wada**] Masaaki Wada, *OPTi*, mathematical software, freely available at http://vivaldi.ics.nara-wu.ac.jp/~wada/OPTi/.

[**Waldhausen**] Friedhelm Waldhausen, "Eine Klasse von dreidimensionalen Mannigfaltigkeiten" I, *Inventiones Mathematicæ* 3 (1967), 308–333; II, *Inventiones Mathematicæ* 4 (1967), 87–117.

[**Weeks$_2$**] Jeffrey Weeks, *SnapPea*, mathematical software, freely available at www.geometrygames.org.

[**Weeks$_4$**] Jeffrey Weeks, "Computation of hyperbolic structures in knot theory", in: *Handbook of knot theory*, 461–480, Elsevier B. V., Amsterdam, 2005.

Index

absolute value, 23, **362**

action of a group, **185**, 185–205, 248

adjacent tiles, **137**, 137–146, **261**

Alexander, James Waddell II (1888–1971), 318

antilinear fractional map, **27**, 27–33, 38, 186, 231–235

antipodal points, **49**, 102

arc length, *see* length

area, 131

  hyperbolic area, **43**, 43–44

  spherical area, 53, 182

$\mathbb{B}^2$, the disk model for the hyperbolic plane, **36**, 36–39, 43, 44, 101, 158–160, 209, 212, 216, 223

$B_d(P, r)$, a ball in a metric space, **4**

ball

  ball in $\mathbb{R}^3$, **5**, 344

  ball in a metric space, **4**, 42–43, 68–79, 240, 262

  ball model for the hyperbolic space, **236**, 240

belong to, **355**

bijection, 5, 90, 185, **357**

bijective, **357**

Bonahon, Francis, iii

Bonnet, Pierre Ossian (1819–1892), 131

boundary

  boundary of a manifold, **348**

  boundary point, **152**

bounded, **62**, 81, **152**, 152–161

Brock, Jeffrey F., 283

Bromberg, Kenneth W., 282

Busemann

  Busemann function, **178**, 183

  Busemann, Herbert (1905–1994), 178

$\mathbb{C}$, the set of all complex numbers, **356**, 361–364

Canary, Richard D., 283

Cannon, James W., 302

canonical tile, **143**, 143–146

Cauchy

  Cauchy sequence, 135, **182**

  Cauchy, Augustin (1789–1857), 182, 211

Cayley, Arthur (1821–1895), 45

center

  center of a horocircle, **172**, 214

  center of a horosphere, 234

Cerf, Jean, 318

circle, **5**, 17–22, 32–33, 38, 48, 85–87, 172, 207–212, 230, 288, 317

great circle, **48**, 129
great circle arc, **49**, 50–51
closed
  closed curve, **49**, **316**
  closed geodesic, **49**, 127–129, 204
  closed subset, 62, 80, **152**,
    152–154, 250–251, 257–258,
    286
compact metric space, **150**,
  150–154, 182, 203, 251, 348
companion knot, **321**
complement, **356**
  knot complement, 293–313,
    319–340, 352–354
complete
  complete geodesic, **21**
  complete metric space, **135**,
    147–155, 174, 182, 183, 203,
    230, 260, 311, 322, 325, 341,
    346, 348
complex number, 6, 11, 23, **361**,
  361–364
  complex conjugate, **6**, 23, 44,
    362
  complex exponential, 242, **364**,
    362–364
composition, **6**, 15, 28, 42, 51, 52,
  185, 186, **357**
cone, **86**
  surface with cone singularities,
    87, 125–126, 136, 192, 205, 342
conjugate, *see* complex conjugate
connected, **64**, 341
continued fraction, **224**
continuous function, **4**, 8, 9, 61, 90,
  151, 302, **360**
convergence, **4**, 135, 147–155, **360**
  limit at infinity, **361**, 360–361
  toward infinity, **361**, 360–361
convex
  euclidean polygon, **62**
  hyperbolic polygon, **82**
  spherical polygon, 82
crossratio, 44–45

differential map, **29**, 29–38, 40,
  49–50, 229–230, 317

dihedral angle, **235**, **259**, 259–265,
  290
dihedron, **235**
Dirichlet
  Dirichlet domain, **198**, 197–201
  Lejeune Dirichlet, Gustav
    (1805–1859), 198
discontinuity domain, 286–287
discontinuous group action, **189**,
  189–197, 203–205, 248, 254,
  260, 268, 280–283, 286, 313
discrete walk, **58**, 58–60, 69–79,
  85–86
disjoint sets, **356**
disk
  disk in the plane, **5**
  disk model for the hyperbolic
    plane, **36**, 36–39, 43, 44, 101,
    158–160, 209, 212, 216, 223
  disk sector, **72**, 86, 205
distance
  distance function, **3**
  euclidean distance, **2**, 1–3, 8, **340**
  hyperbolic distance, **12**, 12–14,
    37, 39, 44, 45, **228**
  signed distance, 218–220, **253**,
    253–254
  spherical distance, **48**, **343**
domino diagram, 225, 291
Dumas, David, 282
Dunne, Edward G., xv

Earle, Clifford J., 280
edge
  edge cycle, **172**, 172–181
  of a hyperbolic polyhedron, **258**,
    258–265
  of a polygon, **61**, 61–83, **257**
element of a set, **355**
elliptic isometry, **41**, **239**, 240
empty set, **355**
Epstein, David B. A., 340
equivalence class, **84**
equivalence relation, **84**
essential surface, 343–346
  essential Klein bottle, **345**
  essential projective plane, **344**
  essential sphere, **344**

essential torus, **344**
euclidean
    euclidean distance, **2**, 3, **340**
    euclidean geodesic, 3, 21
    euclidean isometries, 5–7, **39**
    euclidean length, **2**, 48
    euclidean metric, 67, **340**
    euclidean norm, **33**
    euclidean plane, 1–10
    euclidean polygon, **61**, 61–66,
        183
    euclidean space, **340**, 349
    euclidean surface, **67**, 89–97,
        103–104, 341–343
    euclidean triangle, 125
Euler
    Euler characteristic, **131**
    Euler exponential notation, 6,
        362
    Euler, Leonhard (1707–1783),
        131, 362
exceptional fiber, **347**
exponential
    complex exponential, 242, **364**,
        362–364
exterior point, **152**

face of a polyhedron, **258**, 257–265,
    334–340
Farey
    crooked Farey tessellation, **243**,
        241–247, 265–279, 291–297
    Farey circle packing, **210**,
        207–226
    Farey series, **223**
    Farey sum, **209**
    Farey tessellation, **210**, 207–226
    Farey, John (1766–1826), 210
fiber, **347**
fibration, *see* Seifert fibration
Fields Medal, 322, 350
figure-eight knot, **303**, 293–313,
    352, 353
finite volume, 324
fix, **358**
fixed point, **358**
Ford
    Ford circles, **212**

Ford domain, **331**, 326–340
Ford, Lester Randolph
    (1886–1967), 212, 340
free group action, **192**, **264**,
    280–282, 297, 352
Fuchs, Lazarus Immanuel
    (1833-1902), 284
fuchsian group, **255**, 252–257, 280,
    283–284
    fuchsian group of the first type,
        256
    fuchsian group of the second
        type, **256**
    twisted fuchsian group, **288**
function, **357**
fundamental domain, **192**,
    185–205, **259**, 259–269,
    296–298

Gauss, Johann Carl Friedrich
    (1777–1855), 131
Gauss-Bonnet formula, **131**
generate
    transformation group generated
        by bijections, 135–136, **186**,
        194, 260
genus, 97, 129
geodesic, **21**
    closed geodesic, **49**, 127–129, 204
    complete geodesic, **21**
    euclidean geodesic, 3, 21
    hyperbolic geodesic, 17–22, 230
    spherical geodesic, 49–51
geometric structure, **349**
Gordon, Cameron McA., 340
great circle, **48**, 129
great circle arc, **49**
group, **185**, 185–205
    abstract group, **202**
    fuchsian group, **255**, 252–257,
        280, 283–284
    group action, **185**, 185–205
    group of isometries, **186**
    isometry group, **186**
    kleinian group, **248**, 241–302,
        326–340
    tiling group, **136**, 135–184,
        212–222, 259

transformation group, **185**, 185–205

$\mathbb{H}^2$, the hyperbolic plane, **11**, 11–46
$\mathbb{H}^3$, the 3-dimensional hyperbolic space, **227**, 227–240
Haken, Wolfgang, 350
Hamilton, Richard S., 350
Haros, C., 211
Hatcher, Allen E., 318
Hiatt, Christopher, xvi
homeomorphic, **90**
homeomorphism, 41, **90**, 91, 96–98, 126, 129, 130, 182, 270, 288, 303–311, 317–319, 325
homogeneous, **7**, 15–16, 34–36, 49–50, 128, 129, 230, 349
homothety, **15**, 28, 40, 228, 239, 286
horizontal translation, **15**, 28, 41, 228, 239
horocircle, **172**, 172–181, 214–222, 290
horocycle, **172**
horocyclic isometry, **172**, 172–176, 221–222
horodisk, **178**, 178–180, 256
horosphere, **234**, 234–235, 260–265, 268, 298, 331–332
Howe, Roger E., xv
hyperbolic
　hyperbolic area, **43**, 43–44
　hyperbolic disk, 42
　hyperbolic distance, **12**, 12–14, 37, 39, 44, 45, **228**
　hyperbolic geodesic, 17–22, 230
　hyperbolic isometry, 14–17, **25**, 23–27, 40, **232**, 231–234, 238, 239
　hyperbolic knot, **322**, 322–340
　hyperbolic length, **11**, 17–22, 33–34, **227**, 230
　hyperbolic metric, **12**, 12–14, 37, 39, 44, 45, **82**, **228**
　hyperbolic norm, **33**, 33–36, **228**, 229
　hyperbolic plane, **11**, 11–46, **234**
　hyperbolic polygon, **80**

hyperbolic reflection, 15, **240**
hyperbolic rotation, **40**
hyperbolic space, **227**, 227–240, 349
hyperbolic surface, **82**, 102, 341–343
hyperbolic triangle, 43, 98, 125, 130
hyperbolic volume, **240**, 353

Id_X, the identity map of X, **357**
ideal
　ideal polygon, **116**
　ideal vertex, **116**, **170**, 170–180, **257**, **258**, 257–265
identity map, 185, **357**
image, **357**
imaginary part, 11, **362**
infimum (*pl.* infima), **359**, 358–360
infinite limit, **361**, 360–361
infinity
　limit at infinity, **361**, 360–361
　vertex at infinity, **170**, 170–180, **257**, **258**, 257–265
injective, **357**
interior
　interior of a manifold, **348**
　interior point, **152**
intersection, **356**
inverse map, 185, **357**
inversion
　across a circle, **16**, 16–17, 28, **28**, 42
　across a sphere, **229**, 237
isometric
　isometric extension, **233**, 231–233, 236
　isometric group action, **186**
　locally isometric, **67**
isometry, **5**
　elliptic isometry, **41**, **239**
　euclidean isometries, 5–7, **39**
　group of isometries, **186**
　horocyclic isometry, **172**, 172–176, 221–222
　hyperbolic isometry, 14–17, **25**, 23–27, 40, **232**, 231–234, 238, 239

isometry group, **186**
loxodromic isometry, **40**, **239**
parabolic isometry, **41**, **239**, 327
spherical isometries, 50
isomorphic knots, **317**
isotopic knots, **316**
isotopy, **316**
isotropic, **7**, 34–36, 49–50, 230, 349

Jaco, William H., **321**, 350
jacobian, 238, **317**
Johannson, Klaus, 321, 350

Klein
    Klein bottle, **95**, 93–95, 128, 341
    essential Klein bottle, **345**
    Klein, Felix (1849–1925), 45, 95,
        284
    kleinian group, **248**, 241–302,
        326–340
Kneser, Hellmuth (1898–1973), 350
knot, 316
    companion knot, **321**
    figure-eight knot, **303**
    hyperbolic knot, **322**, 322–340
    knot complement, 293–313,
        319–340, 352–354
    satellite knot, **321**
    torus knot, **319**
    trefoil knot, **319**
    unknot, **317**

length
    euclidean length, **2**, 48
    hyperbolic length, **11**, 17–22,
        33–34, **227**, 230
    length in a metric space, 3, **9**, 87,
        127
    length of a discrete walk, **58**
    length of a sequence, **135**,
        147–155
limit, **4**, 135, 147–155
    limit at infinity, **361**, 360–361
    limit point, **248**
    limit set, **248**, 247–252, 255–256,
        270–279, 299–302
linear fractional map, **27**, 27–33,
        38, 186, 231–235
linear groups, **202**

Listing, Johann Benedict
    (1808–1882), 104
Little, Charles N., 337
locally finite, **134**, 145, **197**, 259,
        261, 333, 334
locally isometric, **67**, 82, 83
loxodromic isometry, **40**, **239**
Luecke, John, 340

manifold, **340**, 340–351
map, **357**
Marden, Albert, 296
maximum (*pl.* maxima), **358**,
        358–360
McMullen, Curtis T., 322
metric, **3**
    euclidean metric, **67**, **340**
    hyperbolic metric, **12**, 12–14, 37,
        39, 44, 45
    metric function, **3**
    metric space, **3**, 3–5, 58–61,
        134–135, 150–155
    path metric, **9**, 63, 81, 82
    product metric, **8**, 344
    quotient metric, **61**, 58–61,
        141–143, 187–189
    spherical metric, **48**, **343**
Milnor, John W., 350
minimum (*pl.* minima), **358**,
        358–360
Minsky, Yair N., 282, 283
Möbius
    Möbius group, **237**
    Möbius strip, **104**, 103–104
    Möbius transformation, **237**
    Möbius, August (1790–1868), 104
Möbius strip, 341
moduli space, **342**
modulus (*pl.* moduli), 23, **362**
Mostow
    Mostow's Rigidity Theorem,
        325, 324–326, 336
    Mostow, George Daniel, 325

N, the set of all positive integers,
        356
nonorientable surface, 104, **129**
norm

euclidean norm, **33**
hyperbolic norm, **33**, 33–36,
    228, 229
normal subgroup, **299**, 299–302

once-punctured torus, **114**,
    114–125, 161–162, 167–169,
    212–214
one-to-one correspondence, **357**
one-to-one map, **357**
onto, **357**
OPTi, 246, 280
orbifold, 342
orbit, **187**, 187–192, 198, 248
orbit space, **187**, 187–192
orientation-preserving, 238, 317
orientation-reversing, 238, 317
Orlik, Peter, 347
orthogonal projection, 40, 253

parabolic isometry, **41**, **239**, 327
parabolic point, **327**
partition, **57**, 57–87, 187–192
    proper partition, **60**
path metric, **9**, 63, 81, 82
Penner, Robert Clark, 340
Perelman, Grigory Y., 350
Perko, Kenneth A., 338
perpendicular bisector, **39**, 198
piecewise differentiable curve, **2**
plane
    euclidean plane, 1–10
    hyperbolic plane, **11**, 11–46, **234**
    projective plane, **102**, 129, 341
        essential projective plane, **344**
Poincaré
    Poincaré Conjecture, 351
    Poincaré's Polygon Theorem,
        174, 169–181
    Poincaré's Polyhedron Theorem,
        260, 257–265
    Poincaré, Jules Henri
        (1854–1912), 283, 351
polygon, 61–87, 89–131, 133–184,
    192–201, 335
    euclidean polygon, **61**, 61–66,
        183
    hyperbolic polygon, **80**, **257**

ideal polygon, **116**
locally finite polygon, **197**
spherical polygon, **82**
polyhedral ball sector, **261**
polyhedron (*pl.* polyhedra), **258**,
    257–265
Prasad, Gopal, 325
preimage, **358**
preserve, **358**
product, 8, **357**
    product metric, **8**, 344
projective
    projective line, **42**
    projective linear groups, **202**
    projective model for the
        hyperbolic plane, **45**
    projective plane, **102**, 129, 341
        essential projective plane, **344**
    projective space, **202**
proper
    proper gluing, **60**
    proper partition, **60**
pseudosphere, **108**, 108–114,
    118–125
Pythagorean triple, **223**

$\mathbb{Q}$, the set of all rational numbers,
    356
quotient
    quotient map, **61**
    quotient metric, **61**, 58–61,
        141–143, 187–189
    quotient space, **60**, 58–61, **187**,
        187–197

$\mathbb{R}$, the set of all real numbers, **356**
$\mathbb{R}^2$, the euclidean plane, 1–10, **357**
$\mathbb{R}^3$, the 3-dimensional euclidean
    space, 4, **357**
real part, 11, **362**
reflection
    euclidean reflection, **6**, 50–52
    hyperbolic reflection, **15**, **240**
regular curve, **316**
respect, **358**
restriction, **358**

Riemann
  Riemann sphere, **27**, 41, 241–252, 276–279
  Riemann, Georg Friedrich Bernhard (1826–1866), 27
Riley, Robert F. (1935–2000), 312, 340
rotation
  euclidean rotation, **6**, 49–52
  hyperbolic rotation, 40
  rotation-reflection, **50**, 51–52

$\mathbb{S}^2$, the 2-dimensional sphere, **47**
$\mathbb{S}^3$, the 3-dimensional sphere, **343**, 351
satellite knot, **321**
Schubert, Horst (1919–2001), 321
Seifert
  Seifert fibration, **347**, 354
  Seifert manifold, **347**
  Seifert, Herbert Karl Johannes (1907–1996), 347
semi-distance function, **3**
semi-metric, **3**, 58–61, 84, 187, 189, 202
sequence, **358**
set, **355**
Shalen, Peter B., 321, 350
shear parameter, **220**, 290
shear-bend parameter, 265, **289**
shearing, **220**
Siebenmann, Laurence C., 322
signed distance, 218–220, **253**, 253–254
simple curve, **316**
SnapPea, 323, 335, 337, 339
sphere, **5**, **47**, 341
  3-dimensional sphere, **343**, 349, 351
  essential sphere, **344**
spherical
  spherical area, 53, 182
  spherical distance, **48**, 343
  spherical geodesic, 49–51
  spherical isometries, 50
  spherical metric, **48**, 84, **343**
  spherical polygon, **82**
  spherical surface, **84**, 341–343

spherical triangle, 52, 125
stabilizer, **190**, 190–192, 204–205
stereographic projection, **41**
subsequence, **150**, 150–155
subset, **355**
supremum (*pl.* suprema), **358**, 358–360
surface, **341**
  euclidean surface, **67**, 89–97, 103–104, 341–343
  hyperbolic surface, **82**, 102, 341–343
  in a 3-dimensional manifold, **344**
  nonorientable surface, 104, **129**
  spherical surface, **84**, 341–343
  surface of genus 2, **97**, 97–102
  surface of genus g, 129
  two-sided surface, **344**
surjective, **357**

tangent line, **276**
tangent map, **29**
tessellation, **133**, 133–184, 207–226, **259**, 259–265
Thurston, William P., 302, 312, 322, 350
tile, **133**, 133–184
tiling group, **136**, 133–184, 212–222, 259
tiling groupoid, **147**
topologically equivalent, **311**, 322, 341, 348
topology
  induce the same topology, **311**, 322, 341, 348
torus (*pl.* tori), 56, 91
  essential torus, **344**
  once-punctured torus, **114**, 114–125, 161–162, 167–169, 212–214
  torus knot, **319**
tractrix, **108**
transformation group, **185**, 185–205
translation
  euclidean translation, **5**
  horizontal translation, **15**, 28, 41, 228, 239

trefoil knot, **319**
triangle
  euclidean triangle, 125
  hyperbolic triangle, 43, 98, 125,
    130
  spherical triangle, 52, 125
  Triangle Inequality, **3**
trivial group, **186**
twisted fuchsian group, **288**
two-sided surface, **344**

unbounded, **62**, 81, 103, 161–162
union, **356**
unknot, **317**

van der Veen, Roland, xvi
vertex (*pl.* vertices)
  of a polyhedron, **258**
  ideal vertex, **116**, **170**, 170–180,
    257, **258**, 257–265
  of a polygon, 61, 80, 82, **257**
  vertex at infinity, **170**, 170–180,
    257, **258**, 257–265
Vogt, Elmar, 347
volume
  finite volume, 324
  hyperbolic volume, **240**, 353

Wada, Masaaki, 246
Waldhausen, Friedhelm, 347, 350
walk, *see* discrete walk
Weeks, Jeffrey R., 323
Wright Sharp, Jennifer, xvi

$\mathbb{Z}$, the set of all integers, **356**
Zieschang, Heiner (1936–2004), 347